Special Inorganic Cements

Modern Concrete Technology Series

Series Editors

Arnon Bentur
National Building Research Institute
Technion–Israel Institute of Technology
Technion City
Haifa 32 000
Israel

Sydney Mindess
Department of Civil Engineering
University of British Columbia
2324 Main Mall
Vancouver
British Columbia
Canada V6T 1W5

Special Inorganic Cements

Ivan Odler

London and New York

First published 2000 by E & FN Spon
11 New Fetter Lane, London EC4P 4EE

Simultaneously published in the USA and Canada
by E & FN Spon
29 West 35th Street, New York, NY 10001

E & FN Spon is an imprint of the Taylor & Francis Group

© 2000 Ivan Odler

Typeset in Times by
J&L Composition Ltd, Filey, North Yorkshire
Printed and bound in Great Britain by
Biddles Ltd, Guildford and King's Lynn

British Library Cataloguing in Publication Data
A catalogue record for this book is available from the British Library

Library of Congress Cataloging in Publication Data

Odler, Ivan, 1930–
    Special inorganic cements/Ivan Odler.
        p.   cm. – (Modern concrete technology)
    Includes bibliographical references and index.
    ISBN 0–419–22790–3 (alk. paper)
    1. Cement. 2. Inorganic compounds.   3. Portland cement.
I. Title.   II. Series: Modern concrete technology series (E & F.N.
Spon); 8.
TA434.035   2000
620.1'35–dc21   99-38936
                        CIP

ISBN 0–419–22790–3

To my wife

Contents

3   Reactive forms of dicalcium silicate and belite cements   55

4   Cements containing calcium sulfoaluminate   69

5   Cements containing the phases $C_{12}A_7$ or $C_{11}A_7.CaF_2$ 88

Introductory remarks

The subject of this book is **special inorganic cements and related cementitious materials**. They include a large variety of products that exhibit cementitious properties, with the notable exception of ordinary Portland cement. The book includes both cements and cementitious materials that are already in commercial use as well as those that are still under development, or are only of scientific interest.

The first 16 sections of the book discuss special inorganic cements from the perspective of their chemical nature and physico-mechanical properties, whereas the remaining 14 sections deal with the selection of inorganic binders for different applications.

The individual phases discussed in the text are designated either by their chemical or mineralogical names or by their chemical formulas. Both conventional chemical formulas and abbreviated chemical symbols and formulas commonly used in cement chemistry are used (see also Table 1.1). The names of the phases most common in inorganic cements are listed in an appendix.

Related references and recommendations for further reading are given at the end of each section. A significant number of them were published in the proceedings of past International Congresses on the Chemistry of Cement, and these are cited in the following way:

- *Seventh International Congress on the Chemistry of Cement*, Paris, 1980: cited as *Proceedings 7th ICCC, Paris.*
- *Eighth International Congress on the Chemistry of Cement*, Rio de Janeiro, 1986: cited as *Proceedings 8th ICCC, Rio de Janeiro.*
- *Ninth International Congress on the Chemistry of Cement*, New Delhi, 1992: cited as *Proceedings 9th ICCC, New Delhi.*
- *Tenth International Congress on the Chemistry of Cement*, Göteborg, 1997: cited as *Proceedings 10th ICCC, Göteborg.*

For some references that are difficult to obtain in their original form, their citation in Chemical Abstracts is also given in the form [volume number/ reference number].

In writing the book, papers and books published up to the end of June 1999 have been taken into consideration.

The author wishes to thank to all those who helped him in writing this book.

Ivan Odler
Winchester, MA, 1999

Glossary of terms commonly used in cement chemistry

Accelerator – additive used to accelerate the setting and hardening of a cement.

Acid attack – a damaging action of acids on concrete.

Activator – a substance that has to be added in small amounts to a latently hydraulic material to make it hydrate.

Air entraining agent – an agent causing an entrapment of air in concrete, thus improving its frost resistance.

Alumina modulus – the ratio $AM = Al_2O_3/Fe_2O_3$ in a Portland cement raw meal or clinker.

Binder – (= inorganic binder) synonymous designation for "Cement".

Blended cement – synonymous designation for "Composite Cement".

Carbonation – reaction of a hardened inorganic binder with carbon dioxide (usually of air).

Cement – (= inorganic cement) a powdered inorganic material which, if mixed with appropriate amounts of water, yields a suspension of plastic consistency that converts spontaneously into a hard solid body at ambient temperatures.

Chemical admixture – a substance added in small amounts to a fresh concrete mix to modify its properties.

Chemical shrinkage – irreversible shrinkage associated with the hydration of a cement paste.

Clinker – (= Portland clinker) product of fusion (heating to temperatures of partial melt formation) of a cement raw meal. In ground form, constituent of inorganic cements.

Clinker mineral – a phase constituting a (Portland) cement clinker.

Coefficient of reflection – a measure of the whiteness of "white" Portland cement.

Composite cement – a cement that contains – in addition to Portland clinker and calcium sulfate – a latent hydraulic or pozzolanic constituent.

Concrete – a material obtained by blending together an inorganic cement, water and concrete aggregate.

Concrete aggregate – a hard, particulate material not reacting in a significant extent with the cement paste, with a maximum size of particles exceeding ≈ 5 mm, used as constituent of concrete mixes.

Consistency – degree of thickness of a fresh cement paste, mortar mix or concrete mix.

Creep – slow deformation of a moist hardened cement paste under the influence of an external force.

Dopant – a foreign ion incorporated in small amounts into the crystalline lattice of a hydraulically reactive phase to preserve at ambient temperature one of its high-temperature modifications and/or to increase its reactivity.

Fluxing agent – an agent added to the raw cement mix to accelerate clinker formation by increasing the amount of melt in the burning process.

Grinding aid – an agent accelerating the cement grinding process.

Hardening – a gradual increase of strength of a cement paste after setting.

Heat curing – process of increasing the hydration rate of a cement paste by exposing it to an elevated temperature (not exceeding 100°C).

Heat of hydration – amount of heat released in the hydration of an inorganic cement.

Hydrate – the reaction product of a hydration process.

Hydration – reaction of an inorganic binder with water yielding hydrates, usually associated with setting and hardening.

Hydraulic binder – inorganic binder which has the capacity to harden both in air and under water.

Hydraulicity – capability of a (hydraulic) material to react with water at ambient or moderately elevated temperatures, yielding hydrates, causing setting and hardening.

Hydraulic material – any powdered material capable of reacting at an ambient or moderately elevated temperature with water to yield a hydrate (or hydrates), causing setting and hardening.

Hydrothermal reaction – a chemical reaction (hydration) taking place in a water vapour saturated atmosphere (in an autoclave) at temperatures exceeding 100°C.

Latently hydraulic material – material which – if ground to a powder – reacts in the presence of small amounts of a suitable activator at ambient or moderately elevated temperatures with water, to yield a hydrate (or hydrates) causing setting and hardening.

Lime saturation factor – the ratio $LSF = CaO/(2.8\ SiO_2 + 1.2\ Al_2O_3 + 0.65\ Fe_2O_3)$ in a Portland cement raw meal or clinker.

Mineralizer – an agent added to the cement raw mix to accelerate the clinker formation in the burning process by its catalytic action.

Modulus of elasticity – the ratio "stress/strain" for a material exposed to an external force.

Mortar – a material obtained by blending together an inorganic binder, water and sand.

Non-hydraulic binder – inorganic binder which hardens only in air, but not under water.

Plasticity – capacity of a fresh cement paste, mortar mix or concrete mix being shaped.

Pore size distribution – distribution of the volume of the pores as function of pore size.

Porosity – volume of space filled with air or pore-solution, per one volume- or mass-unit of a porous body, such as a hardened cement paste.

Pozzolana – (= pozzolanic material) a material with a high content of SiO_2 or $SiO_2 + Al_2O_3$ in amorphous form, which – if sufficiently dispersed – reacts at an ambient or moderately elevated temperature with water and calcium hydroxide to yield a hydrate (or hydrates), causing setting and hardening.

Pozzolanicity – capability of a material to enter, with water and calcium hydroxide, into a pozzolanic reaction.

Setting – spontaneous conversion of a suspension of an inorganic binder in water of plastic consistency into a solid body.

Setting time – time elapsed between mixing and setting.

Silica modulus – the ratio SM = $SiO_2/(Al_2O_3 + Fe_2O_3)$ in a Portland cement raw meal or clinker.

Specific surface area – surface of a mass unit of an inorganic binder or of its hydration products.

Strength – capacity of a hardened material, such as of a hardened cement paste, to resist external stresses without cracking and disintegration.

Through-solution reaction – a chemical reaction, such as a hydration, taking place in a suspension, in which the starting materials dissolve and the reaction product precipitates randomly from the liquid phase.

Topochemical reaction – a chemical reaction, such as a hydration, taking place in a suspension in which the product of reaction occupies the space originally occupied by one of the reactants, rather than being precipitated randomly from the liquid phase.

Unsoundness – excessive expansion, cracking and disintegration of a hardened cement paste due to a delayed expansive reaction of one of its constituents.

1   General characteristics of inorganic cements

Inorganic cements (inorganic binders) are powdered materials that, if allowed to react with a suitable liquid phase (usually water or a water solution of an appropriate reactant), undergo chemical reactions associated – at an appropriate liquid/solid ratio – with the formation of a firm solid structure.

In some cement pastes (that is, suspensions of a binder in a liquid phase) the hardening takes place only in air, and is associated with a loss of free water and/or with a reaction with the CO_2 in the air. Such binders are called **non-hydraulic binders**, as opposed to **hydraulic binders**, in which the hardening may also take place under water, and is associated with a hydration process.

The chemical nature of inorganic cements may vary greatly. In some instances they are materials consisting of a single phase, but more often they contain several phases side by side. Some cements are even blends of two or more constituents, out of which some may be hydraulically active only in combination with others.

Materials or constituents that react spontaneously with water, causing setting and hardening of the original mix, are considered to be **hydraulically reactive**. They exhibit **cementing** or **hydraulic properties**.

Latently hydraulic materials are also able to react hydraulically, but only in the presence of at least small amounts of a suitable activator (such as an alkali hydroxide). Latently hydraulic materials are usually glassy or amorphous, and contain CaO, SiO_2, and Al_2O_3 as their main oxides.

Pozzolanic materials or **pozzolanas** are defined as siliceous or siliceous and aluminous materials that on their own have little or no cementitious value, but will – if present in finely divided form and in the presence of moisture – react chemically with calcium hydroxide at ordinary temperature to yield phases that possess cementing properties. The "pozzolanicity" – that is, the readiness to react with calcium hydroxide – may vary greatly in different pozzolanic materials.

Non-reactive constituents of inorganic binders do not react chemically to any significant extent in the course of hydration. They may, however, alter the rheology of the fresh cement paste and/or some properties of the hardened material by their physico-mechanical action.

Binders that contain latently hydraulic or pozzolanic or non-reactive constituents in combination with Portland clinker (and usually also with calcium sulfate) are called **composite** or **blended cements**.

Most of the compounds exhibiting cementing properties may be envisaged as being composed of simple oxides of different basicity, out of which calcium oxide most often represents the more basic member. For example, tricalcium silicate (Ca_3SiO_5), may be envisaged as a combination of three CaO and one SiO_2.

To express the chemical composition of phases constituting inorganic cements we can use either conventional chemical formulas, or abbreviated formulas in which the individual oxides are expressed by one-letter symbols. Such symbols, common in cement chemistry, are listed in Table 1.1. Very often the chemical composition of the individual compounds is expressed as a sum of oxides. This, however, does not imply that these oxides have a separate existence within the structure of this compound. So, for example, the compound tricalcium silicate may be expressed as Ca_3SiO_5, as $3CaO.SiO_2$, or – in its abbreviated form – as C_3S. The compound tricalcium aluminate trisulfate hydrate may be expressed as $Ca_6Al_2S_3O_{50}H_{64}$, as $6CaO.Al_2O_3.3SO_4.32H_2O$, as $3CaO.Al_2O_3.3CaSO_4.32H_2O$, or by the abbreviated formula $C_3A.3C\bar{S}.32H$ (also as $C_6A\bar{S}_3H_{32}$). It is also not unusual to combine the standard and abbreviated notations within a single chemical formula or within a single chemical equation. So, the compound calcium fluoroaluminate may be expressed as $Ca_{11}Al_{14}O_{32}F_2$, as $11CaO.7Al_2O_3.CaF_2$, or in an abbreviated form as $C_{11}A_7.CaF_2$ (also as $C_{11}A_7\bar{F}$). Note also that in most industrial cements the individual phases are not present as pure chemical compounds, and may contain variable amounts of foreign ions incorporated in their crystalline lattice.

The formation of compounds with cementing properties, just like the chemical reactions causing setting and hardening of cements, may be

Table 1.1 Cement: chemical nomenclature.

Oxide	Standard symbol	Abbreviated symbol
Aluminum oxide	Al_2O_3	A
Calcium oxide	CaO	C
Carbon dioxide	CO_2	$\bar{C}$ or c
Iron oxide	Fe_2O_3	F
Calcium fluoride	CaF_2	$\bar{F}$ or f
Water	H_2O	H
Potassium oxide	K_2O	K
Magnesium oxide	MgO	M
Sodium oxide	Na_2O	N
Phosphorus oxide	P_2O_5	P
Silicon oxide	SiO_2	S
Sulfur oxide	SO_3	$\bar{S}$ or s
Titanium oxide	TiO_2	T

expressed by chemical equations. Here also, either conventional or abbreviated chemical formulas may be employed to characterize the compounds involved.

At appropriate liquid/solid ratios the resultant cementitious mix will exhibit the consistency of a plastic paste or a thick suspension. Its **fluidity** (also called its **consistency** or **workability**) will depend on this ratio, and will improve with increasing amounts of the added liquid. Prior to setting, such a system is called a **fresh cement paste**. Over time the rheological properties of the paste will alter, and its viscosity will increase as a consequence of chemical reactions taking place in the system. Ultimately the paste will lose its plastic properties and will convert to a non-plastic porous solid body. This conversion is called **setting**. The **setting time** – that is, the time between mixing and loss of plasticity – depends on the quality of the cement, on the initial liquid/cement ratio (it lengthens as this ratio increases), on the existing temperature (it shortens as the temperature increases), on the presence of chemical admixtures, and on other factors. The product of this reaction is called **set cement paste**.

As the chemical reactions in the paste will continue even after setting, the hardness and strength of the paste – which are low immediately after setting – will increase, and a **hardened cement paste** will be formed. Ultimately the hardening reaction in the paste comes to an end, owing to the complete consumption of the original binder or insufficient amounts of mixing liquid, and a **mature hardened cement paste** results.

The observed changes in consistency, setting, and hardening are the result of physico-chemical reactions taking place in the cement paste. In many but not all cementitious systems the formation of reaction products is associated with the consumption of the liquid phase (usually water) and thus with the decline of the actual liquid/solid ratio. In the setting/hardening process aggregates of solids are created in which the newly formed reaction products act as a glue, keeping the residua of the non-reacted material together. Initially these aggregates are not interconnected, and the paste preserves its fluid or plastic consistency, even though its viscosity increases. Eventually, however, the solid phases interconnect, and a three-dimensional solid network develops, causing the loss of fluidity or plasticity of the paste. The strength of the material – that is, its ability to resist external mechanical forces – subsequently increases as the chemical reactions continue, and thus the solid network becomes denser and more interconnected.

In cement suspensions with an excessively high liquid/solid ratio the relevant chemical reactions will still take place, but the amount of reaction products formed will be too small to result in the creation of a three-dimensional solid network and thus cause setting. By contrast, if the amount of the liquid phase is too low, it may not be sufficient to wet the surface of the cement particles, to fill the intraparticular space, and to yield a system of paste consistency. Nevertheless, a strength development may also be observed here if the material has been sufficiently compacted.

The chemical reactions causing setting and hardening may be many and varied. In the simplest case, setting and hardening of a mix of plastic consistency occurs as a consequence of the loss of part of the liquid present, for example by evaporation, or by absorption by a porous base. The particles of the binder – initially separated spatially – enter into mutual contact, and bonds develop between them, causing a loss of plasticity and a development of strength. Such bonds may be created as a consequence of recrystallization, in which the area of the existing solid/liquid interface declines and – as a result – the free enthalpy of the system also declines. Examples of such cementitious systems are lime putty (in essence a thick suspension of calcium hydroxide in water) and clay mud, which both lose their plasticity and solidify upon drying.

In another mechanism, hardening occurs as a consequence of a reaction of the cement paste with the carbon dioxide in air, yielding carbonates as products of reaction. Usually the reaction progresses rather slowly, mainly because of the low CO_2 concentration in air. Carbonation often occurs in cementitious systems that have already undergone setting and hardening by another chemical reaction. The carbonation process starts at the paste/air interface, and later becomes controlled by the rate of diffusion of CO_2 into deeper regions of the cementitious body. The rate of carbonation may be accelerated by curing the produced elements in an atmosphere with an elevated CO_2 partial pressure. An example of hardening due to carbonation is the hardening of lime mortars upon long-term exposure to air, resulting in the formation of a calcium carbonate region of increased strength at the surface.

$$Ca(OH)_2 + CO_2 \rightarrow CaCO_3 + H_2O \tag{1.1}$$

In many cements, setting and hardening is based on hydration reactions. In a hydration reaction the initial anhydrous compound reacts with water, and a **hydrate** is formed according to the scheme

$$A + H \rightarrow AH \tag{1.2}$$

An example of such a reaction is the hydration of monocalcium aluminate, the main constituent of high-alumina cement:

$$CaO.Al_2O_3 + 10H_2O \rightarrow CaO.Al_2O_3.10H_2O \tag{1.3}$$

It is also possible for several hydrate phases to be formed simultaneously, as in the hydration of tricalcium aluminate in water:

$$2(3CaO.Al_2O_3) + 27H_2O \rightarrow 2CaO.Al_2O_3.8H_2O + 4CaO.Al_2O_3.19H_2O \tag{1.4}$$

In an even more complex case two or more constituents of the cement may interreact in the course of hydration, as in the formation of ettringite in the hydration of Portland cement:

$$3CaO.Al_2O_3 + 3CaSO_4.2H_2O + 26H_2O \rightarrow 6CaO.Al_2O_3.3SO_3.32H_2O$$

$$(1.5)$$

In some other binder systems a reaction between constituents of different basicity is responsible for setting and hardening, as in calcium phosphate cements for example:

$$CaHPO_4 + Ca_4(PO_4)_2O \rightarrow Ca_5(PO_4)_3OH \tag{1.6}$$

In such instances the liquid phase just acts as a medium for ion transport, and does not participate in the chemical reaction.

Finally, setting and hardening may also be caused by reactions that cannot be included in any of the categories mentioned so far. An example of such a reaction is the setting/hardening reaction in alkali silicate cements:

$$2(Na_2O.nSiO_2) + Na_2SiF_6 \rightarrow (2n+1)SiO_2 + 6NaF \tag{1.7}$$

In the hardening/setting process the constituent(s) of the cement dissolve in the liquid phase, and the reaction product(s) precipitate from it. This occurs because the solubility of the reaction product(s) in the liquid phase is lower than that of the starting material(s). If the rate-controlling step of this process is the precipitation of the reaction product rather than the initial dissolution of the starting constituent, the liquid phase quickly becomes oversaturated with respect to the reaction product, and precipitation occurs randomly in the whole liquid volume. We call this a **through-solution reaction**. When, conversely, the rate-controlling step is the dissolution of one of the starting constituents, the reaction product tends to precipitate right at, or close to, the surface of the solid phase that dissolves most slowly, before oversaturation can be reached in the whole volume of the liquid phase. We call such a process a **topochemical reaction**, as the formation of the end-product is confined to a specific location, rather than occurring randomly throughout the whole volume of the paste.

All setting/hardening reactions of inorganic binders occur at ordinary or only moderately elevated temperatures. All of them are exothermic: that is, they are associated with a liberation of the heat of reaction. The amount of heat liberated in the setting/hardening process may vary greatly in different cements, just like the kinetics of the heat release. Periods of faster and slower rates of heat liberation may alternate. An increased temperature, increased fineness of the binder and an initial agitation of the mix generally increase the rate of reaction.

The volume of reaction products formed equals roughly, but not exactly, the sum of the volumes of the solid and liquid phases entering the reaction. Usually a loss of a fraction of the original volume takes place, and this phenomenon is known as **chemical shrinkage**.

The volume of the hardened paste may decrease further upon losing free water located in the pore system (by evaporation, for example), whereas it may increase if water is taken up from the environment. However, this phenomenon may not be entirely reversible, and may contain a reversible and an irreversible component.

Upon heating, hardened cement pastes undergo chemical reactions (for example, thermal decomposition of the hydrates present and a loss of chemically bound water). These processes may be associated with a decline or even a complete loss of strength, but in some cementitious systems an increase of strength may also be observed.

In spite of many common features, the properties of different inorganic cements can vary greatly, and to obtain satisfactory results it is important to select the right type of binder for every application.

2   Special Portland cements

Portland cements are inorganic binders obtained by grinding to a high fineness Portland clinker alone, or – most commonly – in combination with calcium sulfate, acting as a set regulator. The ASTM standard C219–94 defines Portland cement as "a hydraulic cement produced by pulverizing Portland-cement clinker, and usually containing calcium sulfate." The European standard ENV197–1 requires for Portland cement a clinker content of 95–100% and a content of "minor additional constituents" of no more than 5%, in addition to limited amounts of calcium sulfate.

Portland clinker is a product of burning a raw mix containing the oxides CaO, SiO_2, Al_2O_3, and Fe_2O_3 (plus other oxides in smaller amounts) to temperatures of partial melt formation. Under these conditions calcium oxide, originally present in the form of $CaCO_3$, first converts to free CaO and then reacts with the remaining constituents of the raw mix to yield **clinker minerals**. The most important phase produced in the burning process, and the one characteristic for Portland clinker, is tricalcium silicate ($3CaO.SiO_2$ or Ca_3SiO_5, abbreviation C_3S). The other main phases present in Portland clinker are dicalcium silicate ($2CaO.SiO_2$ or Ca_2SiO_4, abbreviation C_2S), tricalcium aluminate ($3CaO.Al_2O_3$ or $Ca_3Al_2O_6$, abbreviation C_3A), and calcium aluminate ferrite [ferrite phase, $2CaO(Al_2O_3, Fe_2O_3)$, abbreviation $C_2(A,F)$, sometimes also described as C_4AF]. In industrial cements these phases are not present as pure chemical compounds, but contain variable amounts of foreign ions in their crystalline lattices.

2.1   CONSTITUENTS AND COMPOSITION OF PORTLAND CEMENTS

2.1.1   Tricalcium silicate ($3CaO.SiO_2$, abbreviation C_3S)

Tricalcium silicate is an essential constituent of all Portland clinkers. In its crystalline lattice it contains Ca^{2+} cations in combination with SiO_4^{4-} and O^{2-} anions in a ratio of 3:1:1. Upon heating or cooling tricalcium silicate undergoes a series of reversible phase transitions:

$$T_1 \xrightleftharpoons{620\,°C} T_2 \xrightleftharpoons{920\,°C} T_3 \xrightleftharpoons{980\,°C} M_1 \xrightleftharpoons{990\,°C} M_2$$

$$\xrightleftharpoons{1060\,°C} M_3 \xrightleftharpoons{1070\,°C} R$$

(T = triclinic; M = monoclinic; R = rhombohedral)

The crystal structures of the individual tricalcium silicate phases are similar, and differ mainly in the spatial orientations of the SiO_4 tetrahedra.

Tricalcium silicate is thermodynamically stable only between about 1250 °C and 2150 °C. It melts incongruently above 2150 °C, and becomes thermodynamically unstable with respect to dicalcium silicate and calcium oxide below 1250 °C. Thus, to produce a clinker containing tricalcium silicate, the raw mix has to be heated to a temperature that is higher than the lower limit of thermodynamic stability of this phase, and has to be cooled rapidly enough to prevent a noticeable decomposition of tricalcium silicate in the course of cooling. Such a conversion of C_3S to C_2S and C may take place especially at temperatures slightly below 1250 °C, whereas at ambient or only moderately elevated temperatures tricalcium silicate remains preserved for an indefinite time.

In industrial clinkers the existing tricalcium silicate is doped with foreign ions, also present in the raw mix. These dopants prevent a conversion of tricalcium silicate to its T_1 modification, which is the one that is "metastable" at ambient temperature, and stabilize one of the high-temperature modifications of this compound, usually M_1 or M_3. Such a doped form of tricalcium silicate is usually called **alite**.

Upon contact with water, tricalcium silicate undergoes hydration, yielding an amorphous calcium silicate hydrate phase called the **C-S-H phase** (or just C-S-H) and calcium hydroxide as products of hydration. The rate of hydration will depend on the quality and quantity of dopants incorporated within the crystalline lattice, on the cooling rate in the production of the clinker, on the fineness of the cement, and on other factors.

The C-S-H phase is an amorphous or nearly amorphous material of the general formula $CaO_x.SiO_2.H_2O_y$, where both x and y may vary over a wide range. On the nanometer scale the C-S-H phase is structurally related to the crystalline phases 1.4 nm tobermorite and jennite. In cement pastes limited amounts of foreign ions may be incorporated into the C-S-H phase. On the micrometer scale the C-S-H phase appears either as a dense amorphous mass or as a microcrystalline material with an acicular or platelet-like morphology. The material contains pores with radii between about 1 and 10^4 nm, and exhibits a specific surface area exceeding 100 m^2/g.

Calcium hydroxide is formed – besides C-S-H – as the second product of C_3S hydration. This happens because the C/S molar ratio within the C-S-H phase is always distinctly lower than that of the original C_3S. In hydrated tricalcium silicate or Portland cement pastes calcium hydroxide is present in the form of crystals up to about 10–30 μm large, and in this crystalline form is called portlandite.

2.1.2   Dicalcium silicate (2CaO.SiO$_2$, abbreviation C$_2$S)

There exist five polymorphs of dicalcium silicate, designated a, a'_H, a'_L, β, and γ. The structure of dicalcium silicate is built from Ca^{2+} and SiO_4^{4-} ions, and is similar in the individual polymorphs, with the exception of the γ modification, which is somewhat different.

All C$_2$S forms can be produced by solid-state reactions from CaO and SiO$_2$. In a reaction of pure starting compounds the γ modification is the one that is formed in the reaction, whereas an addition of various minor oxides, serving as dopants, is required to produce any of the other modifications.

The γ-C$_2$S modification is the only one that is thermodynamically stable at ordinary temperatures. It is barely hydraulically reactive, and thus is not important for cement chemistry.

The modification that is most reactive and thus the most important in cement chemistry is β-C$_2$S, which is thermodynamically unstable at any temperature. Upon cooling it tends to transform to γ-C$_2$S, unless doped with stabilizing ions. In its doped form β-C$_2$S, also called **belite**, is a regular constituent of a variety of inorganic binders, including Portland cement. High-temperature modifications of dicalcium silicate, stabilized by different dopants, exhibit variable reactivities, and may be found in some special cements.

The products of dicalcium silicate hydration are identical or almost identical to those formed in the hydration of tricalcium silicate; however, the amount of calcium hydroxide formed is distinctly lower. The hydration rate of dicalcium silicate is significantly lower than that of tricalcium silicate, even though it may be influenced to a certain degree by the selection of the dopant ion and the cooling rate. Some highly reactive forms of dicalcium silicate have been synthesized in recent years, but are of theoretical importance only.

2.1.3   Tricalcium aluminate (3CaO.Al$_2$O$_3$, abbreviation C$_3$A)

In Portland cements tricalcium aluminate usually exists in its cubic form; however, in the presence of increased amounts of alkalis in the raw mix an orthorhombic or even a monoclinic modification may be formed instead. The structure of tricalcium aluminate is built from rings of six AlO$_4$ tetrahedra and Ca^{2+} ions. All three modifications of C$_3$A hydrate in a similar way; however, their reactivity may differ, depending on the quality and quantity of the foreign ions incorporated in their crystalline lattices.

In hydration in pure water the hexagonal phases C_2AH_8 and C_4AH_{19} are formed as products of hydration, whereas C_4AH_{19} alone is formed in the presence of calcium hydroxide:

$$2C_3A + 27\,H \rightarrow C_2AH_8 + C_4AH_{19} \qquad (2.1)$$

$$C_3A + CH + 19H \rightarrow C_4AH_{19} \qquad (2.2)$$

C_2AH_8 and C_4AH_{19}, which both belong to the broad group of AFm phases, have a structure that consists of $[Ca_2Al(OH)_6]^+$ layers with OH^- or $[Al(OH)_4]^-$ ions, together with H_2O, in the interlayer region. In addition, C_4AH_{19} contains an additional layer of H_2O between the principal layers; the latter is lost at relative humidities below about 80%, and C_4AH_{19} converts to C_4AH_{13}. Both hexagonal calcium aluminate hydrates are thermodynamically unstable, and tend to convert to the only stable phase, C_3AH_6 [or $Ca_3[Al(OH)_6]_2$], which belongs to the group of hydrogarnets.

The conversion of C_3A to C_4AH_{19} takes place in Portland cement pastes only after the available amounts of calcium sulfate have been used up. In the presence of calcium sulfate, tricalcium aluminate trisulfate hydrate ($C_3A.3C\bar{S}.32H$ or $C_6A\bar{S}_3H_{32}$), also called ettringite, or simply trisulfate, is formed as the product of C_3A hydration. Ettringite belongs to the group of AFt phases. Its nanostructure is based on columns of the composition $[Ca_3Al(OH)_6.12H_2O]^{3-}$ running parallel to the c-axis with SO_4^{2-} ions and H_2O in the intervening channels. Well-developed ettringite crystals exhibit an acicular morphology; however, the material formed in Portland cement pastes immediately after mixing with water is highly dispersed, and tends to cover the cement grains as a thin layer. Because of the more favorable morphology of the hydrated material formed, and a reduced overall hydration rate brought about by the hydrate layer, the period for which the paste preserves its plasticity and remains workable is extended, compared with the conditions existing in the absence of calcium sulfate. For this reason calcium sulfate is a common constituent of Portland cements.

After exhaustion of the available calcium sulfate, the trisulfate phase formed initially may react with the remaining tricalcium aluminate not yet consumed in the hydration and convert to tricalcium aluminate monosulfate hydrate ($C_3A.C\bar{S}.12H$ or $C_4A\bar{S}H_{12}$), commonly called monosulfate. Monosulfate also belongs to the broader group of AFm phases. Its nanostructure consists of $[Ca_2Al(OH)_6]_2^{2-}$ layers with one SO_4^{2-} and six H_2O in the interlayer region. In pure form it crystallizes in the form of thin hexagonal platelets; however, in Portland cement pastes layers of AFm are usually intimately intermixed with layers of the C-S-H phase.

2.1.4 The ferrite phase [calcium aluminate ferrite, $2CaO(Al_2O_3,Fe_2O_3)$, abbreviation $C_2(A,F)$]

The composition of calcium aluminate ferrite (the ferrite phase) in Portland cements may vary between about $C_2(A_{0.7}, F_{0.3})$ and $C_2(A_3, F_{0.7})$, and thus it may be best expressed by the formula $C_2(A,F)$, even though the formula C_4AF is also used occasionally. On the nanometer scale the structure of this phase consists of Ca^{2+} ions, each of them surrounded by seven oxygen neighbors, combined with Al^{3+} and Fe^{3+} ions distributed between octahedral and tetrahedral sites. In industrial clinkers the ferrite phase may also contain significant quantities of foreign ions: the Fe^{3+} is partly substituted by Mg^{2+} and equal amounts of Si^{4+} and/or Ti^{4+} ions.

The products formed in the hydration of the ferrite phase are similar to those formed in the hydration of C_3A: the Fe^{3+} replaces Al^{3+} to a limited degree in the crystalline lattice. The A/F ratio in the hydrates formed is usually higher than that in the original calcium aluminate ferrite, and the fraction of iron that has not been incorporated into any of the hydrate phases remains in the hardened cement paste in the form of amorphous iron oxide, hydroxide or another unspecified iron-containing phase.

2.1.5 Calcium sulfate

In the production of ordinary Portland cement, limited amounts of calcium sulfate, introduced in the form of gypsum ($CaSO_4.2H_2O$) and/or anhydrite ($CaSO_4$), are interground with Portland clinker. In the course of grinding, the added gypsum usually gets dehydrated to hemihydrate or even to anhydrite. The addition of calcium sulfate is needed to control the setting of the resulting cement by making possible the formation of ettringite in the initial hydration of tricalcium aluminate (see section 2.1.3). The amount of sulfate present in Portland cement must be limited to about 3–4% of SO_3, as at higher SO_3 contents excessive amounts of ettringite may be formed in the later stage of hydration, causing expansion and cracking of the resulting concrete.

2.1.6 Free calcium oxide (free lime, CaO)

Free calcium oxide in small amounts (usually below 1%) is a regular constituent of Portland clinker; it originates from the decomposition of calcium carbonate taking place in the burning process. Larger amounts of it may be present if the maximum burning temperature in the production of the clinker is too low, or if the burning time is too short, or if the CaO content in the raw meal exceeds the acceptable range (that is, if the lime saturation factor – see section 2.1.9 – of the raw meal exceeds LSF = 100).

Whereas small amounts of free lime may be considered acceptable, larger amounts may cause expansion, strength loss, and cracking of the hardened paste. This is due to a delayed hydration of free calcium oxide to calcium hydroxide, which takes place topochemically and is associated with an increase of volume and a generation of internal stresses within the paste. Thus excessive amounts of free calcium hydroxide in clinker must be avoided.

2.1.7   Free magnesium oxide (periclase, MgO)

Free MgO, also called periclase, is present in Portland clinkers only at high MgO contents in the raw meal (usually in the form of $MgCO_3$). It may be tolerated if present in amounts of only a few per cent, but may cause concrete damage if present in higher amounts (see also section 2.9).

2.1.8   Alkali sulfates

The following alkali sulfates may be present in Portland clinkers:

- K_2SO_4 (abbreviation $K\bar{S}$; arcanite);
- $K_2Ca_2(SO_4)_3$ (abbreviation $KC_2\bar{S}_3$; calcium langbeinite);
- $K_3Na(SO_4)_2$ (abbreviation $K_3N\bar{S}_4$; aphtitalite).

Upon mixing with water these phases are rapidly dissolved in the liquid phase, and their SO_4^{2-} ions participate in the formation of SO_3-bearing hydrate phases.

2.1.9   Composition of Portland clinker and Portland cement

In ordinary Portland clinker (corresponding to ASTM type I Portland cement) tricalcium silicate is the most abundant phase, present in amounts between about 50% and 70%. Dicalcium silicate usually constitutes 15–30% of the clinker. Typical amounts of tricalcium aluminate are 5–10%, and of the ferrite phase 5– 15%. In special Portland clinkers the individual clinker phases may be present in increased or reduced amounts, or may be absent entirely. Free lime, free MgO and alkali sulfates are minor constituents of Portland clinker that are present only in small amounts or may be absent entirely. Calcium sulfate is only rarely a constituent of Portland clinker, but is interground in limited amounts with it, to obtain Portland cement. In Table 2.1 different types of Portland cement are compared.

The composition of Portland cement raw meals and Portland clinkers can be characterized by the mutual ratio of the main oxides present. The following ratios are those most widely employed (all amounts of oxides are given in wt%):

Table 2.1 Types of Portland cement.

Designation	C_3S	C_2S	C_3A	C_4AF	Other characteristics
Ordinary PC	→	→	→	→	
High-C_3S PC	↑	↓	→	→	
High-C_2S PC	↓	↑	→	→	
High-C_3A PC	→	→	↑	→	
Low-C_3A PC	→	→	↓	→	
Low-iron PC	→	→	↑	↓	
High-iron PC	→	→	↓	↑	
Ferari cement	→	→	↓	↑	A/F = 0.64
Erz cement	→	→	0	0	C_2F present
Kühl cement	→	→	0	↑	
High-MgO PC	→	→	→	→	MgO > 5%
Low-alkali PC	→	→	→	→	Na_2O + K_2O content reduced
Mineralized PC	→	→	→	→	Clinker doped with SO_3 + F
High specific surface area PC	→	→	→	→	Specific surface area ↑
Low-porosity PC	→	→	→	→	No calcium sulfate added Added Na_2CO_3 + lignosulfonate

→ normal value; ↑ increased value; ↓ reduced value; 0 absent.

- lime saturation factor: LSF = 100C/(2.8S + 1.2A + 0.65F);
- silica modulus (or silica ratio): SM = S/(A + F);
- alumina modulus (or alumina ratio): AM = A/F.

In ordinary Portland cement these ratios vary typically in the following ranges: LSF = 90–95; SM = 2.2–2.6; AM = 1.5–2.5. They may deviate significantly from these ranges in special Portland cements.

2.2   THE HYDRATION OF PORTLAND CEMENT

The paste hydration of Portland cement at ambient temperatures is characterized by several stages.

Immediately upon contact of the cement with water, in the **pre-induction period**, a rapid dissolution of ionic species in the liquid phase and the formation of hydrate phases get under way. Alkali sulfates present in the cement dissolve completely within seconds, contributing K^+, Na^+, and SO_4^{2-} ions. Calcium sulfate dissolves until saturation, thus contributing Ca^{2+} and SO_4^{2-} ions. Small amounts of tricalcium silicate (about 1–5%) undergo hydration, and a thin layer of the formed C-S-H phase (first-stage C-S-H) precipitates at the cement grain surface. At the same time Ca^{2+} and OH^- ions, formed simultaneously in the hydration of C_3S, enter the liquid phase. Tricalcium aluminate (about 5–25% of the amount present) dissolves, and reacts with the Ca^{2+} and SO_4^{2-} ions present in the liquid phase, yielding ettringite (AFt phase), which also precipitates on the cement grain surface. The amounts of the C_2S and ferrite phases that hydrate in the pre-induction period are negligible.

This short initial period of rapid hydration is followed by the **induction period** (also called the dormant period), in which the hydration rate slows down significantly. It usually lasts several hours.

In the subsequent **acceleration stage** the hydration accelerates again. The rapid hydration of C_3S is associated with the formation of second-stage C-S-H and the precipitation of crystalline calcium hydroxide (portlandite). A noticeable hydration of C_2S also gets under way. Tricalcium aluminate – and to a lesser extent calcium aluminate ferrite – hydrates, and yields additional amounts of AFt until the whole amount of calcium sulfate present in the original cement has been consumed.

In the **post-acceleration period** the hydration rate starts to slow down gradually, as the amount of cement still present in non-reacted form declines. At the same time the rate of the hydration process becomes diffusion controlled. The C-S-H phase continues to be formed; however, the contribution of C_2S to this process increases gradually.

After the free calcium sulfate has been consumed, the AFt phase that was formed in the earlier stages of hydration reacts with non-reacted C_3A and $C_2(A,F)$ and converts to AFm. This conversion may or may not proceed to full completion, and some AFt may still be present even in fully hydrated pastes.

The main constituent of a mature Portland cement paste is the C-S-H phase. Its mean CaO/SiO_2 molar ratio may vary, depending on the composition of the cement, the water/cement ratio, and the hydration temperature. On the micrometer scale this ratio may vary between about 1.2 and 2.3. The C-S-H phase formed in the hydration of Portland cement also contains limited but variable quantities of foreign ions.

The second most important phase of the hydrated Portland cement paste is calcium hydroxide (portlandite), present in the form of relatively large crystals embedded within the C-S-H phase. The AFm phase is usually intimately intermixed with the C-S-H phase. The amount of AFt declines after reaching a maximum, and this phase is usually absent in mature pastes. As long as it is present, it is distributed within the C-S-H matrix in a microcrystalline form. Some non-hydrated clinker residua may be present even in mature pastes, especially at lower water/cement ratios.

An essential characteristic of any hardened Portland cement paste is the presence of pores. Their radius may vary widely between about 1 and 1000 nm. The overall porosity of the hardened paste declines as the hydration progresses, and increases with increasing water/cement ratio.

2.3 HIGH-C₃S PORTLAND CEMENT

In high-C_3S (high alite) Portland cements the amount of alite in the clinker is increased at the expense of belite. Usually such clinkers also contain moderately increased amounts of tricalcium aluminate. In the production of high-C_3S clinkers the lime saturation factor of the raw meal is increased to values close to or even moderately exceeding LSF = 100.

To burn such raw meals effectively, and to reduce the free lime content in the clinker to acceptable levels, the raw meal must be finely ground and well homogenized. The addition of small amounts of a mineralizer may also be required. The energy consumption in producing such clinkers is increased above that of ordinary Portland cement, mainly because of the presence of higher proportions of calcium carbonate in the raw meal, which must be decarbonated in a reaction that is highly endothermic (1782 kJ per 1 kg of $CaCO_3$).

High-C_3S clinkers are employed in the production of high early strength Portland cement (corresponding to ASTM Type III cement). To produce such cement, the clinker, in combination with appropriate amounts of calcium sulfate, must be ground to a specific surface area higher than that common in ordinary Portland cement: that is, to 400–500 m^2/kg (Blaine), rather than the 300–350 m^2/kg (Blaine) typical for ordinary Portland cement.

2.4   PORTLAND CEMENT WITH ELEVATED C$_2$S CONTENT

Some Portland cements may contain elevated amounts of belite (C$_2$S), at the expense of alite (C$_3$S). To produce such clinkers, the lime saturation factor of the raw meal must be reduced, typically to values between about LSF = 80 and LSF = 90, instead of the LSF > 90 that exists in ordinary Portland cement. (Cements made from clinkers with even lower lime saturation factors, which contain no or only very limited amounts of alite, are called belite cements, and are discussed in section 3.2.)

Raw meals for high-belite clinkers are relatively easy to burn, and the consumption of energy is reduced, mainly because of a reduced CaCO$_3$ content. Compared with ordinary Portland cement clinkers, it is possible to reduce the burning temperature by up to about 100 °C and along with it the NO$_x$ emission. At the same time the grindability of the clinker is improved (Ludwig and Pohlman, 1986).

In the formed clinker the M$_1$ modification of C$_3$S prevails. Belite is present mostly in its β modification, but some α- and α'-C$_2$S may also be formed. At high cooling rates the amount of these high-temperature modifications increases at the expense of β-C$_2$S.

Owing to the lower reactivity of belite, the overall rate of hydration, and along with it the strength development up to about 90–180 days, is slowed down with increasing belite and decreasing alite contents in the cement (Bei and Ludwig, 1990). At the same time the final strength of a Portland cement with an elevated C$_2$S content may exceed that of an ordinary Portland cement, because more C-S-H and less portlandite is formed in the hydration of dicalcium than of tricalcium silicate.

The main difference in the composition of cement pastes made from cements with increased belite and reduced alite contents is their lower calcium hydroxide content. This may affect positively the resistance of such hardened pastes to chemical corrosion. At the same time the depth of carbonation increases with declining C$_3$S content in cement (Kelham and Moir, 1992).

The amount of heat liberated in the hydration declines with increasing dicalcium silicate and decreasing tricalcium silicate contents in cement. At the same time the rate at which the heat is released is slowed down (Yoshida, 1992). The rate of heat release may be reduced even further by coarser grinding of the cement. A comparison of strength development and hydration heat liberation reveals that the strength per unit quantity of heat of hydration generally increases with increasing C$_2$S and decreasing C$_3$S content in the clinker (Yoshida and Igarashi, 1992).

It has been reported (Albats and Shein, 1997) that the strength of Portland cement with an elevated belite content may be increased by using a very high heating rate in the clinker burning process (as high as 20 °C/min) and by simultaneously increasing the alumina modulus (MA =

A/F) of the raw mix. The increase of short-term strength was explained by a lowering of the C/S molar ratio in the alite formed (from 3.07 to 2.84), which resulted in an increase of the overall alite content in clinker at the expense of belite. Furthermore, under such conditions a fraction of the C_3A present decomposes to $C_{12}A_7+C$, and the released free lime reacts with belite to yield alite. In parallel, $C_{12}A_7$ reacts with the aluminoferrite phase, increasing its A/F ratio and thus its reactivity:

$$C_{12}A_7 + 2C + 7C_4AF \rightarrow 7C_6A_2F \qquad (2.3)$$

Garbacik (1997) reported the possibility of producing a clinker with an elevated belite content, which exhibits favorable strength properties, by burning at maximum temperatures as low as 1250–1300 °C a raw mix containing chalky and marly starting constituents and very fine crystalline reactive forms of silicates. The alite and belite phases formed under these conditions exhibit very small dimensions of their crystals, and are exceptionally reactive.

Clinkers with increased C_2S and reduced C_3S contents are employed in the production of cements with reduced hydration heat evolution (see section 19), such as moderate heat of hardening cement (corresponding to ASTM Type II cement) and low-heat cement (corresponding to ASTM Type IV cement). To slow down the rate of heat evolution, cements employed for this purpose are usually ground to a relatively low specific surface area. Cements of this type are used in applications in which a reduced release of hydration heat is required, as in the construction of dams and other bulk concrete structures (Kelham and Moir, 1992; Sone *et al.*, 1992). High-C_2S clinkers, especially those with a reduced C_3A content, are also constituents of some oil well cements (see section 27). Low-C_3S clinkers are less suitable for the production of blended cements, owing to the reduced amount of free calcium hydroxide produced in the hydration of such clinkers.

Belite-rich clinkers (with C_2S contents above 55%) have also been developed in which, in addition to C_3S, the amounts of C_3A and C_4AF have also been reduced significantly, to less than 3% and 8% respectively (Okamura *et al.*, 1998). Cements made from such clinkers usually exhibit excellent fluidity; however, this depends greatly on the amount and form of calcium sulfate present.

2.5   HIGH-C_3A PORTLAND CEMENT

High-C_3A Portland cement contains elevated amounts of tricalcium aluminate in its clinker, compared with ordinary Portland cement. Such clinkers may be produced by increasing the Al_2O_3 content in the raw meal and thus lowering its silica modulus [SM = S/(A + F)], and by increasing the alumina modulus (AM = A/F). High-C_3A clinkers are used for producing type S expansive cements. To produce such cement the clinker must be ground with relatively high amounts of calcium sulfate, to permit the formation of high amounts of expansive ettringite in the course of hydration. (For more detail of type S expansive cement see section 21.3.3.)

2.6   LOW-C$_3$A PORTLAND CEMENT

A Portland cement with a reduced C$_3$A content may be produced from raw meals with a very low content of Al$_2$O$_3$ (Stark *et al.*, 1997b). The iron oxide content in the mix is usually also reduced, and the silica modulus [SM = S/(A + F)] is increased. The calcium sulfate content in the raw meal may be kept relatively low, as only a small amount of AFt may be formed in the hydration.

Cements produced from low-C$_3$A clinkers exhibit particularly good frost resistance. In hardened cement pastes made from ordinary Portland cement the AFm phase formed in hydration at ambient temperatures tends to convert to the AFt phase at temperatures near or below the freezing point. This formation of secondary ettringite, brought about by a reversion of the thermodynamic equilibrium, enhances the concrete damage caused by the formation of ice in the pore system of the cement paste at subfreezing temperatures (Stark *et al.*, 1997a). This effect may be reduced or prevented by reducing the C$_3$A content in the clinker.

2.7   LOW-IRON (WHITE) PORTLAND CEMENT

In low-iron Portland cements the content of Fe$_2$O$_3$ in the clinker is reduced, usually to amounts below 0.5%. Consequently, the aluminate modulus (AM = A/F) increases to values of AM = 10 and above. Such clinkers contain no or only insignificant amounts of the calcium aluminoferrite phase, and virtually all the Fe$_2$O$_3$ present is incorporated into the crystalline lattices of other clinker minerals, mainly in the C$_3$A and glass phases (Boikova *et al.*, 1986). All the Al$_2$O$_3$ (except for a small fraction incorporated in the crystalline lattices of calcium silicates) is present in the form of tricalcium aluminate.

As the typical gray color of ordinary Portland cement is due to the presence of the ferrite phase, low-iron clinkers may be used in the production of white Portland cement, if the Fe$_2$O$_3$ content is kept sufficiently low, and if appropriate measures are taken to eliminate or reduce the effect of residual iron on the color of the resultant clinker.

Low-iron clinkers are produced from raw meals made from starting materials that contain only negligible amounts of Fe$_2$O$_3$, such as limestone with a high CaCO$_3$ content, or chalk, in combination with high-purity kaolin, bentonite, or china clay. Such raw meals generally exhibit poor burnability, a fact that has to be taken into consideration if a low-iron clinker is produced. The addition of small amounts of a mineralizer, such as fluorspar (CaF_2) or cryolite (sodium aluminum fluoride), may be necessary to facilitate clinker formation (Blanco-Varela *et al.*, 1996; Blanco-Varela, 1997; Puertas *et al.*, 1997a, 1997b, 1998). Higher burning temperatures (1550–1600 °C) may also be needed to achieve sufficient clinkering. To avoid iron contamination from the fuel, natural gas or heating oil, rather than coal, must be used as the heat source.

To minimize the effect of residual iron on the color of the produced cement, it is necessary to burn the clinker under slightly reducing conditions, and to cool it rapidly. In a reducing atmosphere trivalent iron (Fe^{3+}) present in the raw meal is reduced to its bivalent form (Fe^{2+}), which is not involved in producing the gray coloration of the produced cement. Unlike Fe^{3+}, Fe^{2+} cannot participate in the formation of the ferrite phase, but instead replaces Ca^{2+} in the clinker minerals that are present. Rapid cooling of the clinker is necessary to prevent a noticeable decomposition of the C_3S phase, which tends to be enhanced under reducing conditions. Under such cooling conditions a small fraction of the clinker remains present in the form of a glass phase.

The clinker produced usually has a low alkali content, which is due to the more intensive alkali volatilization caused by the higher burning temperature employed. It also may contain increased amounts of free lime, owing to the relatively poor burnability of the raw meal. The reactivities of the alite and belite phases may be relatively low, owing to the reduced amounts of foreign ions incorporated into their crystalline lattices.

To prevent contamination of the cement with iron, the produced clinker, together with "white" gypsum, must be ground in a grinding mill with ceramic grinding elements or pebbles, rather than with the more efficient steel balls. Limited amounts of titanium oxide may also be added to the mix to be ground, to act as a brightening agent and to increase the whiteness of the final product.

The "whiteness" of the clinker, as defined by the coefficient of reflection, depends on the presence of minor constituents in the crystalline lattices of the individual clinker phases (Frigione and Zenone, 1985; Gaidzhurov *et al.*, 1994). At equal amounts of the foreign ion, this coefficient declines in the following order:

Under oxidizing conditions:

- for C_2S and C_3S: $\quad Ti^{4+} > Mn^{4+} > Fe^{3+}$
- for C_3A: $\quad\quad\quad\quad\ Fe^{3+} > Ti^{4+} > Mn^{4+}$

Under reducing conditions:

- for C_2S: $\quad\quad Mn^{2+} > Ti^{2+} > Ti^{3+}$
- for C_3S: $\quad\quad Mn^{2+} > Ti^{3+} > Fe^{2+}$
- for C_3A: $\quad\quad Ti^{3+} > Mn^{2+} > Fe^{2+\ or\ 3+}$

Another factor that affects the coefficient of reflection is the cooling rate of the clinker, which influences the composition of the formed glass phase. It has been reported that the highest whiteness is obtained if the composition of the glass phase is close to eutectic (Gaidzhurov and Zhubar, 1994). To achieve this, a fast cooling rate of the clinker is necessary.

A "super white" cement with an especially high degree of whiteness may be produced by a sol-gel route in the system $CaO-Al_2O_3-SiO_2$ containing a fluoride-based mineralizer (Chatterjee *et al.*, 1990).

Muntean (1992) produced a white Portland cement with a high belite content by burning a raw meal consisting of limestone and bentonite or tuff at 1300–1350 °C. The whiteness of the resultant cement could be significantly improved by adding 5% of calcium, magnesium or barium sulfate to the raw mix.

Owing to the presence of elevated amounts of C_3A in low-iron clinkers it is advisable to introduce the SO_3 required for set control in the form of gypsum, rather than anhydrite, as it has a faster dissolution rate and thus may be more effective in preventing flash setting.

Portland cements with very low iron contents are used mainly as "white cements" in applications where a white rather than a gray color is desired for decorative purposes. Compared with similar gray cements, concrete mixes made with white cements exhibit a shorter setting time and higher tensile and compressive strengths as well as higher slumps (Bilal, 1995). At the same time low-iron Portland cements are particularly prone to sulfate attack, owing to their high C_3A contents, and must not be used in applications where the concrete may come into contact with water containing more than insignificant quantities of sulfate ions (Bilal, 1995).

A white cement with a reduced C_3A content (that is, below 5% rather than above the 10% that exists in conventional white cements) may be produced by adding a combination of fluorspar (CaF_2) and anhydrite ($CaSO_4$) to the raw meal, to act as a mineralizer (Blanco-Varela *et al.*, 1996; Blanco-Varela, 1997; Puertas *et al.*, 1997a, 1998). This is important to overcome the particularly poor burnability of such raw meals. The clinker produced contains distinct amounts of fluorellastidite ($3C_2S.3CaSO_4.CaF_2$) and limited amounts of the phases $C_4A_3\bar{S}$ and $C_{11}A_7.CaF_2$. White cements with such low C_3A contents exhibit a high resistance to sulfate and seawater attack.

2.8   HIGH-IRON PORTLAND CEMENT

In high-iron Portland cements the content of Fe_2O_3 is increased above the levels that are common in ordinary Portland cement. Often the amount of Al_2O_3 is reduced at the same time. As a consequence of both measures the content of the calcium aluminoferrite phase is increased and that of the tricalcium aluminate phase is reduced. In clinkers with a very low alumina modulus the latter phase may even be absent completely.

The Al_2O_3/Fe_2O_3 molar ratio in the compound $4CaO.Al_2O_3.FE_2O_3$ (abbreviation C_4AF) corresponds to an alumina modulus of $AM = 0.64$. Thus no C_3A is supposed to be present in clinkers in which AM is equal to or lower than this value. However, in industrial clinkers the Al_2O_3/Fe_2O_3 molar ratio in the ferrite phase may differ from $A/F = 1.0$, and in high-iron clinkers it is usually lower than this value. Consequently some C_3A may be present in the clinkers even at an alumina modulus lower than $AM = 0.64$. In clinkers with alumina moduli even lower than that corresponding to the end-member of the $C_2A_xF_{(1-x)}$ solid solution series, the phase C_2F (dicalcium ferrite) may also be present. Historically a cement with an alumina modulus of exactly $AM = 0.64$ is called **Ferrari cement**. A cement with an extremely high iron oxide (around 8%) and very low aluminum oxide (around 2%) contents, produced from a raw meal in which clay has been replaced by iron ore, is called **iron ore cement** or **Erz cement**. A cement with a low alumina modulus ($AM = 1.0–1.5$) and simultaneously a reduced silica modulus (to $SM = 1.0–1.5$) is called **Kühl cement**.

To prepare a raw meal for the production of high-iron Portland clinker, additional Fe_2O_3 in the form of a suitable starting material, such as iron ore, has to be added to the raw meal to increase its overall Fe_2O_3 content. Such raw meals usually exhibit good burnability, as the Fe_2O_3 present also acts as a mineralizer. In the burning of clinkers with very low alumina contents a decline of the tricalcium silicate and an increase of the dicalcium silicate content has been observed at high burning temperatures (Stark *et al.*, 1997b), which was attributed to the reaction

$$C_3S + C_2(A,F) \rightarrow C_2S + C_3(A,F) \tag{2.4}$$

High-iron clinkers contain well-developed acicular or prismatic crystals of the calcium aluminoferrite phase, as well as dicalcium ferrite. The A/F molar ratio in the aluminoferrite phase may be as low as 0.11 (Stark *et al.*, 1997b).

In the hydration of high-iron Portland cement the calcium aluminate ferrite phase reacts with the calcium sulfate present to yield the AFt (ettringite) phase. As the $C/(A + F)$ ratio in $C_2(A,F)$ is lower than in the formed AFt phase, some calcium hydroxide formed in the hydration of C_3S and C_2S is also consumed in the formation of AFt. The formation of ettringite in high-iron cements progresses more slowly than in cements with Fe_2O_3 contents corresponding to ordinary or low-iron Portland cements (Chen and Odler, 1992b). In the later course of hydration, after all the free calcium sulfate has been consumed, the primary formed AFt tends to convert to AFm. However, this conversion progresses more slowly than in ordinary or low-iron cements, and significant amounts of AFt may be present even in mature high-iron Portland cement pastes (Chen and Odler, 1992b).

The AFt and AFm phases formed in the hydration of high-iron Portland cements contain significant amounts of Fe^{3+} incorporated in their crystalline lattices, substituting for Al^{3+}. However, as the A/F ratio in the formed AFt and AFm is consistently higher than in the original ferrite phase, a fraction of the original Fe^{3+} is accommodated in the form of a poorly crystallized hydrogarnet phase $[C_3(A,F)H_{6-2x}S_x]$ (Gollop and Taylor, 1995). Thus in a mature paste the Al_2O_3 is distributed between ettringite, C-S-H, a hydrotalcite phase, and monosulfate, whereas Fe_2O_3 is largely confined in the hydrogarnet phase with lesser amounts in ettringite and monosulfate.

Compared with other types of Portland cement, high-iron cement exhibits a significantly improved, yet not absolute, resistance to sulfates. Any sulfate expansion in concrete structures made from high-iron cement progresses very slowly, and becomes significant only after years of service (Kalousek *et al.*, 1972; Schoner and Wierig, 1989). This improved sulfate resistance is due mainly to the presence of reduced amounts of monosulfate in the hardened paste and thus to a reduced formation of additional AFt (ettringite), which is formed in a reaction of monosulfate and sulfate ions migrating into the paste from an outside source. There are also indications that an AFt phase with a high Fe^{3+} content, as formed in high-iron Portland cement pastes, has an altered crystal morphology and exhibits a reduced expansivity, as compared with a low-Fe^{3+} AFt phase formed in pastes made from ordinary Portland cement (Regourd *et al.*, 1980).

If a concrete surface is in contact with water that contains distinct amounts of SO_4^{2-} in combination with Na^+ or K^+ the sulfates migrate into the cement paste, and a zone is formed in which the amount of ettringite is increased and any monosulfate has disappeared. The newly formed ettringite is intimately mixed with C-S-H at or below the micrometer scale. In high-iron Portland cement pastes the amount of newly formed ettringite is low and, unlike ordinary Portland cement pastes, it causes no, or barely any, disruption of the structure. The ettringite-enriched zone is followed by one in which gypsum, formed in a reaction of SO_4^{2-} with free calcium hydroxide of the paste, is precipitated. In high-iron cement pastes any crack formation is confined to this zone, and is associated with gypsum crystallization (Gollop and Taylor, 1995).

In contact with $MgSO_4$ the reaction of the sulfate ions with the constituents of the cement paste is essentially identical. Parallel to it, however, a decalcification of the C-S-H phase occurs, resulting in a distinct strength loss of the material. Brucite $[Mg(OH)_2]$ is precipitated in the surface zone, in addition to gypsum. Because of these additional reactions, high-iron Portland cement does not exhibit any improved resistance to $MgSO_4$ attack, contrary to the action of other soluble sulfates.

The strength development and most other properties of high-iron

Portland cement differ only slightly from those of ordinary Portland cement. Its color, however, is distinctly darker.

High-iron Portland cements tend to fulfill the existing requirements for cement with increased resistance to sulfates. They are used mainly in applications in which contact of the resulting concrete with water or soils having an elevated sulfate content must be expected. High-iron clinkers are also constituents of some oil well cements.

2.9   HIGH-MgO PORTLAND CEMENT

In the course of clinker burning, magnesium oxide – originally present in the raw meal constituents – is incorporated into the crystalline lattices of the clinker minerals formed. In all clinker minerals, distinct but limited amounts of Ca^{2+} may be replaced by Mg^{2+} ions. In addition, in the ferrite phase some Al^{3+} may be substituted by Mg^{2+} in combination with Si^{4+} or Ti^{4+} ions:

$$2Al^{3+} \rightarrow Mg^{2+} + Si^{4+} \text{ (or } Ti^{4+}) \tag{2.5}$$

The amount of MgO taken up by a clinker mineral generally increases with the amount of this oxide in the raw meal up to a maximum value, above which no additional Mg^{2+} can be accommodated in the crystalline lattice. The maximum amount of MgO that may be taken up by different clinker phases varies. About 2–3% of MgO can be incorporated into the alite and belite phases: this value is somewhat higher in the tricalcium aluminate phase and even higher in the ferrite phase, which under extreme conditions may accommodate up to about 8–10% of MgO. The residual amount of MgO, not taken up by the clinker phases, remains in the clinker in the form of free magnesium oxide, called **periclase**. Table 2.2 shows the distribution of MgO in five different clinkers. The following contents of MgO were found in the individual clinker phases of three clinkers with overall MgO contents of about 7 wt%:

- alite: 1.64–1.84 wt%;
- belite: 0.58–1.00 wt%;
- aluminate phase: 0.50–0.60 wt%;
- aluminoferrite phase: 2.60–2.85 wt%.

Periclase is present in Portland clinker in the form of relatively large (up to 20 μm) cubic crystals.

Another effect of high MgO contents is the appearance of α'-C_2S (bregidite) as well as β-C_2S (larnite) in the clinker; the former phase is able to accommodate somewhat higher amounts of MgO than the latter (Luginina *et al.*, 1992).

Table 2.2 Distribution of MgO in Portland clinker (in wt. %).

Clinker	Total MgO	Periclase	MgO in alite and belite	MgO in ferrite phase	Rest MgO	CaO free
A	1.52	0.00	0.86	0.23	0.43	0.26
B	5.05	2.53	0.92	0.87	0.73	0.74
C	6.01	3.14	0.74	1.27	0.86	0.32
D	8.00	4.19	0.73	1.54	1.59	2.15
E	11.00	6.90	0.74	1.77	1.59	1.96

Source: Hu (1997)

In the hydration of cement, Mg^{2+} ions present in the crystalline lattices of clinker minerals are incorporated into the structure of the formed hydrate phases. Free MgO, present as periclase, also hydrates, yielding hexagonal magnesium hydroxide [$Mg(OH)_2$], called **brucite**:

$$MgO + H_2O \rightarrow Mg(OH)_2 \tag{2.6}$$

Magnesium oxide hydrates very slowly, and it may be 20 years or more before the process is completed (Rehsi and Garg, 1992). The reaction is topochemical in nature, which means that water migrates towards magnesium oxide, and magnesium hydroxide is formed in its place. As a consequence, the distribution of the magnesium hydroxide formed within the cement paste will reflect that of the original oxide. However, as the mass of the hydroxide is greater than that of the oxide (by 44.4%), and its density is lower ($D_{MgO} = 3590$ kg/m^3; $D_{Mg(OH)_2} = OH_2$ 2360 kg/m^3), the conversion of the oxide to hydroxide is associated with an increase of the solid volume (by 119.5%) (Chatterji, 1995). Most of the oxide becomes converted to hydroxide only when the remaining cement constituents have been largely hydrated, and a firm structure has already been developed within the paste. Thus the conversion of MgO to $Mg(OH)_2$ creates stresses within the hardened paste, which may cause expansion and eventually crack formation and disintegration of the material (Ramachandran, 1980). Because of this possibility, periclase is an undesirable constituent of Portland clinker. In this respect free magnesium oxide behaves similarly to free calcium oxide; however, the hydration of magnesium oxide progresses much more slowly, and thus any problems associated with its presence will become apparent only after years of service.

The amount of magnesium oxide that may be present in ordinary Portland cement without causing expansion, cracking, and strength loss (due to the formation of microcracks within the structure) lies at around 4–5%. Most of it will be incorporated into the crystalline lattices of clinker minerals; however, a limited amount of free magnesium oxide may also be present without causing problems. Virtually all Portland cement specifications set a limit for the total amount of MgO that is allowed to be present in the clinker or cement. Such limits vary between 2.5% and 6.5% in different specifications.

To determine experimentally whether a hardened paste made from a cement with a given amount of MgO will preserve its volume stability indefinitely, accelerated testing has to be performed. In such tests the test specimens are exposed to an autoclave treatment, and the expansion that takes place under these conditions is assessed.

In high-MgO Portland cements, the MgO content in the clinker is increased above values commonly considered safe. Such cements are produced so that starting materials with high MgO contents, such as dolomitic limestone, can be used. To obtain a cement with an acceptable performance even at these elevated MgO levels, one or a combination of several measures must be taken in the production process:

- The Fe_2O_3 content in the raw meal, and thus the amount of the ferrite phase in the clinker, must be increased. As the capacity of the ferrite phase to accommodate Mg^{2+} ions is greatest of all clinker minerals, the fraction of MgO incorporated into the crystalline lattices of clinker minerals is thus increased. At the same time the amount of free periclase (the only form of MgO that may cause problems) is reduced.
- In the burning of clinker the rate of cooling must be speeded up as much as possible. Cooling of the clinker with water, rather than in air, has been found to be particularly effective (Sharma *et al.*, 1992). Under these conditions the average size of periclase crystals that crystallize from the melt in the course of cooling is reduced, while their number is increased. In the hydration of such a cement the stresses caused by the conversion of periclase to brucite are more evenly distributed within the hardened paste, and the formation of cracks due to the presence of large MgO crystals may be prevented or reduced. Some reduction of the size of the formed periclase crystals may also be achieved by finer grinding of the raw meal.
- Excessive expansion and crack formation of high-MgO cement may also be prevented or reduced by grinding the high-MgO clinker together with fly ash as well as gypsum (Ali and Mullick, 1998). Less effective are some natural pozzolanic materials, limestone or granulated blast furnace slag (Sharma *et al.*, 1992; Hu, 1997). The amount of these materials that has to be used to achieve effective protection against cracking ranges up to 30%. Table 2.3 shows the effect of added fly ash and granulated blast furnace slag on the expansion of cements with elevated MgO contents.

The mechanism by which the added pozzolanic material prevents excessive expansion is not obvious. It has been suggested that the silica present

Table 2.3 Effect of fly ash and granulated blastfurnace slag on expansion of cements with high MgO contents.

Clinker	MgO total in clinker (wt%)	MgO free in clinker (wt%)	MgO free in cement (wt%)	Fly ash (wt%)	Slag (wt%)	Expansion 4h 100°C (%)
1	8.00	4.14	3.97	0	–	0.131 D
			3.35	13	–	0.064 D
			2.94	25	–	0.080
			2.32	40	–	0.066
			3.35	–	13	0.070 D
			2.73	–	30	0.039
2	11.00	6.90	6.62	0	–	0.119 D
			5.73	13	–	0.110 D
			4.90	25	–	0.084 D
			3.86	40	–	0.064 D
			5.73	–	13	0.087 D
			4.55	–	30	0.044

D = disintegration of test specimen

Source: Hu (1997)

in the additive reacts with periclase to yield magnesium silicates; however, the presence of such phases could not be confirmed experimentally (Ananenko, 1986; Rehsi and Garg, 1992; Sharma *et al.*, 1992). It has also been suggested that the increased amount of C-S-H formed in the presence of the pozzolanic additive makes the cement paste firmer and hence better able to resist expansive forces (Sharma *et al.*, 1992). There are indications, however, that the main reason for the beneficial effect of fly ash is a reduction of the amount of periclase that becomes hydrated under these conditions (Ananenko, 1986). Also, the overall amount of free MgO in blended cements made from high-MgO clinker is lower than in pure Portland cement made from the same clinker.

The amount of MgO in clinker that may be tolerated if the above measures are effectively applied may reach 10% or even 12%. This makes it possible to utilize starting materials high in MgO that are not suitable for conventional cement production (Hu, 1997).

The hydration of high-MgO Portland cements does not differ significantly from that of cements with "normal" MgO contents. Investigations on samples hydrated for 20 years revealed that even after this time the conversion of MgO to $Mg(OH)_2$ was not fully completed, either in the presence or in the absence of fly ash (Rehsi and Garg, 1985, 1992). The hardened pastes also contained significant amounts of ettringite, due apparently to the high iron content in such cements, as under these conditions a complete conversion of primary formed ettringite to monosulfate is prevented. The amount of portlandite in pastes with added fly ash is reduced, owing to the pozzolanic action of this constituent.

The physico-mechanical properties of high-MgO Portland cements, in which appropriate measures have been undertaken to eliminate the detrimental effect of the periclase that is present, do not differ significantly from those of cements with "normal" MgO contents. Such cements may be used in the same applications without any limitations. According to the European standard ENV197–1, cements with added amounts of fly ash exceeding 6% qualify as Portland fly ash cements rather than as Portland cements.

2.10   LOW-ALKALI PORTLAND CEMENT

The alkali oxides Na_2O and K_2O, as constituents of clinker phases, are regular constituents of Portland cement, and may be considered acceptable in amounts up to about 2%. In general, increased amounts of alkali tend to increase the early strength and decrease the late strength of the cement. The setting time is shortened, especially in the presence of high amounts of potassium sulfate in cement, owing to the formation of syngenite $[CaK_2(SO_4)_2.H_2O]$. The consistency of the resultant concrete becomes stiffer (Gouda, 1986). Also, the amount of free lime in the clinker increases with increasing amount of alkali in the raw meal (Slota and Lewandowska-Kanas, 1986).

Alkali metals are present in clinker preferentially in the form of sulfates. The rest are incorporated into the crystalline lattices of clinker minerals. Sodium tends to be preferentially taken up by the calcium aluminate phase, whereas potassium is mainly incorporated into the crystalline lattice of dicalcium silicate, which becomes stabilized in its α' form. In either case, in the course of hydration Na^+ and K^+ ions enter the liquid phase together with the balancing anions, which eventually become incorporated into the hydration products, being replaced with equivalent quantities of OH^- ions:

$$K_2SO_4 + Ca(OH)_2 + 2H_2O \rightarrow CaSO_4.2H_2O + 2K^+ + 2OH^- \quad (2.7)$$

$$3CaSO_4.2H_2O + 3CaO.Al_2O_3 + 26H_2O \rightarrow Ca_6Al_2S_3O_{18}.32H_2O$$
$$\text{(ettringite)} \quad (2.8)$$

The hydroxides present in the liquid phase, which constitutes the pore solution, are mainly responsible for its high pH. In steel-reinforced concrete the high pH of the pore solution may be considered beneficial, as it protects the reinforcement from corrosion.

In concrete mixes made out of Portland cement with a high alkali content and an aggregate that contains SiO_2 in a reactive form (such as opal, flint, chert, some silicate minerals, volcanic glasses, and even sufficiently strained or microcrystalline quartz) an interaction between the alkaline hydroxide present in the pore solution and the SiO_2 of the aggregate may take place, resulting in an expansion of the material and eventual cracking. Such a process is called **alkali-silica (or silicate) reaction**. Alkaline hydroxides of the pore solution may also react with some dolomitic rocks containing – in addition to dolomite – clay minerals, and the resulting **dedolomitization reaction** may also cause expansion and cracking. A necessary condition for such reactions to occur is a sufficiently high hydroxide content in the pore solution caused by a high alkali content of the clinker. (For more details see Chapter 22.)

To avoid expansion and concrete damage even in the presence of alkali-sensitive aggregates the content of equivalent NaO (defined as $Na_2O_e = Na_2O + 0.66K_2O$) in the concrete mix should not exceed 4 or even 3 kg/m^3. To meet this criterion a special Portland cement with particularly low alkali content must be employed. In such a cement the equivalent Na_2O content should not exceed 0.6%.

Low-alkali Portland cement may be produced by using starting materials with sufficiently low alkali contents; however, in most instances such materials are not available. In such situations some of the alkali metals present in the starting materials must be removed in the course of manufacture.

In the course of clinker burning a part of the alkalis present in the material become volatilized in the sintering zone, in which temperatures of the raw mix of around 1450 °C are reached. The volatility of alkalis is especially high if they can volatize in the form of sulfates: of those, potassium sulfate exhibits a significantly greater volatility than sodium sulfate. The volatilized constituents are swept back with the hot gases, and if the kiln is equipped with a preheater, most of the volatilized sulfates become reabsorbed by the cold raw meal introduced into the system, and reenter the kiln. The fraction of alkali metals that leave the system with the hot gases as gas or dust is normally negligible, and thus the alkali content of the produced clinker corresponds roughly to that of the raw meal.

If a reduction of the alkali content in the clinker is required, part of the hot gases carrying the alkaline compounds must be diverted though a bypass as they are leaving the rotary kiln and before they enter the preheater or precalciner. The dust of the bypass gases, rich in alkalis, is then filtered out and disposed of. The fraction of hot gases that must be diverted through the bypass system depends on the quantities of alkalis that have to be removed: in most instances a diversion of 10–20% of the gases through the bypass is sufficient to reduce the alkali content of the clinker to an acceptable level. The process is associated with a loss of thermal energy, as heat cannot be recovered from the hot gases that have been diverted.

To increase the amounts of alkalis undergoing volatilization in the sintering zone, and thus to reduce the alkali content of the produced clinker even further, small amounts of calcium chloride may be added to the raw meal (Puscasu and Dimitrescu, 1994). Under these conditions additional amounts of alkalis escape in the form of potassium and sodium chloride, which are substances with an even greater tendency to volatilize than the corresponding sulfates.

To reduce the adverse effect of alkalis on the free lime content and ultimate strength, it has been recommended that gypsum should be added to the raw meal in amounts of up to 3% (Slota and Lewandowska-Kanas, 1986). Under these conditions a significant fraction of the alkalis becomes bound in the form of sulfates. In this way the alite and belite crystals become better developed, and the C_3A phase tends to crystallize in its cubic form.

The physico-mechanical properties of low-alkali Portland cements do not differ significantly from corresponding cements with "normal" alkali contents. Their marketing and use make sense only in regions with deposits of rocks that may be susceptible to alkali-silica or dedolomitization reactions, if employed as concrete aggregates.

2.11   MINERALIZED PORTLAND CEMENT

In the production of mineralized Portland cements small amounts of "minor" constituents are added to the raw meal to enhance the manufacturing process and/or the properties of the produced cement.

The constituent most widely used is calcium fluoride in the form of fluorspar. If added in amounts of about 0.2–1.0%, calcium fluoride accelerates the clinkering process significantly and makes it possible either to increase the rate of clinker burning or to reduce the burning temperature (Klemm *et al.*, 1979; Odler and Abdul-Maula, 1980; Blanco-Varela *et al.*, 1984; Surana and Jeshi, 1990). Data on the effect of calcium fluoride on cement strength are not uniform, and both an increase and a decline of strength due to CaF_2 additions to the raw meal have been reported. At higher fluorine contents the phase $11CaO.7Al_2O_3.CaF_2$ may also be formed in the clinker. Cements containing this phase are discussed in Chapter 5.

Calcium sulfate added to the raw meal in amounts of up to about 2–3% SO_3 (per weight of clinker) makes it possible to reduce the burning temperature to below 1350 °C (Odler and Zhang, 1996, 1997). The resulting clinker is highly porous and easy to grind. Most of the introduced SO_3 is incorporated in the crystalline lattices of the clinker minerals, especially dicalcium silicate, resulting in a significant increase of reactivity of this phase. A fraction of the added calcium sulfate may also react with the alkali metals of the raw meal, yielding calcium langbeinite [$Ca_2K_2(SO_4)_3$] (Borgholm *et al.*, 1995; Michaud and Sunderman, 1997). Because of its good solubility, this phase may serve as a source of the sulfate ions needed for ettringite formation, and it ensures proper setting of the resulting cement, even if it is produced by grinding the clinker alone or with significantly reduced amounts of calcium sulfate. Higher amounts of sulfates in the raw meal slow down or even prevent the formation of the C_3S phase (Strunge, 1985; Saada *et al.*, 1988; Cheong *et al.*, 1992, 1994, 1997; Su *et al.*, 1992).

The formation of the clinker may be particularly effectively enhanced and the reactivity of the resulting cement increased by simultaneously adding to the raw meal calcium sulfate and calcium fluoride (Tong and Lin, 1986; Gimenez *et al.*, 1991; Borgholm *et al.*, 1995; Blanco-Varela *et al.*, 1996; Odler and Zhang, 1996; Blanco-Varela, 1997; Chen and Lou, 1997; Puertas *et al.*, 1997a, 1997b, 1998). In such mixes the formation of C_3S takes place at around 1120 °C, owing to the decomposition of the intermediate phase $3C_3S.CaF_2$. At higher temperatures the intermediate phase $3C_3S.3CaSO_4.CaF_2$ (fluorellastidite) may be formed (Gimenez *et al.*, 1991; Borgholm *et al.*, 1995). This phase prevents the formation of C_2S, which results in enhanced C_3S formation (Chen and Lou, 1997).

2.12   HIGH SPECIFIC SURFACE AREA PORTLAND CEMENT

The fineness of cement may be conveniently characterized by the Blaine method, even though the values obtained by this procedure are systematically lower than those found by the more accurate, but also more complicated, BET method. The specific surface area of ordinary Portland cement typically ranges between 280 and 350 m^2/kg (Blaine), whereas high early strength cement may be ground up to a specific surface area of 450–500 m^2/kg (Blaine). This higher surface area contributes to an accelerated hydration of the cement, but also increases its water requirement.

The various constituents of Portland clinker differ in their grindability. In general, the grindability declines in the order calcium sulfate > alite > aluminate > ferrite > belite. As a consequence, different cement particle size fractions may vary in their phase composition.

The highest surface areas attainable by mechanical grinding lie at around 1200 m^2/kg (Blaine) (Yoshida and Okabayashi, 1992), but it is not practical to produce cements with such an extremely high fineness. Apart from the technical difficulties and the high costs associated with comminution to such a high fineness, such cements would exhibit extremely high water requirements and excessive reactivity, and would not be suitable for practical use. To suppress the excessive reactivity of cements with extremely high fineness it has been suggested that a retarder, such as sodium citrate, should be added to the cement/concrete mix (Yoshida and Okabayashi, 1992). The amount of retarder that is required will depend on the composition of the clinker employed and the fineness of the ground material. It has also been suggested that the cement should be allowed to undergo a partial prehydration, and that it should be used in combination with a superplasticizer (Fujita *et al.*, 1993). In this way compressive strengths of concrete of up to 90 MPa after one day and well over 100 MPa after 28 days may be attained.

In addition to the overall fineness, the particle size distribution also has a significant effect on cement properties, and these may be controlled – to a degree – by classification. In general, the fluidity of the cement paste is reduced both when there are too many fine particles and when there are too few. An optimum particle size distribution exists at which the fluidity of the produced mix is optimal (Uchida *et al.*, 1992). The optimum particle size distribution depends on many factors, among which a crucial role is played by the quantity of very fine particles: that is, those with diameters of about 10 μm or less. For more detail on the subject the reader should turn to the original literature (Frigione and Mara, 1976; Sumner *et al.*, 1986).

High surface area cements possessing good fluidity and favorable strength development may be produced by removing fine clinker particles from the original, finely ground cement and substituting them with finely ground silica, limestone, fly ash, or blast furnace slag (Uchida *et al.*, 1992). Nehdi *et al.* (1998) recommended replacing the fine cement fraction with silica fume in combination with a superplasticizer, to obtain high surface area mixes of acceptable fluidity.

The particle size distribution also determines the packing density of the cement particles. Within reasonable limits a wider distribution affects this parameter favorably. However, the degree of hydration increases with a more homogeneous distribution (Wang and Zhang, 1997). As to the effect on strength, the influence of packing density is greater at shorter hydration times, whereas the degree of hydration becomes more important at later ages.

2.13 LOW SPECIFIC SURFACE AREA PORTLAND CEMENT

Low specific surface area Portland cement is ground to a specific surface area (Blaine) below 300 m^2/kg or even below 250 m^2/kg (Blaine) with reduced amounts of calcium sulfate. It is not intended for general use, but for special applications in which the presence of fine particles is not desired and their amount should be kept as low as possible. Such applications include:

- *Production of fiber cement*: In the production of fiber cements (such as asbestos cement) by the Hatschek process a water suspension of cement and fibers is filtered through a filter cloth, to remove the excessive water and to produce a dense cement/fiber conglomerate, which – after setting and hardening – yields the final product. The cement employed must be coarsely ground, to retain as much as possible of it at the filter and to minimize the fraction of cement that passes the filter cloth together with the excess water.

- *Production of concrete elements by a centrifugation process*: Some prefabricated concrete products, such as concrete pipes or utility poles, may be produced by centrifuging a fresh concrete mix made with excessive amounts of water. In the course of centrifuging, a fraction of the water separates from the rest of the mix, whereas the residual material becomes compacted. To guarantee effective segregation and to minimize the amount of cement that remains suspended in the separated water, the cement must be coarsely ground.
- *Underwater concreting*: In underwater concreting the cement used must be coarsely ground to minimize the amount that becomes lost in the surrounding water.

2.14 LIMESTONE-MODIFIED PORTLAND CEMENT

Limited amounts of limestone, interground with Portland clinker and calcium sulfate, may beneficially affect some properties of the resultant cement. This effect is achieved mainly by physical action, as calcite ($CaCO_3$) – the main or sole constituent of limestone – participates only marginally in the hydration process.

The limestone used must have a high $CaCO_3$ content, and must not contain clay minerals, to avoid an adverse effect on the rheology and frost resistance of the resulting cement.

If ground together with Portland clinker, limestone will predominate in the fine fractions, as it is significantly softer than the clinker. In a fresh paste made from such a cement, the fine limestone particles will fill the spaces between the coarser particles of clinker, and consequently the amount of water needed to attain a given consistency will decline, up to an optimum limestone addition. This amount of limestone may vary in different cements, up to as much as 25 wt%.

A limited substitution of Portland clinker by limestone may result in a moderate increase of strength of the resultant cement (at a constant consistency of the fresh mix). This is brought about mainly by the water-reducing effect of limestone.

The added limestone also exhibits an accelerating effect on the hydration process, and mainly on the hydration of alite, as its particles can act as nuclei for the crystallization of portlandite. This effect may also positively affect the short-term strength of the cement.

In cements with a sufficiently high tricalcium aluminate content the calcium carbonate of limestone may react with this phase, yielding calcium carboaluminate hydrate ($C_4A\bar{C}H_{11}$). Up to about 6 wt% of calcium carbonate may be consumed in this reaction within 28 days (Ingram and Daugherty,

1992). Whether the formation of calcium carboaluminate also contributes to a strength increase of the hardened cement paste remains uncertain.

At smaller limestone additions the texture of the hardened paste does not differ significantly from that of plain Portland cement. At higher limestone additions, however, limestone may appear in the cement matrix in the form of crystalline particles evenly distributed within the paste.

As tricalcium aluminate may react to a limited extent with calcium carbonate, the limestone that is present may partially replace the gypsum present in the cement to control setting. As, however, such substitution may adversely affect the mechanical properties of the hardened paste, it should be avoided, except in cases of a limited availability of gypsum.

In assessing the durability of concrete made with limestone-modified Portland cement one has to bear in mind that – at a constant water/cement ratio – the porosity of the paste, and along with it its permeability, increases with increasing limestone content in cement. Nevertheless, adequate durability may be preserved, as long as the amount of added limestone is not excessive (Tezuka *et al.*, 1992).

The European specification ENV197–1 permits an addition of maximum 5 wt% of limestone in Portland cement. A Portland limestone cement may contain up to 20 wt% (notation II/A-L) or even up to 35 wt% (notation II/B-L) of limestone. No addition of limestone is permitted in the American standards (C150–95 and C595M-95).

2.15 PORTLAND CEMENTS MODIFIED WITH CHEMICAL AGENTS

Some chemical agents, introduced in small amounts, can facilitate the production process, or modify the properties of the resulting cement or those of the fresh concrete mix and hardened concrete. They include

- agents facilitating cement grinding;
- agents that improve handling and storage of the cement;
- modifiers of fresh concrete mix rheology;
- hydrophobic agents;
- accelerators and retarders of setting and hardening;
- air-entraining agents;
- expansive additives;
- agents inhibiting corrosion of the reinforcement in steel-reinforced concrete;
- agents that reduce the permeability of the hardened concrete;
- coloring agents.

In most instances these agents are intermixed with the fresh concrete mix as chemical admixtures; however, it is also possible to interblend or intergrind some of them with the cement in the course of its manufacture.

Grinding aids are agents employed to intensify the cement-grinding process so that a high fineness of the material can be achieved. They are typically volatile organic compounds, whose vapor, in the course of grinding, is adsorbed on the newly created surface of the cement particles. In this way they prevent agglomeration of the particles and enhance the progress of comminution. Triethanol amine and ethylene glycol are examples of such agents. In addition to their effect on grinding, grinding aids also prevent "packing" of the cement, which may take place during long-term storage under warm and humid climatic conditions, and may impede the free flow and handling of the cement.

In a hydrophobic Portland cement (Bensted, 1992, 1993) the cement particles are coated with a suitable **hydrophobic agent**, such as oleic, lauric or stearic acid or pentachlorophenol, in amounts of about 0.2–0.3%. This is achieved by grinding together Portland clinker, gypsum and the agent. Such a measure gives the cement a large degree of protection against deterioration upon storage. Hydrophobic cement is favored in situations where poor storage conditions, such as high air humidity, may impede cement quality. Other benefits of hydrophobic Portland cement include improved plasticity of the fresh concrete mix and water repellency of the hardened concrete. Lower grades of hydrophobic Portland cement are designated water-repellent and waterproofed Portland cements. A more extensive mixing of the concrete mix is required if hydrophobic rather than ordinary Portland cement is employed. The setting time is usually extended significantly.

An air-entrained Portland cement contains small amounts (between about 0.02% and 0.05%) of an **air-entraining agent**. These are organic compounds whose molecules consist of a long non-polar chain coupled with a polar group on one end. Because of this structural feature, upon dissolving in water they tend to concentrate at the liquid/air interface, with the polar group in the liquid phase and the non-polar chain in air. At the same time cations associated with the molecule dissociate and enter the liquid phase. In the presence of such an "anionic surfactant," air-filled, electrically charged, spherical pores, with diameters of 10–250 μm, are produced in the cement paste in the course of mixing. The presence of such pores improves the workability of the resultant fresh concrete mix and increases significantly the frost resistance of the hardened concrete. Cements with an added air-entraining agent are used mainly in road construction in regions where the hardened concrete may be exposed to freezing temperatures.

In a superplasticized cement a suitable sulfonated naphthalene-based **superplasticizer** is interground with the binder (Rosetti *et al.*, 1992; Bouzoubaa *et al.*, 1998). Under these conditions the rheological properties of the produced concrete are more uniform than in cases in which the superplasticizer is added to the concrete mix in the course of mixing. The added superplasticizer also acts as a grinding aid.

2.16 GYPSUM-FREE PORTLAND CEMENTS

Most Portland clinkers ground to cement fineness do not yield pastes of plastic consistency, if mixed with the amounts of water commonly used in producing cement pastes. Instead, an instant setting, accompanied by intensive heat development, occurs. The resultant product is non-flowable, and exhibits limited strength immediately after setting. The phenomenon is called **quick** or **flash setting**. It is caused by an excessively fast initial hydration of the C_3A phase of the clinker, and by the formation of platelet-like crystals of the C_4AH_x phase (an AFm phase) at the cement grain surface, which obstruct the free flow of the paste. The subsequent strength development of such pastes is usually reduced, owing to incomplete compaction and to a weakening of the texture of the paste by the platelets of the calcium aluminate hydrate. The phenomenon is more pronounced in clinkers with distinct amount of tricalcium aluminate than in those in which this phase is absent or present only in reduced amounts. Cements of this kind cannot be employed for the production of plastic concrete or mortar mixes; however, they may be employed in applications in which instant setting is required. A spraying technique has to be applied in such instances, to attain an acceptable compaction of the mix.

To obtain a cement that exhibits **normal setting**, calcium sulfate in the form of gypsum ($CaSO_4.2H_2O$) and/or anhydrite ($CaSO_4$), in amounts corresponding to about 1.5–3.5% SO_3, has to be interground with the clinker in the production of Portland cement. After mixing with water, such a cement yields a paste of plastic consistency, which sets within a few hours. The action of calcium sulfate may be explained by the formation of ettringite ($C_3A.3C\bar{S}.32H$), instead of C_4AH_x, after mixing the cement with water. The improvement of the rheology of the system under these conditions is due to a slow-down of the overall hydration rate and to a more favorable morphology of the reaction product of C_3A hydration: this precipitates on the cement grain surface in the form of a layer of fine ettringite crystals, which does not interfere with the flow of the cement suspension.

2.16.1 Low-porosity cement

It has long been recognized that, even in the absence of calcium sulfate, flash setting of cement may be prevented and a cement suspension of flowing consistency obtained by combining finely ground Portland clinker with an alkali metal carbonate or hydrogen carbonate (Na_2CO_3, K_2CO_3, $NaHCO_3$, $KHCO_3$), and with calcium or sodium lignosulfonate or a sulfonated lignin (Brunauer *et al.*, 1972–1973; Hanna, 1977; Hanna and Taha, 1977; Diamond and Gomez-Toledo, 1978; Odler *et al.*, 1978; Collepardi *et al.*, 1980, 1982; Skvara *et al.*, 1981; Paori *et al.*, 1982; Skvara and Rybinova, 1985; Kokanyev and Kuznetsova, 1992; Hrazdira, 1994;

Satava and Tyle, 1994; Bojadjieva and Glavchev, 1995; Opoczky and Papo, 1997). The amounts of additive that have to be introduced to obtain optimum rheology vary in different clinkers, but range typically between about 0.25% and 1.0% of each carbonate and lignosulfonate. It is remarkable that flash setting may be prevented and adequately flowable pastes may be obtained *only* by a combination of carbonate and lignosulfonate, whereas either of them alone is ineffective in this respect. The two additives may be interground with the clinker, interblended with the preground material, or added to the mixing water; however, the way in which they are introduced will affect their performance, especially their effect on setting time. In general, the setting time is extended if the two additives are dissolved in the mixing water prior to mixing with the ground clinker.

Some investigators have suggested grinding the clinker to a high fineness using suitable grinding aids (Brunauer *et al.*, 1972–1973; Sanitsky *et al.*, 1989; Sanitsky and Sobol, 1992). Such a measure affects the strength development positively, yet has little effect on the rheology of the fresh mix. For optimum flow, the quantity of the additives has to be increased under these conditions.

With appropriate amounts of added carbonate and lignosulfonate the flow properties of the paste are significantly improved, compared with pastes made from ordinary cement based on the same clinker (at equal fineness and water/cement ratio). Freely flowing suspensions may be produced at water/cement ratios as low as $w/c = 0.25$ or less, if mixed intensively enough. The rheological characteristics of such "gypsum-free" cement pastes, as assessed visually, are not identical with those of ordinary Portland cement. They possess peculiar plastic-elastic properties, which cannot be readily characterized in terms of normal suspension behavior. They exhibit a distinct thixotropy, and – if not agitated – appear rather sticky and rubber-like. More diluted suspensions – if studied with standard equipment – can be approximately described as dilatant, or Bingham-like, with a very low plastic yield stress (Diamond and Gomez-Toledo, 1978; Odler *et al.*, 1978). It has been reported that the rheological properties of concrete mixes made with "low-porosity" cements may be further improved by adding a superplasticizer to the system (Peukert, 1997).

The time for which the suspension preserves its flowability may vary between a few minutes and several hours, and may be controlled by the amount of added additives (Hanna, 1977; Odler *et al.*, 1978; Satava and Tyle, 1994). In general, the setting time is extended with an increasing amount of lignosulfonate, whereas the amount of carbonate has little effect on this behavior. Unlike the behavior of ordinary Portland cement pastes, the transition from the plastic to the set state occurs very suddenly.

At higher lignosulfate additions the paste exhibits only a very low strength immediately after setting, and – at ordinary temperatures – no significant strength increase takes place for a period lasting typically several hours or even a few days, before a rapid strength development gets under way. The length of this "induction period" may vary in different clinkers, and generally is extended with increasing lignosulfonate addition. However, it may be significantly shortened by a moderate increase of temperature (Kokanyev and Kuznetsova, 1992).

Instead of lignosulfonates, other sulfonated polyelectrolytes – in combination with an inorganic salt (usually Na_2CO_3) – may also be used to control the rheology of the mix. It has also been reported that the retardation of strength development is less pronounced if a sulfonated polyphenolate (a sulfonated phenol-formaldehyde polycondensate) is employed instead of lignosulfonate (Skvara and Rybinova, 1985; Satava and Tyle, 1994; Skvara, 1995). The ultimate strength of gypsum-free, carbonate + lignosulfonate modified Portland cement pastes is generally higher than that of ordinary Portland cement pastes of equal flow properties, yet lower if compared on an equal w/c basis.

For optimum performance the clinker that is used should contain at least 55% C_3S and not more than 6% C_3A (Kokanyev and Kuznetsova, 1992). The overall calcium silicate content should be over 80% (Peukert, 1997).

The mechanism of the set-controlling and liquefying action of the carbonate + lignosulfonate combination is not obvious. It is assumed that it is brought about by a retardation of the initial hydration and an effective dispersion of the cement particles (Odler *et al.*, 1978; Collepardi *et al.*, 1980, 1982). In the hydration process hexagonal calcium aluminate hydrates, rather than ettringite, are formed as the hydration of C_3A gets under way, which tend to convert into the cubic C_3AH_6 in the later stage of hydration (Skvara and Rybinova, 1985; Sanitsky and Sobol, 1992; Skvara, 1995). Some carboaluminates are probably also formed. The hydration of tricalcium silicate becomes distinctly retarded, and along with it the formation of calcium hydroxide (Odler *et al.*, 1978; Skvara and Rybinova, 1985; Skvara, 1986; Sanitsky and Sobol, 1992). The formed C-S-H phase has a very high C/S ratio: 2.7 ± 0.1, as determined by EDAX analysis (Skvara, 1986). Its degree of polycondensation is higher than in ordinary Portland cement pastes (Opoczky and Papo, 1997). Large portlandite crystals, common in hardened ordinary Portland cement pastes, are virtually absent, suggesting a high degree of dispersion of this phase (Skvara *et al.*, 1981; Skvara and Rybinova, 1985; Skvara, 1986, 1995). Owing to the very low water/cement ratios that may be used in making pastes containing a combination of carbonate and lignosulfonate, the hardened pastes typically exhibit very low porosities: hence the designation "low-porosity cement."

The addition of an alkali carbonate + lignosulfonate combination to ordinary Portland cement containing calcium sulfate also causes a liquefying effect. This effect is greater than that of the lignosulfate alone, yet less pronounced than that existing in sulfate-free cements (Odler *et al.*, 1978;

Skvara, 1996). Under these condition an initial liquefaction is followed by a rapid increase of paste viscosity.

Low-porosity cements may be considered for high-strength concretes, or systems in which a very low porosity of the hardened concrete/mortar is essential (for applications in highly corrosive environments, for example, or for radioactive or toxic waste disposal). It has also been reported that these cements may be successfully employed and exhibit an acceptable strength development even at subfreezing temperatures (Skvara *et al.*, 1981; Sanitsky *et al.*, 1989; Sanitsky and Sobol, 1992). This is due mainly to the distinctly lower freezing temperature of the liquid phase existing in such cement pastes.

In handling mixes made with low-porosity cement, their particular rheology and setting behavior must be taken into consideration. So far, low-porosity cements have been produced and applied on an industrial scale only in the former Czechoslovakia (Skvara *et al.*, 1981).

2.17   SPECIAL APPROACHES IN PORTLAND CEMENT MANUFACTURE

In conventional production of Portland clinker a preground and adequately homogenized blend of starting materials with the desired oxide composition, commonly called **raw meal**, is fired in a rotary kiln at a maximum temperature of about 1450–1500 °C. Coal or – less often – mineral oil or natural gas serve as sources of thermal energy. The non-volatile inorganic constituents of the fuel combine with the raw meal and eventually become constituents of the resulting clinker. In most production units a preheater is installed in front of the rotary kiln in which the raw meal is preheated by the hot gases leaving the kiln, for better fuel economy. In some units a precalciner is also installed between the rotary kiln and preheater, in which the $CaCO_3$ present in the raw meal is decomposed to CaO + CO_2 before it enters the kiln. The hot clinker leaving the kiln is cooled down to acceptable temperatures, and finally is ground, together with added gypsum or anhydrite, to the desired fineness.

In the production of the raw meal some non-conventional starting materials, instead of or in addition to limestone, clay, marl, or quartz, are sometimes used. They may include fly ash (Odler and Zhang, 1997), blast furnace slag (Blanco-Varela *et al.*, 1988), and a variety of other industrial by-products or wastes. Out of these, some may also contain variable amounts of organic residues that supply part of the required thermal energy. Additional amounts of energy may be saved if the starting mix contains distinct amounts of decarbonated CaO, as is the case in some industrial slags or CaO-rich ashes.

It is possible to intensify the clinker mineral formation process by adding to the raw meal small amounts of a mineralizer/fluxing agent

(Gouda, 1980; Odler and Abdul-Maula, 1980a, 1980b; Odler, 1991; Surana and Jeshi, 1990). Of these, calcium fluoride and other fluorine compounds are the most effective and the most widely used (Klemm *et al.*, 1979; Gouda, 1980; Odler and Abdul-Maula, 1980a, 1980b; Blanco-Varela *et al.*, 1984; Surana and Jeshi, 1990; Ayed *et al.*, 1992). Other known mineralizers/fluxes include ZnO (Tsuboi *et al.*, 1972; Knöfel, 1978; Gouda, 1980; Odler and Abdul-Maula, 1980b; Baeker *et al.*, 1983; Kakali and Parissakis, 1995), MgO (Tsuboi, 1972; Gouda, 1980), and CuO (Odler and Abdul-Maula, 1980b; Kakali *et al.*, 1996). A mineralizer/flux considered especially effective is a combination of CaF_2 and $CaSO_4$ (Tong and Lin, 1986; Borgholm *et al.*, 1995; Blanco-Varela *et al.*, 1996; Odler and Zhang, 1996), or even a combination of CaF_2, $CaSO_4$, and ZnO (Huang *et al.*, 1992; Lu *et al.*, 1992). It has to be stressed, however, that mineralizers/fluxing agents added to the raw meal not only accelerate the formation of Portland clinker, but also influence its structure and the properties of the resultant cement (Odler and Abdul-Maula, 1980a, 1980b; Page *et al.*, 1986; Borgholm *et al.*, 1995; Odler and Zhang, 1996).

To intensify the formation of clinker without the use of mineralizers/fluxes it has been suggested that extremely high heating rates should be used in the burning process, up to 800 °C/min (Hu, 1992; Huang *et al.*, 1992; Ji and Xu, 1992; Lu *et al.*, 1992), making it possible to lower the maximum burning temperature by up to 200 °C. It remains questionable, however, whether such heating rates may be realized under industrial conditions, and whether they would result in a reduction of fuel consumption or other economic benefits.

Barbanyagre *et al.* (1992) suggested producing Portland clinker in a two-step firing process. In the first step – preliminary firing – only a part of the oxides participating in clinker formation is brought to reaction; the rest of the raw meal components are introduced only prior to the main firing stage. In one of the variants that were found meaningful, first a blend with a very low CaO content is fired to below 1200 °C, melting completely at this temperature. Subsequently, the rest of the CaO is allowed to combine with this melt. In a modification of this approach a raw mix with a high $CaCO_3$ content is fed into the cold end of the rotary kiln, while a high-silicate, low-basicity mix is blown into the combustion zone of the kiln from its lower end (Barbanyagre, 1997). The practicability and economy of such a procedure must be questioned, however.

In another approach it has been suggested that the raw mix should be compacted mechanically as it leaves the preheater and enters the rotary kiln (Sulimenko and Albats, 1997).

In a melted Portland cement the raw meal is heated to temperatures of complete melting – that is, to 1700–1800 °C – prior to air or water cooling (Khadilkar *et al.*, 1992). The microstructural features of clinkers produced in this way differ distinctly from those of clinkers produced by conventional technology. Owing to a rather high free lime content in the clinker, aeration (exposure to humid air) of the resultant cement may be needed to eliminate the negative effect of this constituent on volume stability and strength. The burning temperature in the production of melted cement may be reduced by reducing the lime saturation factor, by increasing the iron content in the raw meal, and by adding fluxes – such as CaF_2 – to the mix. The produced clinker is inhomogeneous, with large nests of alite, belite or liquid phase present (Khadilkar *et al.*, 1997). Melted Portland cement clinkers are difficult to grind. There is no technical or economic justification for an industrial production of melted Portland cement.

On a laboratory or pilot scale, Portland clinker has been produced: in a fluidized bed; by irradiation of the raw meal with accelerated electrons (Yegorov *et al.*, 1982; Abramson *et al.*, 1992; Handoo *et al.*, 1992); by microwave processing (Yi *et al.*, 1996; Fang *et al.*, 1997); in a sol-gel process (Chatterjee *et al.*, 1990; Varadarajan *et al.*, 1992); and by a variety of other approaches. In all instances the chosen process influenced the structure of the clinker and the properties of the resultant cement.

Technologies have been developed in which Portland clinker is produced simultaneously with another product (such as pig iron, alumina, ammonium sulfate, sulfuric acid, phosphates, and potassium salts), but most of these technologies are not used commercially at present. In the combined Portland cement + sulfuric acid technology (Müller-Kühne process) a raw meal rich in sulfur is burnt in a rotary kiln. The SO_2 formed, which escapes with the flue gas, is oxidized catalytically to SO_3 and used for sulfuric acid production, whereas the clinker leaving the kiln is ground to Portland cement. The technology is especially suited for processing and recycling sulfur-rich industrial waste products such as waste sulfuric acid, acid sludges, and SO_3-rich ashes. A plant using this technology has been operational in the former East Germany (Anonymous, 1996).

As the last step in Portland cement manufacture, Portland clinker, in combination with limited amounts of gypsum and/or anhydrite, has to be ground to a fine powder. A ball mill is the standard equipment used for this purpose. It was reported recently (Goldstein, 1997) that the power consumption in the grinding process may be reduced if – in place of a single clinker – a blend of two clinkers is used, of which one has a higher and the other a lower lime saturation factor.

In recent years high-pressure roller mills have been introduced into practical use in increasing numbers, mainly because of their lower energy consumption. A cement ground in a high-pressure roller mill differs distinctly from one ground in a ball mill. In general, its water requirement is increased and the setting time is shortened. The reasons for the differences are as follows (Wolter and Dreizler, 1988; Rosemann *et al.*, 1989; Ellenbrock *et al.*, 1990; Odler and Chen, 1990, 1993; Thorman and Schmitz, 1991; Chen and Odler, 1992a).

If the cement is ground in a high-pressure roller mill rather than in a ball mill, decomposition of the calcium sulfate dihydrate (gypsum) to the more rapidly soluble hemihydrate does not take place, as insufficient heat is generated in the comminution process, and the temperature stays too low for the decomposition of the dihydrate. In addition, unlike ball mill comminution, the gypsum is not spread over the cement particle surface, and the individual gypsum particles remain spatially separated from those of the clinker. This happens because, in a high-pressure roller mill, comminution takes place as a consequence of static pressure exerted on the clinker, rather than by a rubbing action as in ball mill grinding. For both reasons, after mixing with water, the tricalcium aluminate undergoing hydration does not have enough sulfate ions available for conversion to ettringite, and converts partially or completely to monosulfate instead. Owing to the platelet-like morphology of this phase the flow of the cement paste is impaired, compared with the situation that exists if a smooth, dense ettringite layer forms on the particle surface. Moreover, the monosulfate layer developed at the clinker surface is not as effective in slowing down the hydration rate of tricalcium aluminate as that of ettringite. Other factors that adversely affect the rheological properties of cement ground in a high-pressure roller mill are the narrower particle-size distribution of the cement and the less favorable shape factor and aspect ratio of the cement particles that are formed. There are also indications that the reactivity of the clinker phases, especially that of C_3A, increases with high-pressure roller mill comminution.

To eliminate these unfavorable effects of high-pressure roller mill grinding on cement quality, this grinding technique is usually employed in combination with ball mill grinding, rather than as the sole comminution technology.

The possibility of increasing the strength properties of existing cements ground in a conventional mill by mechanical activation was reported by Sekulic *et al.* (1980, 1997). Such activation may be achieved by an additional grinding of the binder in a vibro-mill for several minutes. In addition to increased strength, the specific surface area of the cement also increases, but only moderately. X-ray investigations revealed a decline of the maximum peak intensity of the clinker phases, indicating the production of defects in their crystalline lattices.

2.18   SPECIAL APPROACHES IN CEMENT PROCESSING

Standard processing of cements consists in mixing them with water and suitable fillers/aggregates to obtain a fresh concrete/mortar mix of plastic consistency. The resulting mix is placed into appropriate molds or forms, compacted, and allowed to set and harden at ambient temperature. Under special conditions, however, other approaches may also be used.

One factor that may affect the rheological properties of the mix and the strength development is the method of mixing. In general, more intensive or prolonged mixing improves the distribution of the cement particles within the cement paste and thus reduces its tendency to segregate. It may even increase the strength of the hardened material. Strength increases at 28 days of up to 23% have been reported where an "ultramixer" rather than a normal mixer was employed in the mixing of plain cement pastes (Rejeb, 1995, 1996).

An improvement of concrete strength was also reported where the mixing water was introduced in two portions added to the mixer at separate times (Tamimi and Al-Najeim, 1994). Under such conditions the portlandite crystals that adhere to the aggregate surface exhibit a more ordered structure, with the *c*-axis lying perpendicular to the interface.

A favorable effect on ultimate strength was also found if the coarse aggregate was added to a cement paste or grout that had been premixed first (Rejeb, 1966).

It has also been claimed that the plasticity and mobility of the mortar and the strength of reinforced concrete may be increased by a preceding "magnetic treatment" of the mixing water (Afanaseva, 1992).

Yu *et al.* (1998) reported an increase of concrete strength (by at least 5% and by as much as 50% after one day) in concrete mixes that were produced with "electron water" or were cured in such water. This form of water was obtained by exposing normal water in a tank to an electric current of an appropriate electric voltage. The authors claim that under such conditions the average cluster size of water is reduced and – as a result – the cement hydration is improved.

To obtain high strength, low permeability, and good corrosion resistance of the hardened cement paste, the water/cement ratio of the mix should be as low as possible. At the same time, however, the workability of the fresh paste deteriorates, and compacting of the mix becomes gradually more difficult as the water/cement ratio decreases. Special methods of compaction have been suggested to overcome this dilemma. To compact mixes with a low water/cement ratio, pressing rather than vibration is widely employed. Especially favorable strength results are obtained if hot-pressing is employed, in which a cement mix of very low water/cement ratio is exposed to high pressure (up to 350 MPa) and heated to high temperatures (up to 250 °C) simultaneously (Roy *et al.*, 1972; Roy and Gouda, 1973; Pashchenko *et al.*, 1992; Roy, 1992). Tensile strengths of up to about 70 MPa and compressive strengths of up to 650 MPa may be achieved in this way. Owing to the very low water/cement ratio, complete hydration cannot take place in such systems, and they have to be considered as microcomposites in which up to 50 wt% of the original cement remains non-hydrated.

In another approach a cement paste produced with a normal water/cement ratio is exposed to uniaxial pressure after the paste has achieved its initial set. In this way a portion of the mixing water is removed, and the water/solid ratio may be reduced to about 0.1. The residual water allows a continuation of the hydration process, in which a very low-porosity, high-strength material, called pore-reduced cement (PRC), is produced (Macphee, 1990; Macphee *et al.*, 1992; Geslin, 1995). Compressive and tensile strength improvements of up to three and six times respectively may be achieved by this approach.

To accelerate the hardening process heat-curing is widely employed, in which the hydration process takes place at elevated temperatures. Several possibilities for heating the concrete mix are available:

- Heating one or more of the individual concrete mix constituents prior to mixing, placing and compacting the fresh mix. The difficulty with this approach is the accelerated hydration rate of the resultant mix, which causes a drastic shortening of the time available for its processing, even at temperatures that are only moderately elevated.
- Heating the walls of the mold in which the fresh concrete mix has been placed.
- Keeping the concrete mix, after it has been placed into a mold and compacted, in a confined space at elevated temperature and high humidity (steam curing).
- Exposing the surface of the fresh concrete mix to infrared radiation. Such an approach is limited to applications in which thin concrete elements are produced.
- Heating the concrete body with an alternating electric current, using the steel form and the steel reinforcement, embedded into the concrete mix, as opposite electrodes. The use of this approach is very limited, owing to high costs, difficulties with attaining a relatively uniform distribution of temperature within the concrete body, and safety considerations.
- Heating the concrete mix in a microwave field. This is the most progressive approach, which provides the most uniform temperature distribution within the concrete body (Sohn and Lynn Johnson, 1999), and yields optimum quality of the hardened concrete. The method is also economical, as the heating is virtually confined to the concrete mix.

In applying heat curing one has to realize that the nature of the hydration products becomes altered at elevated temperatures (Odler *et al.*, 1987; Lewis *et al.*, 1995). In general, the C-S-H phase that is formed attains a lower bound water content and an increased C/S ratio as the hydration temperature goes up. At the same time the quantities of foreign ions (Al^{3+}, Fe^{3+}, SO_4^{2-}) incorporated into the C-S-H phase increase. Above about 70 °C the AFt phase becomes unstable and the AFm phase is formed instead, unless the Al^{3+}, Fe^{3+}, and SO_4^{2-} ions have been completely taken up by the C-S-H phase. All these changes result in a moderate decline of ultimate strength, especially at curing temperatures approaching 100 °C. In steam-curing experiments performed at 65 °C it was observed that optimum strength values were obtained in cements that contained anhydrite rather than gypsum, and at SO_3 contents between 4 and 5 wt% (Zhang *et al.*, 1996). Another factor that may adversely affect the quality of the hardened concrete is macrocracks, which may be formed as a result of thermal gradients that develop in the material.

In some concrete mixes heat-cured at temperatures above about 70 °C expansion and cracking may be observed, which start after an induction period of one year or more and are associated with a formation of ettringite in the hardened concrete. The damage in concrete is characterized by the formation of gaps around larger aggregate particles, usually – but not always – filled with ettringite. It is not obvious whether the gap formation occurs by the crystallization pressure of ettringite that crystallizes in the vicinity of the aggregate surface, or whether ettringite crystallizes in the gaps after they have been formed by expansion of the hardened cement paste (secondary ettringite).

The delayed ettringite expansion may be prevented either by reducing the temperature of heat-curing, or by selecting an appropriate binder in the production of the concrete mix. To prevent this phenomenon it has been suggested that only Portland cements should be used in which the SO_3/Al_2O_3 molar ratio does exceed 0.55 or which do not contain the C_3A phase (Heinz and Ludwig, 1986, 1997). In laboratory experiments performed on plain cement pastes heat-cured at 90 °C it was found that a deleterious expansion occurred only in a cement that possessed simultaneously high SO_3 and C_3A contents (5.0 and 10.2 wt% respectively) (Odler and Chen, 1995). In another study (Fu and Beaudoin, 1996) it was observed that concrete samples made from ASTM Type III cement – if heat-treated at 90 °C – expanded significantly, whereas similar samples made from Type I and Type V cements did not. The increased tendency of Type III cement to expand was attributed to its higher SO_3 and C_3A contents. Another possibility for preventing delayed ettringite expansion consists in employing blended cements containing granulated blast furnace slag, or fly ash or natural pozzolana, instead of plain Portland cement (Heinz and Ludwig, 1986, 1997).

The final properties of hardened cement may be modified by carbonation that takes place upon curing in an atmosphere of CO_2, especially at high humidity. The rate of carbonation depends on the partial pressure of CO_2, and fairly high CO_2 concentrations are required to attain an acceptable reaction rate. As there is always a steep gradient in the degree of carbonation, depending on the distance from the surface, the procedure is suitable only for thin-wall concrete/mortar elements.

In the course of carbonation the CO_2 reacts with calcium hydroxide and the C-S-H phase of the hydrated cement paste, yielding calcium carbonate and a partially decomposed C-S-H phase with a reduced C/S ratio and a higher degree of SiO_4 polymerization, which eventually converts to amorphous SiO_2. A concurrent carbonation of the AFm phase also takes place (Thomas *et al.*, 1993). The carbonation results in a decline of porosity and along with it an increase of strength (Matsusato *et al.*, 1992; Honda and Shizawa, 1994); however, at a high degree of carbonation a decline of flexural strength may also be observed (Honda and Shizawa, 1994).

An additional increase of strength may be achieved by impregnating a hardened cement paste with a silicic ester solution, followed by aging at room temperature. Under these condition a silica gel is precipitated within the pores after the solvent has been evaporated, resulting in a decrease of porosity and an increase of strength (Atzeni *et al.*, 1992):

$$Si(O\text{-}Et)_4 + 4H_2O \rightarrow 4Et\text{-}OH + H_4SiO_4$$

$$H_4SiO_4 \rightarrow SiO_2 + 2H_2O$$

REFERENCES

Abramson, I.G. *et al.* (1980) Clinker formation in high energy flux of accelerated electrons, in *Proceedings 7th ICCC, Paris*, Vol. 2, pp. 26–28.

Abramson, I.G. *et al.* (1992) Development of radiation thermal technology of cement manufacture, in *Proceedings 9th ICCC, New Delhi*, Vol. 2, pp. 171–176.

Afanaseva, V.F. (1992) Magnetic water treatment in the production of precast reinforced concrete (in Russian). *Beton Zhelezobeton (Moscow)* (12) 26–28 [ref. CA 121/162688].

Albats, B.S., and Shein, A.L. (1997) High-quality Portland cement produced from low-base raw mix with low energy consumption, in *Proceedings 10th ICCC, Göteborg*, paper 1i015.

Ali, M.M., and Mullick, A.K. (1998) Volume stabilization of high MgO cement: effect of curing conditions and fly ash addition. *Cement and Concrete Research* **28**, 1585–1594.

Aluno-Rosetti, V., and Curcio, F. (1997) A contribution to the knowledge of the properties of Portland-limestone cement concretes with respect to the requirements of European and Italian design code, in *Proceedings 10th ICCC, Göteborg*, paper 3v026.

Ananenko, N.F. (1986) Magnesium oxide in Portland cement, in *Proceedings 8th ICCC, Rio de Janeiro*, Vol. 6, pp. 143–148.

Anonymous (1996) The WSZ Schwefelsäure und Zement GmbH – unique or a model for the future. *ZKG Internat.* **49**, A33–34.

Atzeni, C., Massida, L., and Sanna, U. (1992) Densifying of cement pastes by chemical deposition of silica in the pores, in *Proceedings 9th ICCC, New Delhi*, Vol.3, pp. 419–424.

Ayed, F.A., Castanetan, R., and Sorrentino, F.P. (1992) Thermal behaviour of mineralized Portland cement raw meal, in *Proceedings 9th ICCC, New Delhi*, Vol. 2, pp. 287–293.

Baeker, C., Lampe, F., and Worzala, H. (1983) Effect of mineralizers on alite formation (in German). *Silikattechnik (Berlin)* **34**, 81–82.

Barbanyagre, V.D. (1997) Clinker formation in high silicate low temperature melt, in *Proceedings 10th ICCC, Göteborg*, paper 1i011.

Barbanyagre, V.D., Shamshurov, V.M., and Timoshenko, T.I. (1992) Peculiarities of the process of cement clinker formation by changing the interaction sequence of reacting components, in *Proceedings 9th ICCC, New Delhi*, Vol. 2, pp. 196–200.

Bei, R., and Ludwig, U. (1990) Hydraulic reactivity of belite-rich cements (in German). *Zement-Kalk-Gips* **43**, 506–510.

Ben Besset, M., Nixon, P.J., and Hardcastle, J. (1988) The effect of differences in the composition of Portland cement on the properties of hardened concrete. *Magazine of Concrete Research* **40** (143), 106–110.

Bensted, J. (1992–1993) Hydrophobic Portland cement. *World Cement* **23**, 30–31: **24**, 54–55.

Bertrandy, A., and Poitevin, P. (1991) Limestone filler for concrete, fresh research and practice, in *Proceedings International Conference on Blended Cements in Constructions, Sheffield*, pp. 16–31.

Bilal, S.H. (1995) Chemical and physical properties of white cement concrete. *Advances in Cement-based Materials* **21**, 161–167.

Blanco-Varela, M.T. (1997) CaF_2 and $CaSO_4$ in white cement clinker production. *Advances in Cement Research* **9**, 105–113.

Blanco-Varela, M.T., Palomo, P., and Vazquer, T. (1984) Effect of fluorspar in the formation of clinker phases. *Cement and Concrete Research* **14**, 397–406.

Blanco-Varela, M.T. *et al.* (1988) Reactivity and burnability of raw mixes with crystallized blast furnace slag. *Zement-Kalk-Gips* **41**, 398–402, 628–636.

Blanco-Varela, M.T. *et al.* (1996) Modelling of the burnability of white cement raw mixes made with CaF_2 and $CaSO_4$. *Cement and Concrete Research* **26**, 457–464.

Boikova, A.I., Fomicheva, O.I., and Zubekhin, A.P. (1986) Chemical composition of phases and phase relations in the unbleached and bleached clinkers, in *Proceedings 8th ICCC, Rio de Janeiro*, Vol. 2, pp. 243–248.

Boikova, A.I., Fomicheva, O.I., and Luginina, I.G. (1997) Phases in high magnesium clinkers, in *Proceedings 10th ICCC, Göteborg*, paper 1i057.

Bojadjieva, C., and Glavchev, I. (1995) Investigations on the influence of some plasticizers on gypsum-free cement paste. *Cement and Concrete Research* **25**, 685–684.

Borgholm, H.E., Herfort, D., and Rasmussen, S. (1995) A new blended cement based on mineralized clinker. *World Cement* **26**, 27–32.

Bouzoubaa, N., Zhang, M.H., and Malhotra, V.M. (1998) Superplasticized Portland cement: production and compressive strength of mortars and concrete. *Cement and Concrete Research* **28**, 1783–1796.

Brunauer, S. *et al.* (1972–1973) Hardened Portland cement pastes of low porosity I-VII. *Cement and Concrete Research* **2**, 313–330; 331–348; 463–480; 577–589; 731–743; and **3**, 129–147; 279–293.

Chatterjee, A.K. *et al.* (1990) Development of sol-gel technology for cement manufacture, in *Proceedings Symposium Advances in Cementitious Materials, Gaithersburgh*, pp. 640–660.

Chatterji, S. (1995) Mechanism of expansion of concrete due to the presence of dead burnt CaO and MgO. *Cement and Concrete Research* **25**, 51–56.

Chen, W., and Lou, Z. (1997) Studies of the intermediate mineral formation pattern and the improvement of C_3S formation at low temperature in the presence of F and SO_3 (in Chinese). *Cuisuanyan Xuebo* **25**, 651–656 [ref: CA 128/274182].

Chen, Y., and Odler, I. (1992a) Effect of the grinding technique on the shape of cement particles, in *Proceedings 14th International Conference on Cement Microscopy, Costa Mesa*, pp. 22–28.

Chen, Y., and Odler, I. (1992b) The progress of Portland cement hydration: effect of clinker composition, in *Proceedings 9th ICCC, New Delhi*, Vol. 4, pp. 24–30.

Cheong, H.M. *et al.* (1992) Effect of sulfate on the reaction of C_3S formation, in *Proceedings 9th ICCC, New Delhi*, Vol. 2, pp. 335–341.

Cheong, H.M., Choi, S.H., and Han, K.S. (1994) Prevention mechanism of $3CaO.SiO_2$ formation in clinker with excess of SO_3. *International Ceramics Monographs 1994*, I (Proceedings International Ceramics Conference), pp. 141–146 [ref. CA 121/63994].

Cheong, H.M., Choi, S.H., and Han K.S. (1997) Effect of alkalis and SO_3 on tricalcium silicate formation and microstructure, in *Proceedings 10th ICCC, Göteborg*, paper 1i049.

Cochet, G., and Sorrentino, F. (1993) Limestone filler cement: properties and uses, in *Mineral Admixtures in Cement and Concrete* (ed. S.N. Ghosh), ABI Books, New Delhi, pp. 212–295.

Collepardi, M. *et al.* (1980, 1982) Combined effect of lignosulfonate and carbonate on pure Portland clinker compounds hydration. *Cement and Concrete Research* **10**, 455–461; **12**, 271–277, 425–435.

Coswami, G. *et al.* (1990) Phase formation during sintering of magnesium and fluorine containing raw mix in cement rotary kiln. *Zement-Kalk-Gips* **43**, 253–256.

Diamond, S., and Gomez-Toledo, C. (1978) Consistency, setting and strength gain characteristics of a "low porosity" Portland cement paste. *Cement and Concrete Research* **8**, 613–622.

Ellenbrock, H.G., Sprung, S., and Kuhlman, K. (1990) Particle size distribution and cement properties III, Effect of grinding technology (in German). *Zement-Kalk-Gips* **43**, 13–19.

Fang, Y. *et al.* (1997) Cement clinker formation from fly ash by microwave processing, in *Proceedings 10th ICCC, Göteborg*, paper 1i010.

Frigione, G., and Mara, S. (1976) Relationship between particle size distribution and compressive strength. *Cement and Concrete Research* **6**, 113–128.

Frigione, G., and Zenone, P. (1985) Influences on the whiteness of cement. *Il Cemento* **82**, 95–106.

Fu, Y., and Beaudoin, J.J. (1996) Microcracking as precursor to delayed ettringite formation in cement systems. *Cement and Concrete Research* **26**, 1493–1498.

Fu, Z. (1992) Characteristics of masonry cement produced at 975 °C in a fluidized bed boiler, in *Proceedings 9th ICCC, New Delhi*, Vol. 3, pp. 58–64.

Führ, C., Müller, A and Knöfel, D. (1992) Durability of mortars and concretes with varying lime standards, in *Proceedings 9th ICCC, New Delhi*, Vol. 5, pp. 26–31.

Fujita, K., Yoshimori, K., and Yamazaki, Y. (1993) Strength development of prehydrated cement (in Japanese). *Semento Konkurito Ronbunshu* **47**, 100–105 [ref. CA 120/329692].

Gaidzhurov, P.P., and Zubar, T.G. (1994) Concentration dependence of glass-phase parameters on whiteness of white Portland cement clinker (in Russian). *Izvestiya Vysshikh Uchebnykh Zavedenii, Khimiya I Khimicheskaya Tekhnologiya* **37**, 73–75 [ref. CA 121/261973].

Gaidzhurov, P.P. *et al.* (1994) Effect of the alumina modulus and 3rd transition metal oxides on properties of white Portland cement (in Russian). *Izvestiya Vysshikh Uchebnykh Zavedenii, Khimiya I Khimicheskaya Tekhnologiya* 37, 102–106 [ref. CA 120/225155].

Garbacik, A. (1997) Properties of low temperature Portland clinker, in *Proceedings 10th ICCC, Göteborg*, paper 1i018.

Geslin, N.M. *et al.* (1995) Durability and microstructure of pore reduced cements. *Materials Research Society Symposium Proceedings* 370, 237–244.

Gimenez, S. *et al.* (1991) Production of cement requiring low energy expenditure. *Zement-Kalk-Gips* 44, 12–15.

Goldstein, I.Ya. (1997) Energy saving technology of cement production by means of combined grinding of clinkers of variable composition, in *Proceedings 10th ICCC, Göteborg*, paper 1i013.

Gollop, R.S., and Taylor, H.F.W. (1992, 1994, 1995) Microstructural and microanalytical studies of sulfate attack I-III. *Cement and Concrete Research* 22, 1027–1035; 24, 1347–1353; 25, 1581–1590.

Gouda, G.R. (1980) Fluxes conserve energy in cement manufacture. *Rock Products* 83 (4), 52–56.

Gouda, G.R. (1986) Microstructure and properties of high-alkali clinker, in *Proceedings 8th ICCC, Rio de Janeiro*, Vol. 2, pp. 234–239.

Handoo, S.K. *et al.* (1992) Evaluation of radiation synthesized clinker, in *Proceedings 9th ICCC, New Delhi*, Vol. 2, pp. 177–182.

Hanna, K.M. (1977) Application of experience with low porosity cement pastes and mortar (in German). *Zement-Kalk-Gips* 30, 140–142.

Hanna, K.M., and Taha, A. (1977) Rheological properties of low-porosity cement pastes (in German). *Zement-Kalk-Gips* 30, 293–295.

Heinz, D., and Ludwig, U. (1986) Mechanism of subsequent ettringite formation in mortars and concretes after heat treatment, in *Proceedings 8th ICCC, Rio de Janeiro*, Vol. 5, pp. 189–194.

Heinz, D., and Ludwig, U. (1997) Mechanism of secondary ettringite formation in mortars and concretes subjected to heat treatment, in *Concrete Durability, Katherine and Bryant Mather International Conference* (ed. J.M. Scanlon), American Concrete Institute SP-100, pp. 189–194.

Honda, M., and Shizawa, M. (1994) Influence of carbonation on the mechanical properties and texture of hardened mortar. *Ceramics Transactions* 401, 203–212.

Hrazdira, J. (1994) Role of superplasticizers in gypsumless Portland cements. American Concrete Institute SP-148, pp. 404–416.

Hu, G. (1997) The application of high magnesium oxide content limestone in cement industry, in *Proceedings 10th ICCC, Göteborg*, paper 1i023.

Hu, S. (1992) Studies on rapid burning temperature, time and raw meal modulus range in Portland cement, in *Proceedings 9th ICCC, New Delhi*, Vol.2, pp. 81–87.

Huang, W., Lu, Z., and Xu, G. (1992) The study on the formation of Portland cement clinker in low temperature and rapid heating-up process, in *Proceedings 9th ICCC, New Delhi*, Vol. 2, pp. 164–170.

Ingram, K.D., and Daugherty, K.E. (1992) Limestone additions to Portland cement: uptake, chemistry and effect, in *Proceedings 9th ICCC, New Delhi*, Vol. 3, pp. 180–186.

Ingram, K.D. *et al.* (1990) *Carbonate addition to cement.* ASTM STP 1064, pp. 14–23.

Ji, S.X., and Xu, F. (1992) Research on forming mechanism of Portland cement clinker under different burning conditions, in *Proceedings 9th ICCC, New Delhi*, Vol. 2, pp. 36–42.

Kakali, G., and Parissakis, G. (1995) Investigations on the effect of Zn oxide on the formation of Portland cement clinker. *Cement and Concrete Research* **25**, 79–85.

Kakali, G., Parissakis, G., and Pouras, D. (1996) A study on the burnability and the phase formation of PC clinker containing Cu oxide. *Cement and Concrete Research* **26**, 1473–1478.

Kalousek, G., Porter, L.C., and Benton, E.J. (1972) Concrete for long time service in sulfate environment. *Cement and Concrete Research* **2**, 79–89.

Kelham, S., and Moir, G.K. (1992) The effect of C₃S level on concrete properties, in *Proceedings 9th ICCC, New Delhi*, Vol. 5, pp. 3–8.

Khadilkar, S.A. *et al.* (1992) Clinkerization through melting – a new approach, in *Proceedings 9th ICCC, New Delhi*, Vol. 2, pp. 190–295.

Khadilkar, S.A. *et al.* (1997) Melting process for clinkerization – technical reexamination, in *Proceedings 10th ICCC, Göteborg*, paper 1i008.

Kim, S., Tsurumi, T., and Diamon, M. (1992) Carbonation of jet cement pastes (in Japanese). *Semento Konkurito Ronbunshu* **46**, 586–591 [ref. CA 119/102017].

Klemm, W.A., Jawed, I., and Holub, K.J. (1979) Effect of calcium fluoride mineralization on silicate and melt formation in Portland cement clinker. *Cement and Concrete Research* **9**, 489–496.

Knöfel, D. (1978) Modifying some properties of Portland cement clinker and Portland cement by means of ZnO and ZnS (in German). *TIZ Fachberichte* **107**, 324–327.

Kokanyev, N.F., and Kuznetsova, T.V. (1992) Properties of gypsum-free Portland cement (in Russian). *Tsement (Moscow)* (1), 54–59.

Kuznetsova, T.V. *et al.* (1992) The structure and properties of low CaO contained clinker, in *Proceedings 9th ICCC, New Delhi*, Vol. 2, pp. 88–93.

Livesen, P. (1991) Performance of limestone-filled cement, in *Proceedings International Conference on Blended Cements in Construction, Sheffield*, pp. 1–15.

Lewis, M.C., Scrivener, K.L., and Kelham, S. (1995) Heat curing and delayed ettringite formation. *Materials Research Society Symposium Proceedings* **370**, 67–76.

Louginina, I.G. *et al.* (1986) Hardening of Portland cement with elevated content of magnesium oxide and other magnesium compounds, in *Proceedings 8th ICCC, Rio de Janeiro*, Vol. 6, pp. 164–169.

Lu, Z., Huang, W., and Xu, G. (1992) The effect of the composite mineralizers of CaF₂, CaSO₄ and ZnO on the formation of Portland cement clinker in a rapid heating-up burning, in *Proceedings 9th ICCC, New Delhi*, Vol. 2, pp. 386–392.

Ludwig, U., and Pohlman, R. (1986) Investigations on the production of low lime Portland cements, in *Proceedings 8th ICCC, Rio de Janeiro*, Vol. 2, pp. 363–371.

Luginina, I.G., Boykova, A.I., and Formichova, O.I. (1992) Effect of increased magnesium oxide content and mixture basicity upon clinker phases, in *Proceedings 9th ICCC, New Delhi*, Vol. 2, pp. 114–117.

Macphee, D.E. (1990) PRC-pore reduced cement: high density cement pastes following fluid extraction. *Advances in Cement Research* **3**, 135–142.

Macphee, D.E. *et al.* (1992) Microstructural development in pore reduced cement (PRC). *Materials Research Society Symposium Proceedings* **245**, 303–308.

Matsusato, H., Funato, M., and Yamazaki, Y. (1992) Strength and microstructure of carbonated hardened cement (in Japanese). *Semento Konkurito Ronbunshu* **46**, 592–597 [ref. CA 119/102018].

Michaud, V., and Sunderman, R. (1997) Sulfate solubility in high SO_3 clinkers, in *Proceedings 10th ICCC, Göteborg*, paper 2ii011.

Muntean, M. (1992) White belitic Portland cement, in *Proceedings 9th ICCC, New Delhi*, Vol. 3, pp. 45–50.

Nagaoka, S., Mizukoshi, M., and Ono, S. (1994) Studies on the relation between the content of C_2S in cement and the strength development of concrete. *Semento Konkurito Ronbunshu* **48**, 130–135 [ref. CA 121/151166].

Nehdi, M., Mindess, S., and Aitcin, P.C. (1998) Rheology of high performance concrete: effect of ultrafine particles. *Cement and Concrete Research* **28**, 687–697.

Nikoforov, Yu.V., and Zozulya, R.A. (1992) Long-term hardening of cements with higher content of magnesium oxide, in *Proceedings 9th ICCC, New Delhi*, Vol. 4, pp. 64–68.

Odler, I. (1991) Improving energy efficiency in Portland clinker manufacturing, in *Cement and Concrete Science and Technology* (ed. S. N. Ghosh), ABI Books, New Delhi, pp. 174–200.

Odler, I., and Abdul-Maula, S. (1980a) Structure and properties of Portland cement clinker doped with CaF_2. *Journal of the American Ceramic Society* **63**, 654–659.

Odler, I., and Abdul-Maula, S. (1980b) Effect of mineralizers on the burning of Portland clinker (in German). *Zement-Kalk-Gips* **33**, 132–136, 278–282.

Odler, I., and Chen, Y. (1990) Effect of grinding in high pressure roller mill on properties of Portland cement (in German). *Zement-Kalk-Gips* **43**, 188–191.

Odler, I., and Chen, Y. (1993) Influence of the method of comminution on the properties of cement, in *Proceedings Congress on Cement Technology, Düsseldorf*, pp. 78–92.

Odler, I., and Chen, Y. (1995) Effect of cement composition on the expansion of heat-cured cement pastes. *Cement and Concrete Research* **25**, 853–862.

Odler, I., and Schmidt, O. (1980) Structure and properties of Portland cement clinker doped with zinc oxide. *Journal of the American Ceramic Society* **63**, 13–16.

Odler, I., and Wonnemann, R. (1983) Effect of alkalis on Portland cement hydration. *Cement and Concrete Research* **13**, 771–777.

Odler, I., and Zhang, H. (1996) Investigations on high SO_3 Portland cement clinkers. *World Cement* **27**, 73–77.

Odler, I., and Zhang, H. (1997) Possibilities of utilizing high SO_3 fluidized bed ash in the production of Portland clinker and cement, in *Proceedings 10th ICCC, Göteborg*, paper 2ii63.

Odler, I., Duckstein, U., and Becker, Th. (1978) On the combined effect of water soluble lignosulfonates and carbonates on Portland cement and clinker pastes. *Cement and Concrete Research* **8**, 469–480, 525–538.

Odler, I., Abdul-Maula, S., and Lu, Z. (1987) Effect of hydration temperature on cement paste structure. *Materials Research Society Symposium Proceedings* **85**, 139–144.

Okamura, T., Harada, H., and Daimon, M. (1998) Influence of calcium sulfate in belite-rich cement on the change in fluidity of mortar with time. *Cement and Concrete Research* **12**, 1297–1308.

Opoczky, L., and Papo, A. (1997) Rheological and hydration properties of a high strength gypsum free cement, in *Proceedings 10th ICCC, Göteborg*, paper 2ii004.

Page, C.H., Ghosh, D., and Chatterjee, H.K. (1986) Influence of mineralizers on the constitution and properties of Portland clinker, in *Proceedings 8th ICCC, Rio de Janeiro*, Vol. 2, pp. 152–157.

Paori, M., Baldini, G., and Collepardi, M. (1982) Combined effect of lignosulfonate and carbonate on pure Portland clinker compounds hydration. *Cement and Concrete Research* **12**, 271–277, 425–435.

Pashchenko, A.A., Chystiakov, V.V., and Myasnikova, Ye.A. (1992) High strength hot-pressed cement paste, in *Proceedings 9th ICCC, New Delhi*, Vol. 3, pp. 361–366.

Peukert, S. (1997) Cement for HPC concrete, in *Proceedings 10th ICCC, Göteborg*, paper 2ii034.

Piasta, W. *et al.* (1997) Influence of limestone powder filler on microstructure and mechanical properties of concrete under sulphate attack, in *Proceedings 10th ICCC, Göteborg*, paper 4iv 018.

Puertas, F. *et al.* (1997,1998) Behaviour of a new white cement fabricated with raw materials containing CaF_2 and $CaSO_4.2H_2O$. *ZKG International* **50**, 232–239; **51**, 94–100.

Puertas, F. *et al.* (1997) Burnability of mineralized white cement: optimization of raw meal, in *Proceedings 10th ICCC, Göteborg*, paper 1i034.

Puscasu, D., and Dimitrescu, C. (1994) Manufacture of low-alkali cement clinker batch at Devia cement plant, Bulgaria (in Romanian). *Materiale de Constructii (Bucharest)* **24**, 171–175 [ref. CA 122/15689].

Ramachandran, V.S. (1980) Unsoundness of cements containing MgO and CaO. *Il Cemento* **77**, 159–168.

Regourd, M., Hornain, H., and Matureux, B. (1980) Microstructure of concrete in aggressive environments, in *Durability of Building Materials and Components*, ASTM, Philadelphia, pp. 253–268.

Rehsi, S.S., and Garg, S. K. (1985) Long-term study on stability of high magnesia cement containing fly ash. *Durability of Building Materials* **21**, 265–273.

Rehsi, S.S., and Garg, S.K. (1992) 20 year study on hydration of high magnesia cement with fly ash at ambient temperature, in *Proceedings 9th ICCC, New Delhi*, Vol. 4, pp. 684–690.

Rejeb, S.K. (1995) Technique of multistep concrete mixing. *Materials and Structures* **28**, 230–234.

Rejeb, S.K. (1996) Improving compressive strength of concrete by a two-step mixing method. *Cement and Concrete Research* **26**, 585–592.

Rosemann, H. *et al.* (1989) Investigation on the use of high pressure roller mill in fine grinding of cement (in German). *Zement-Kalk-Gips* **42**, 165–169.

Rosetti, V.A., Curcio, F., and Cussino, L. (1992) Production performances and utilization of special superplasticized cement, in *Proceedings 9th ICCC, New Delhi*, Vol. 3, pp. 38–44.

Roy, D.M. (1992) Advanced cement systems, including CBC, DSP, MDF, in *Proceedings 9th ICCC, New Delhi*, Vol. 1, pp. 257–380.

Roy, D.M., and Gouda, G.R. (1973) High strength generation in cement pastes. *Cement and Concrete Research* **3**, 807–820.

Roy, D.M., Gouda, G.R., and Brabovski, A. (1972) Very high strength cement paste prepared by pressing and other high pressure techniques. *Cement and Concrete Research* **2**, 349–366.

Saada, Y., Seidel, G., and Müller, W. (1988) Effect of elevated sulfate contents in raw meal on burnability and hydration properties (in German). *Silikattechnik (Berlin)* **39**, 115–117.

Sane, T., Fujiyama, O., and Tanimura, M. (1992) Properties of concrete using low heat cement containing a large amount of belite. *Semento Konkurito Ronbunshu* **46**, 392–397.

Sanitsky, M., and Sobol, Ch.S. (1992) Gypsum-free rapid hardening and mixed Portland cement, in *Proceedings 9th ICCC, New Delhi*, Vol. 3, pp. 438–443.

Sanitsky, M. *et al.* (1989) Effective rapid hardening gypsum-free Portland cements (in Russian). *Cement (Moscow)* (8), 6–17.

Satava, V., and Tyle, P. (1994) Rheological behaviour of the system Portland clinker – Na_2CO_3 – sulfonated polyphenolate – H_2O. *Ceramics-Silikaty (Prague)* **38**, 9–15.

Schoner, W., and Wierig, J. (1989) Long term trials with concretes made with aggregates of high sulfate content (in German). *Zement-Kalk-Gips* **42**, 476–480.

Sekulic, Z., Popov, S., and Milosevic, S. (1980) Comminution and mechanical activation of Portland cement in different mill types. *Ceramics-Silikaty (Prague)* **42**, 25–28.

Sekulic, Z., Stefanovic, M., and Zivanovic, B. (1997) Mechanical activation of ordinary Portland cement and cement with additives, in *Proceedings 10th ICCC, Göteborg*, paper 3iii028.

Sharma, K.M. *et al.* (1992) Volume stability of high magnesia cements, in *Proceedings 9th ICCC, New Delhi*, Vol. 5, pp. 614–620.

Singh, N.B., and Singh, A.K. (1989) Effect of Melment on hydration of white Portland cement. *Cement and Concrete Research* **19**, 547–553.

Skvara, F. (1986) Microstructure of hardened pastes of gypsum-free Portland and slag cements, in *Proceedings 8th ICCC, Rio de Janeiro*, Vol. 3, pp. 356–362.

Skvara, F. (1995) Gypsum-free Portland cement pastes of low water-to-cement ratio. *Materials Research Society Symposium Proceedings* **370**, 153–158.

Skvara, F. (1996) Properties of gypsum-free Portland cement pastes with a low water-to-cement ratio. *Ceram-Silik. (Prague)* **40**, 36–40.

Skvara, F., and Rybinova, M. (1985) The system ground clinker – polyphenol sulfonate – alkali metal salts. *Cement and Concrete Research* **15**, 1013–1021.

Skvara, F. *et al.* (1981) Portland cement without gypsum (Czech). *Silikaty (Prague)* **25**, 251–261.

Slota, R.J., and Lewandowska-Kanas, A. (1986) Modification of clinker phase composition and strength properties of a high-alkali cement clinker by gypsum additions, in *Proceedings 8th ICCC, Rio de Janeiro*, Vol. 2, pp. 205–210.

Sohn, D., and Lynn Johnson, D. (1999) Microwave curing effects on the 28–day strength of cementitious materials. *Cement and Concrete Research* **29**, 241–247.

Sone, T., Fujiyama, O., and Tanimura, M. (1992) Properties of concrete using low-heat cement containing a large amount of belite (in Japanese). *Semento Konkurito Ronbunshu* **46**, 392–397 [ref. CA 119/102051].

Stark, J., Eckard, A., and Ludwig, H.M. (1997) Influence of C_3A content on frost and scaling resistance. *RILEM Proceedings* **34**, 100–110.

Stark, J., Osokin, P., and Potapova, N. (1997) Technology and properties of low-aluminate clinkers and cements, in *Proceedings 10th ICCC, Göteborg*, paper 1i005.

Strunge, J., Knöfel, D., and Dreizler, I. (1985, 1990) Effect of alkalis and sulfur on properties of cement (in German). *Zement-Kalk-Gips* **38**, 150–158; **43**, 199–208.

Su, D. *et al.* (1992) Effect of SO_3 on mineral formation and properties of clinker, in *Proceedings 9th ICCC, New Delhi*, Vol. 2, pp. 322–328.

Sulimenko, L.M., and Albats, B. (1997) Clinkering of pre-formed mixes, in *Proceedings 10th ICCC, Göteborg*, paper 1i007.

Sumner, M., Hepher, N., and Moir, G. (1986) The influence of a narrow cement particle size distribution on cement paste and concrete water demand, in *Proceedings 8th ICCC, Rio de Janeiro*, Vol. 2, pp. 310–315.

Surana, M.S., and Jeshi, S.N. (1990) Use of mineralizers and fluxes on improved clinkerization and conservation of energy. *Zement-Kalk-Gips* **43**, 43–45.

Tamimi, A.K., and Al-Najeim, H. (1994) XRD and SEM investigations of the cement paste – aggregate interface for concrete produced by both the conventional mixing and a new mixing technique, in *Proceedings 16th International Conference on Cement Microscopy*, pp. 354–368.

Tezuka, Y. *et al.* (1992) Durability aspects of cements with high limestone filler content, in *Proceedings 9th ICCC, New Delhi*. Vol. 5, pp. 53–59.

Thomas, S. *et al.* (1993) MAS NMR studies on partially carbonated Portland cement and tricalcium silicate pastes. *Journal of the American Ceramic Society* **76**, 1998–2004.

Thorman, P., and Schmitz, Th. (1991) Effect of grinding in high pressure roller mill on cement properties (in German). *Zement-Kalk-Gips* **45**, 188–193.

Tong, D., and Lin, Z. (1986) The role of CaF_2 and $CaSO_4$ in cement clinkering, in *Proceedings 8th ICCC, Rio de Janeiro*, Vol. 2, pp. 117–121.

Tsuboi, T.T. *et al.* (1972) The effect of MgO, SO_3 and ZnO on the sintering of Portland cement clinker (in German). *Zement-Kalk-Gips* **25**, 426–430.

Uchida, K., Fukubayashi, Y., and Yanashite, S. (1992) Properties of special cement with adjusted particle size distribution, in *Proceedings 9th ICCC, New Delhi*, Vol. 3, pp. 23–29.

Varadajaran, V. *et al.* (1992) Experience in the use of industrial by-products sludge for cement manufacture by sol-gel technology, in *Proceedings 9th ICCC, New Delhi*, Vol. 2, pp. 278–284.

Wang, A., and Zhang, C. (1997) Study on the influence of the particle size distribution on the properties of cement. *Cement and Concrete Research* **27**, 685–695.

Wang, X., Xiang, Y., and Yang, S. (1991) Formulation process of Portland cement clinker with high sulfate combined mineralizer (in Chinese). *Wuhan Congye Daxue Xuebao* **13**, 43–49 [ref. CA 118/153077].

Wolter, A., and Dreizler, I. (1988) Effect of high pressure roller mill on cement properties (in German). *Zement-Kalk-Gips* **41**, 64–70.

Yegorov, G.G. *et al.* (1982) Investigation of clinker formation process in flows of accelerated electrons. *Tsement (Moscow)* (1), 14–18.

Yi, E., Roy, D.M., and Roy, R. (1996) Microwave clinkering of ordinary and colored Portland cement. *Cement and Concrete Research* **26**, 41–47.

Yo, S., and Guan, H. (1991) Effect of magnesium oxide in the production of white cement clinker (in Chinese). *Wuhan Gongye Daxue Xuebao* **13**, 10–14 [ref. CA 118/153076].

Yoshida, K., and Igarashi, H. (1992) Strength development and heat of hydration of low heat cement, in *Proceedings 9th ICCC, New Delhi*, Vol.3, pp. 16–22.

Yoshida, K., and Okabayashi, S. (1992) Physical properties and early hydration reaction of finely ground cement, in *Proceedings 9th ICCC, New Delhi*, Vol. 4, pp. 104–110.

Yu, Q. *et al.* (1998) Effect of electron water curing and electron charging curing on concrete strength. *Cement and Concrete Research* **28**, 1201–1208.

Zhang, Z., Lin, Z., and Tong, D. (1996) Influence of the type of calcium sulfate on the strength and hydration of Portland cement under initial steam-curing condition. *Cement and Concrete Research* **26**, 1505–1511.

3 Reactive forms of dicalcium silicate and belite cements

The two most important constituents of Portland cement are **alite**, a form of tricalcium silicate, and **belite**, a form of dicalcium silicate. In their hydration both calcium silicates yield – in addition to calcium hydroxide – a nearly amorphous calcium silicate hydrate phase (C-S-H phase), and this hydration product is mainly responsible for the strength and other physico-mechanical properties of the hardened cement paste.

In the production of Portland clinker, calcium carbonate serves as the source of the CaO that is needed. Thus, as a first step, calcium carbonate must be decomposed to CaO and CO_2, before C_2S or C_3S can be formed in a subsequent reaction between CaO and SiO_2:

$$CaCO_3 = CaO + CO_2 \ (+178.2 \text{ J/mol}) \tag{3.1}$$

$$2CaO + SiO_2 = Ca_2SiO_4 \ (-126.3 \text{ J/mol}) \tag{3.2}$$

$$CaO + SiO_2 = Ca_3SiO_5 \ (-112.9 \text{ J/Mol}) \tag{3.3}$$

The thermal decomposition of calcium carbonate is a highly endothermic reaction, and even though the formations of both dicalcium and tricalcium silicate from CaO and SiO_2 are exothermic, the overall formation process of these compounds from $CaCO_3$ and SiO_2 remains endothermic. A comparison of both reactions reveals that the overall chemical enthalpy of C_3S formation is distinctly higher than that of C_2S (1268 J/kg for C_2S and 1770 J/kg for C_3S). Thus the potential exists to reduce the energy consumption in the production of Portland clinker by increasing its belite content at the expense of alite.

Another difference between dicalcium and tricalcium silicate is the temperature at which both compounds may be synthesized. Whereas dicalcium silicate is readily formed from CaO and SiO_2 at temperatures as low as 1000–1100 °C, tricalcium silicate becomes thermodynamically stable only at about 1250 °C, and is formed at a reasonable rate at even higher temperatures. This makes it necessary to employ burning temperatures of up to 1450–1500 °C in the production of Portland clinker, and this fact is mainly responsible for the significant thermal losses accompanying the process. These losses include heat losses due to radiation and convection, heat lost in the clinker, heat lost in the exit gases and dust, and heat lost in the air from the cooler.

In efforts to reduce the energy consumption in the production of cement, the possibilities of reducing the alite content in Portland clinker, or even of eliminating this phase completely, have been extensively studied. Cements of this type are called "belitic cements," "belite cements," "belite-rich cements" or "low-CaO cements." They belong in the category of low-energy cements (see section 16), so called as they may be produced with a lower consumption of energy than ordinary Portland cement. Note, however, that only those low-CaO cements that, in addition to belite, still contain at least some alite may be considered "Portland cements"; cements that are completely devoid of this phase are not "Portland cements," by definition.

Even though the intrinsic bond properties of fully hydrated C_2S are superior to those of C_3S (because of a higher amount of C-S-H formed per unit weight of original calcium silicate), the main problem with belite cements is the fact that dicalcium silicate hydrates much more slowly than tricalcium silicate. Consequently, the strength development in such cements is rather slow, which may be a handicap in many applications. Thus the possibilities of activating the belite phase in belite cements have been studied extensively, and several solutions have been suggested for coping with this problem. In parallel studies, the possibilities of producing more reactive forms of dicalcium silicate have also been explored, and progress achieved in this area will also be discussed in this chapter. Unfortunately, however, these highly reactive products are scarcely of practical interest at present.

3.1   REACTIVE FORMS OF DICALCIUM SILICATE

3.1.1   *Doped forms of dicalcium silicate*

There are known five polymorphs of dicalcium silicate existing at ordinary pressure, designated α, α'_H, α'_L, β, and γ. Their transformation temperatures are as follows:

$$\alpha \xleftrightarrow{\;1425\,°C\;} \alpha'_H \xleftrightarrow{\;1160\,°C\;} \alpha'_L \xleftrightarrow{\;630-680\,°C\;} \beta \xleftrightarrow{\;<500\,°C\;} \gamma$$

$$\alpha'_L \xrightarrow{\;780-860\,°C\;} \gamma$$

The structure of all of them is built from Ca^{2+} and SiO_4^{4-} ions. The structural arrangement of these ions is similar in the first four modifications; all of them are derived from a-C_2S by a progressive decrease of symmetry brought about by changes in the orientation of the SiO_4^{4-} tetrahedra and small alterations in the positions of the Ca^{2+} ions. The structure of γ-C_2S is somewhat different, and this is helpful in efforts to stabilize at ambient temperature (other than the γ modification) by doping the crystalline lattice with small amounts of foreign ions.

Out of the individual polymorphs of dicalcium silicate, the γ form is the one that is thermodynamically stable at ambient temperature. The a-, a'_H- and a'_L-dicalcium silicates are high-temperature modifications, and β-C_2S is a form that is thermodynamically unstable at any temperature.

If produced by burning a CaO (or CaCO) + SiO_2 blend at a high temperature, the C_2S modification that is stable at this temperature and is formed initially in a solid-state reaction tends to convert gradually to γ-C_2S upon cooling. The phase transitions occur in the sequence $a \rightarrow a'_H \rightarrow a'_L \rightarrow \beta \rightarrow \gamma$. However, the transition is usually incomplete, and a mixture of the β and γ modifications is obtained as a result of the cooling process. If the crystallites of β-C_2S are very small, the $\beta \rightarrow \gamma$ transformation may even be prevented completely, and β-C_2S remains preserved in a "metastable" state. Nearly pure β-C_2S may be produced by reheating the initially formed γ-C_2S to temperatures above 1000 °C and rapid quenching (Asaga, 1986). The reactivity of the β-C_2S that is produced may vary in different preparations, and may be affected by the heating rate, maximum burning temperature, rate of cooling, and other factors.

All high-temperature polymorphs of dicalcium silicate, just like its β form, may be stabilized by incorporating suitable foreign ions into their crystalline lattices. The amount of the foreign ion needed for stabilization generally increases along the sequence from a- to β-C_2S. It is not uncommon, however, for the product produced by burning a C_2S raw meal with an added dopant to contain – in addition to its main constituent – also minor amounts of another C_2S modification.

In addition to stabilizing a particular dicalcium silicate polymorph, the foreign compound added as dopant to the starting CaO (or $CaCO_3$) + SiO_2 mix also affects the rate at which C_2S is formed, and the temperature that must be employed in the synthesis of this substance (Feng, 1986; Weisweiler *et al.*, 1986; Rajczyk, 1997). If $CaCO_3$ is used as the CaO source, the dopant may also lower the temperature at which the thermal dissociation of the carbonate gets under way, as well as the activation energy and enthalpy of the reaction (Ahluwalia and Mathur, 1986; Saraswat *et al.*, 1986; Mathur, 1992). Table 3.1 summarizes data on the effect of different dopants on the temperature of $CaCO_3$ decomposition and the activation energy and reaction enthalpy of dicalcium silicate synthesis.

The dopant ion replaces Ca^{2+} or Si^{4+} in the crystalline lattice of C_2S, thus stabilizing the particular C_2S modification. Such a replacement may cause distinct deformations and distortions of the crystalline lattice; the

Table 3.1 Effect of selected dopants on the temperature of $CaCO_3$ decomposition (T_i), activation energy (E_a) and reaction enthalpy (ΔH) of C_2S formation.

Dopant		$T_i(°C)$	E_a (kJ/mol)	$\Delta H(kJ/mol)$
None		740	247	160
Cr_2O_3	0.5%	680	216	142
	1.0%	680	202	139
TiO_2	0.5%	660	172	160
	1.0%	650	161	163
MnO_2	0.5%	670	200	152
	1.0%	670	183	140
Li_2CO_3	0.5%	620	200	135
	1.0%	620	187	120
Na_2O	0.5%	660	216	150
	1.0%	660	200	148
K_2O	0.5%	690	210	148
	1.0%	690	195	140

Sources: Ahluvalia and Mathur (1986), Saraswat et al. (1986), Mathur et al. (1992)

crystal-chemical environment for both Ca and Si atoms changes, as well as the electron binding energies (Yang and Song, 1992). In the case of dopant ions having different charges than Ca^{2+} and Si^{4+}, vacancies in the crystalline lattice may be created. In positron annihilation studies it was found that doping with monovalent cations, such as Na^+, creates oxygen vacancies, whereas doping with P^{5+}, which replaces Si^{4+}, leads to vacancies of Ca^{2+} (Feng, 1986). All these factors may affect the reactivity of the dicalcium silicate that is formed.

The most common procedure for making doped forms of dicalcium silicate consists in burning a well-compacted stoichiometric blend of CaO (or $CaCO_3$) and SiO_2, which has been finely ground together with the dopant, to high temperatures. In most instances it is the β form of C_2S that is obtained, but high-temperature forms may be also produced, if appropriate dopants, in appropriate amounts, are employed. The following ions and their combinations are among those found to be able to stabilize β-C_2S and/or high-temperature polymorphs of dicalcium silicate: Al^{3+}, B^{3+}, Ba^{2+}, Cr^{3+}, Fe^{3+}, K^+, Li^+, Mg^{2+}, Mn^{4+}, Na^+, PO_4^{3-}, SO_4^{2-}, Ti^{4+} (Pritt and Daugherty, 1976; Jelenic et al., 1978; Kantro and Weise, 1979; Matkovic et al., 1980, 1986; Jelenic and Bajzak, 1981; Fierens and Torloq, 1983; Ahluwalia and Mathu, 1986; Asaga, 1986; Choubine and Khnykine, 1986; Feng et al., 1986; Saraswat et al., 1986; Weisweler et al., 1986; Gawlicky, 1992; Mathur et al., 1992; Prakash et al., 1992; Yang and Song, 1992; Rajczyk, 1997).

The reactivity and strength development of dicalcium silicate may be affected by the quality and quantity of the dopant. It is believed that the presence of dopants in the crystalline lattice generally increases the

Table 3.2 Effect of different dopants on C$_2$S reactivity.

Dopant	Effectiveness		Method	Reference
B$_2$O$_3$ 0.6%	10 d: −25%	100 d: −31%	XRD	Asaga (1986)
BaSO$_4$ 2%	1 h: +38%	24 h: +18%	Cal	Yang and Song (1992)
Fe$_2$O$_3$ 0.7%	3 d: −12%	28 d: +115%	C. str.	Prakash et al. (1992)
K$_2$O 2%	1 h: +8%	24 h: +9%	Cal	Yang (1992)
MgO 3%	1 h: +12%	24 h: +30%	Cal	Yang (1992)

XRD = X-ray diffraction; Cal = calorimetry; C. str. = compressive strength

reactivity of dicalcium silicate by introducing defects or distorting the structure of the crystalline lattice. However, there are only a few publications in the literature in which the reactivity of the doped material is compared directly with that of non-doped C$_2$S. Table 3.2 summarizes data from experiments in which the reactivity of doped C$_2$S was compared directly with that of a non-doped preparation produced in a similar way. For better comparison of different investigators' data, the differences found are expressed as a percentage of the value found in the control sample.

Data on the effect of different dopants on the hydration rate and strength published by different investigators vary widely. This may be because the rate at which a dicalcium silicate preparation reacts with water will depend not only on the dopant selected, but also on its quantity and on the conditions of sample preparation. So it was reported that the rate of reaction will increase with increasing burning temperature (Fierens and Tierloq, 1983), and will decline with repeated firing.

A direct side-by-side comparison of the reactivities of different C$_2$S polymorphs is impossible, as they must be prepared with different dopants or at least with different additions of the same dopant. Nevertheless, there are indications that the α and α' modifications are more reactive than the β modification.

In conclusion it may be stated that doping with appropriate foreign ions may help to increase the reactivity of dicalcium silicate; however, such effects are rather limited, and generally the rate at which the doped material will react with water will remain below that of tricalcium silicate. The resultant reactivity will also depend on the amount of dopant used, and on the conditions of sample preparation. The rate of C$_2$S hydration may be accelerated, to a limited extent, by adding suitable accelerators, such as CaCl$_2$, Ca(NO$_3$)$_2$, K$_2$CO$_3$ or calcium acetate (El-Didamony *et al.*, 1996). These act by forming insoluble compounds with Ca(OH)$_2$ or catalytically.

3.1.2 Mechano-chemically activated forms of dicalcium silicate

One factor that increases the initial rate of reaction of inorganic binders with water is their fineness, as at this stage the rate of hydration increases proportionately with the surface of the solid component. Thus the potential exists to increase the initial reactivity of dicalcium silicate by fine grinding. Unfortunately, however, there are limits on how far a compact solid material can be ground by conventional mechanical grinding; a specific surface area of about 1000 m^2/kg (Blaine) seems to be the upper limit in most instances.

Upon prolonged grinding the surface area of the ground material increases up to a maximum value, and may start to decrease as the grinding continues. This is due to the microplasticity of extremely fine particles, which causes their gradual agglomeration under the influence of mechanical forces.

In addition to an increase of the specific surface area, prolonged grinding may result in a mechano-chemical activation of the material, and may also increase its chemical reactivity. In experiments in which B_2O_3-doped β-C_2S was ground in a vibration mill for up to 70 hours, the specific surface area increased gradually up to 30 hours and subsequently decreased. In samples exposed to extreme mechano-chemical activation the crystallinity of the material decreased. The crystallite size decreased continually, and changes were observed in the crystal unit cell dimensions. Up to an optimum degree of grinding the reactivity of the dicalcium silicate increased and along with it its compressive strength (Luhul *et al.*, 1992). The following values were reported for samples ground for different periods of time:

Grinding time (h)	Particles $d < 5\,\mu m$ (%)	Compressive strength	
		7 d	28 d
		(MPa)	
1	30.9	1.1	12.1
5	44.6	3.7	28.6
30	71.4	14.3	58.5
70	63.3	6.8	42.9

It is obvious that the observed dramatic increase of the strength values cannot be explained merely by the increased fineness of the material, and that the mechano-chemical activation taking place in the course of grinding was essential for obtaining these high strength values.

The disadvantages of mechano-chemical activation include the high energy costs associated with prolonged grinding, and the increased water requirement of the finely ground binder. It must also be expected that the reactivity of the material will gradually decline upon prolonged storage.

3.1.3   Reactive forms of dicalcium silicate produced by low-temperature procedures

As well as high-temperature sintering of a blend of CaO and SiO_2, methods have been developed that make it possible to synthesize highly reactive forms of dicalcium silicate by reducing the required burning temperature.

Roy and Oyefesobi (1977) produced β-C_2S with an extremely high specific surface area by spraying a $Ca(NO_3)_2$ solution, together with colloidal SiO_2, into a vertical tube furnace preheated to 750–950 °C. Alternatively, a solution containing both starting materials in an appropriate stoichiometric ratio was first dried at 70 °C and then calcined at 760 °C. The products obtained reacted about 10 times faster with water than β-C_2S produced at high temperatures by solid phase reactions. In a similar approach, a highly reactive β-C_2S was produced by drying and calcining at low temperature a gel obtained by hydrolysis of an ethanolic solution of calcium nitrate and tetraethoxy silane (Chatterjee, 1996).

Yang *et al.* (1982, 1986, 1992) produced a highly reactive form of β-C_2S by a hydrothermal treatment at 100 °C of a moistened blend of CaO and aerosil, or silica gel, and by subsequent firing at 850–1300 °C. The product exhibited a greatly disordered structure, with Ca^{2+} ions irregularly distributed in the crystal lattice sites. The specific surface area (BET) of the product ranged between 9 and 22 m^2/g, and decreased with increasing burning temperature. The hydraulic reactivity of this form of dicalcium silicate was found to be very high, and similar to that of tricalcium silicate.

Kurdowski *et al.* (1997) produced a highly reactive form of dicalcium silicate by shaking a stoichiometric suspension of calcium hydroxide and amorphous silica at ambient temperature for one month, drying the product obtained, and burning it at 800 or 900 °C. Singh *et al.* (1997) produced active dicalcium silicate in a similar way, by using calcium nitrate and silica gel as starting materials, and by burning the product at temperatures between 600 and 1000 °C.

Ishida and co-workers (Ishida *et al.*, 1992; Sasaki *et al.*, 1993; Okada, 1994) studied the possibility of producing reactive forms of C_2S by a low-temperature dehydration of calcium silicate hydrates with a C/S ratio of 2.0. These are α-dicalcium silicate hydrate $[Ca_2(HSiO_4)(OH)]$, dellaite $[Ca_6(Si_2O_7)(SiO_4)(OH)_2]$, and hillebrandite $[Ca_2(SiO_3)(OH)_2]$. Out of these, the last may be produced conveniently in a pure form from CaO and quartz in an autoclave process at 250 °C. From it a highly reactive form of β-C_2S may be produced by thermal decomposition at 600 °C. The product produced in this way had a specific surface area of about 10 m^2/g and at 80 °C hydrated completely within 24 hours. The specific surface area and reactivity of the β-C_2S formed also depended on the temperature used in the thermal decomposition of hillebrandite, and declined as this temperature was increased. The C-S-H phase formed in the hydration of this form of β-C_2S had a C/S ratio higher than is common in the hydration of a material formed by solid-state reactions at high temperatures, i.e. 1.9–2.0.

Komljenovic *et al.* (1997) produced hydrothermally (at 200–275 °C) α-C_2SH (in some samples together with C_3SH_2), which was subsequently converted to a reactive β-C_2S by burning at 1000 °C.

3.1.4   Polymerized dicalcium silicate pastes

Lu *et al.* (1990) produced specimens by compacting under high pressure (250 MPa) a reactive β-C_2S mixed with 15% of water. Prior to testing, the specimens were stored either under water or in solutions of Fe^{2+}, Ni^{2+}, or a silane compound having the formula $CH_2=CH-Si(OC_2H_5)_3$. The C-S-H phase formed in the hydration process in samples stored in such solutions exhibited an increased degree of SiO_4 polymerization compared with specimens stored in plain water. At the same time the strength of the material was increased (by up to 53%), and this effect was attributed to a cross-linking action of the agents employed.

3.2   BELITE CEMENTS

Belite (or belitic) cements are produced by grinding belitic clinkers with limited amounts of calcium sulfate (gypsum or anhydrite). Such clinkers contain belite (dicalcium silicate) as their sole or main calcium silicate phase. In addition, they contain tricalcium aluminate and the ferritic phase. Alite (tricalcium silicate) may also be present in some belitic clinkers, but only in very limited amounts. They differ both from ordinary Portland clinker and from 'Portland clinker with an elevated C_2S content' (see section 2.4) by having a lower CaO content, which results in a lime saturation factor of not more than LSF = 80.

Research on belite cements was spurred by efforts to reduce energy consumption in the manufacture of cement, as clinker of this type may be produced from raw meals with reduced calcium carbonate contents and at a reduced burning temperature. Typically, a reduction of the lime saturation factor from LSF = 100 to LSF = 80 will reduce the amount of $CaCO_3$ in the raw meal by about 120 kg per tonne of clinker, which corresponds to a reduction of the chemical enthalpy of clinker formation by about 220 kJ/t.

Table 3.3 summarizes the required burning temperature, the bulk density, and the alite and belite contents of clinkers produced from identical starting materials with different lime saturation factors. Also given are the compressive strengths of cements made from these clinkers with identical sulfate additions and ground to the same specific surface area (Ludwig and Pöhlman, 1986). The burning temperature required to reduce the free lime content to acceptable levels declined with decreasing lime saturation factor, indicating the possibility of reducing the thermal losses in the burning process. At the same time a decrease of the bulk density of the clinker, and along with it an improvement of its grindability, was also observed. Unfortunately, however, these advantages are balanced by a reduction in the strength properties of the resultant cements, especially after shorter hydration times.

To obtain strengths that are still acceptable, even though reduced, it

Table 3.3 Properties of cements with different lime saturation factors.

LSF	Burning temp (°C)	Bulk dens. (kg/L)	Alite (%)	Belite (%)	Compressive strength (MPa)	
					7 d	180 d
75	1280	1.8	0	80	–	15
80	1280	2.2	12	68	5	25
85	1290	2.2	32	48	18	47
90	1300	2.2	46	34	25	55
95	1330	2.3	56	24	30	59
100	1380	2.7	80	0	42	64

appears that the amount of calcium carbonate in the raw meal should be reduced only to levels at which some alite can still be formed in the burning process. It was found that such an approach distinctly increases the strengths obtained, compared with clinkers in which this phase is completely absent. As an alternative solution it was suggested that belite cements with acceptable strength properties could be produced by blending an alite-free clinker, made with a very low lime saturation factor, with limited amounts of another clinker, whose composition corresponds to ordinary Portland cement and which may serve as a source of the required alite (Ludwig and Pöhlman, 1986).

To improve the strength development of belite cements it has also been recommended that small amounts of BaO should be added to the raw meal (Rajczyk and Nocun-Wczelik, 1992; Rajczyk *et al.*, 1992; Rajczyk, 1997). This oxide acts as a dopant for C_2S and increases its reactivity.

In addition to the C_3S/C_2S ratio, the compressive strength of belite cements also depends on the Al_2O_3/Fe_2O_3 ratio in the interstitial phase (Ikabata, 1997), though to a lesser extent. Up to about 28 days the strength of the cement tends to increase with an increasing value of this ratio.

Belite cements may be used in applications in which high early strength is not essential. They also meet the requirements for cements with low hydration temperature.

3.3 ACTIVATED BELITE CEMENTS

Activated belite (or belitic) cements differ from "normal" belite cements by the presence of a belite phase whose reactivity has been enhanced by specific measures. The main purpose of such activation is to improve the strength development of the resultant cement, especially at shorter hydration times.

It has been found that very effective activation may be achieved by combining an increased alkali content of the raw meal and a very rapid cooling of the clinker formed, in particular in the temperature range 1300–900 °C (Stark *et al.*, 1979, 1986; Gies and Knöfel, 1986; Lampe,

1986; Milke, 1992). In this way the α'-C_2S phase, which is more reactive than β-C_2S and exists in the resultant clinker at high temperatures, is at least partially preserved in the cooled product. Moreover, defects are introduced into the crystalline lattice, which increase the reactivity of the dicalcium silicate present even further. To effectively prevent the α'-$C_2S \rightarrow \beta$-C_2S conversion, cooling rates of up to 1000 K/min were found to be necessary. About 5% of alkali oxides and other foreign ions are typically present in the stabilized α'-C_2S phase.

Another dopant that may be considered for increasing the reactivity of belite cements is SO_4^{2-} ions (Gies and Knöfel, 1987; Stark et al., 1987). Under these conditions the final clinker contains the β-C_2S rather than the α'-C_2S phase, even if high cooling rates have been employed; however, the SO_4^{2-}-doped form of β-dicalcium silicate is much more reactive than its SO_4^{2-}-free counterpart. The doped C_2S typically contains about 3% each of SO_3 and Al_2O_3 in its crystalline lattice. If an SO_3-doped belitic clinker that also contains some alite is to be produced, the amount of SO_3 in the raw meal must not exceed about 2–3% (related to the final clinker), as higher amounts of sulfates hamper, or even prevent, the formation of the latter phase. In cements made from sulfate-doped clinkers the amount of calcium sulfate interground with the clinker must be reduced, to keep the overall SO_3 content within an acceptable range.

The reactivity of belite cements may also be increased by adding BaO to the raw meal (Gawlicky, 1992; Rajczyk and Nocun-Wczelik, 1992; Rajczyk et al., 1992; Rajczyk, 1997). Up to about 1.5% of BaO may be incorporated in belite, thus increasing its reactivity. Usually barium monoaluminate is also formed in the burning process under these conditions.

Kuznetsova et al. (1992) produced activated belite cements by introducing into the raw meal industrial waste products containing phosphorus, chromium, and titanium, and by employing cooling rates of up to 5000 K/min. Such clinkers contained dicalcium silicate mainly in its β form, but α and α' modifications were also present. By adding carbon to the raw mix, regions with local temperatures above 2100 °C were obtained before the bulk of limestone present was decarbonated. This resulted in a non-stoichiometry of the produced clinker minerals and along with it an increase of reactivity.

Cements made from doped and rapidly cooled belitic clinkers by grinding them with appropriate amounts of gypsum exhibit more favorable strength development than "ordinary" belite cements. At an equal fineness the 28-day strength of such cements is similar to that of ordinary Portland cement; however, the strengths at shorter times still tend to be somewhat lower. Strengths after one year may be higher, owing to the higher amount of C-S-H and lower amount of calcium hydroxide formed in the hydration. The hydration heat development of such cements is lower, whereas most other properties are similar to those of ordinary Portland cement (Matsunaga et al., 1987).

The composition of the hardened paste made from both "normal" and "activated" belite cement is similar to that of ordinary Portland cement, except for a distinctly lower portlandite [Ca(OH)$_2$] content. If alkali activated, the alkali content of the paste is increased. The reduced portlandite content does not adversely affect the carbonation behavior of concrete structures made with belite cements (Fuhr and Knöfel, 1992; Fuhr *et al.*, 1992). By contrast, in alkali-activated belite cements the higher alkali content may be critical, should the cement be used in combination with an alkali-sensitive aggregate, even though it has been reported that the alkali content in the pore fluid of such cement pastes is not higher than that common in ordinary Portland cement (Fuhr and Knöfel, 1992).

Activated belite cement, just like non-activated cement, may be produced from raw meals with a lower CaCO$_3$ content, thus lowering the emission of CO$_2$ in the production process. The NO$_x$ emission will also be reduced, owing to the lower burning temperature. The clinker may be burnt in a conventional rotary kiln; however, if an alkali-doped clinker is produced, the alkali salts that are present may cause ring formation in the kiln or precipitation of deposits in the preheater.

A serious technical problem shows the necessity for extremely fast cooling in the production of "activated" belitic clinkers. Under laboratory conditions very high cooling rates may be achieved by quenching the clinker in a suitable liquid (such as water); however, such an approach may not be practical under large-scale conditions. In conventional industrial coolers, as used in today's cement plants, the cooling rate ranges between 150 and 300 K/min, which is short of the required rate of 1000 K/min. Thus new coolers must be developed that are able to meet this requirement. Owing to the size of the clinker granules the hot clinker will have to be comminuted first, to achieve the required cooling rate in the whole material.

The extremely rapid cooling will necessarily degrade the overall thermal economy of the burning process. It has been estimated that the heat losses due to cooling will increase from 450 kJ/kg, typical for a conventionally cooled clinker, to 650 kJ/kg in rapidly cooled clinkers, thus practically eliminating the energy savings brought about by the lower lime saturation factor and reduced burning temperature (Locher, 1986).

The texture of belitic clinker is usually rather fine, with a medium size of both alite and belite crystals of around 10 μm. The clinker is rather porous, and exhibits very good grindability.

A typical activated belite cement meets the requirements of specifications for ordinary and low-heat Portland cements, and appears to be suitable for applications in which moderate early strength development may be acceptable. A white belite cement may be produced by lowering the Fe$_2$O$_3$ content in the raw meal to near-zero values (Muntean *et al.*, 1992).

REFERENCES

Ahluwalia, S.A., and Mathur, V.M. (1986) Thermal studies on the effect of some transition metal oxides on the kinetics of formation and stabilization of β-dicalcium silicate, in *Proceedings 8th ICCC, Rio de Janeiro*, Vol. 2, pp. 40–45.

Asaga, K. (1986) Effect of thermal history and additives on the hydration of β-Ca_2SiO_4, in *Proceedings 8th ICCC, Rio de Janeiro*, Vol. 3, pp. 148–153.

Chatterjee, A.K. (1996) High belite cements – present status and future technological options: Parts I + II. *Cement and Concrete Research* **26**, 1213–1237.

Choubine, V., and Khnykine, Y. (1998) Modification of belite clinkers (in French), in *Proceedings 8th ICCC, Rio de Janeiro*, Vol. 2, pp. 253–255.

El-Didamony, H. *et al.* (1996) Hydration characteristics of β-C_2S in the presence of some accelerators. *Cement and Concrete Research* **26**, 1179–1187.

Feng, X.-J. *et al.* (1986) Investigations on doped β-C_2S with positron annihilation, in *Proceedings 8th ICCC, Rio de Janeiro*, Vol. 2, pp. 128–133.

Fierens, P., and Tirloq, J. (1983) Nature and concentration effect of stabilizing elements of beta dicalcium silicate on its hydration rate. *Cement and Concrete Research* **13**, 267–276.

Fuhr, C., and Knöfel D. (1992) Activated belite-rich cements – ready for practical use? Investigation of pore fluids and durability, in *Proceedings 9th ICCC, New Delhi*, Vol. 5, pp. 32–38.

Fuhr, C., Müller, A., and Knöfel, D. (1992) Durability of mortars and concretes with varying lime standards, in *Proceedings 9th ICCC, New Delhi*, Vol. 5, pp. 26–31.

Gawlicky, M. (1992) Studies of barium addition on the β-Ca_2SiO_4 hydration, in *Proceedings 9th ICCC, New Delhi*, Vol. 4, pp. 449–453.

Gies, A., and Knöfel, D. (1986) Influence of alkalis on the composition of belite-rich clinkers and the technological properties of the resulting cements. *Cement and Concrete Research* **16**, 411–422.

Gies, A., and Knöfel, D. (1987) Influence of sulfur on the composition of belite-rich cement clinkers and technological properties of resulting cements. *Cement and Concrete Research* **17**, 317–328.

Ikabata, T., and Takemura, H. (1997) Influence of the composition of interstitial phases on the hydration properties of belite cements, in *Proceedings 10th ICCC, Göteborg*, paper 2ii007.

Ishida H. *et al.* (1992) Hydration of β-C_2S prepared at 600 °C from hillebrandite: C-S-H with Ca/Si = 1.9–2.0, in *Proceedings 9th ICCC, New Delhi*, Vol. 4, pp. 76–82.

Jelenic, I., and Bajzak, A. (1981) On the hydration kinetics of α'- and β-modifications of dicalcium silicate. *Cement and Concrete Research* **11**, 467–471.

Jelenic, I., Bajzak, A., and Bujan, M. (1978) Hydration of B_2O_3-stabilized α'- and β-modifications of dicalcium silicate. *Cement and Concrete Research* **8**, 173–180.

Kantro, D.L., and Weise, C.H. (1979) Hydration of various beta-dicalcium silicate preparation. *Journal of the American Ceramic Society* **62**, 621–626.

Komljenovic, M.M., Zivanovic, B.M., and Rosic, A. (1997) β-dicalcium silicate synthesis by liquid state reaction, in *Proceedings 10th ICCC Göteborg*, paper 1i017.

Kurdowski, W., and Wieja, K. (1992) Belite cement with increased brownmillerite (in Polish). *Cement-Wapno-Gips* **45**, 45–50.

Kurdowski, W., Diszak, S., and Trybalska, B. (1997) Belite produced by means of low temperature synthesis. *Cement and Concrete Research* **27**, 51–62.

Kuznetsova, T.V., Osokin, A.P., and Akimova, V.G. (1992) The structure and properties of low-CaO contained clinker, in *Proceedings 9th ICCC, New Delhi*, Vol. 2, pp. 88–93.

Lampe, F., Altrichter, B., and Müller, A. (1986) On the relationship between minor constituents in belite, cooling rate, crystal modification and hydraulic properties (in German). *Silicattechnik (Berlin)* **37**, 238–243.

Locher, F.W. (1986) Low energy clinker, in *Proceedings 8th ICCC, Rio de Janeiro*, Vol. 1, pp. 57–67.

Lu, P., Zhao, J., and Shen, W. (1990) A study of low porosity polymerized β-C_2S paste. *Materials Research Society Symposium Proceedings* **179**, 235–241.

Ludwig, U., and Pöhlman, R. (1986) Investigations on the production of low lime Portland cement, in *Proceedings 8th ICCC, Rio de Janeiro*, Vol. 2, pp. 363–371.

Luhul, Z.J., Fan, Y., and Yang, J. (1992) The mechano-chemical activation of belite, in *Proceedings 9th ICCC, New Delhi*, Vol. 3, pp. 51–57.

Mathur, V.K., Gupta, R.S., and Ahluwalia, S.C. (1992) Lithium as intensifier in the formation of dicalcium silicate phase, in *Proceedings 9th ICCC, New Delhi*, Vol. 2, pp. 406–412.

Milke, I., Müller, A., and Stark, J. (1992) Active belite cement, in *Proceedings 9th ICCC, New Delhi*, Vol. 2, pp. 399–405.

Matkovic, B. *et al.* (1980) Reactivity of belite stabilized by $Ca_5(PO_4)_3OH$, in *Proceedings 7th ICCC, Paris*, Vol. 2, pp. 1/189–1/194.

Matkovic, B. *et al.* (1986) Dicalcium silicate doped with phosphates, in *Proceedings 8th ICCC, Rio de Janeiro*, Vol. 2, pp. 276–280.

Matsunaga, A., Ito, T., and Chikamatsu, R. (1997) Study on strength and adiabatic temperature rise of high strength concrete using low-heat Portland cement (in Japanese). *Semento Konkurito Ronbushu* **51**, 370–375 [ref. CA 128/220689].

Muntean, M. *et al.* (1992) White belitic cement, in *Proceedings 9th ICCC, New Delhi*, Vol. 3, pp. 45–50.

Okada, Y. (1994) Characterization of C-S-H from high reactive β-dicalcium silicate prepared from hillebrandite. *Journal of the American Ceramic Society* **77**, 1313–1318.

Prakash, R., Ahluwalia, S.C., and Sharma, J.M. (1992) Microstructural studies on pure and iron doped unhydrated and hydrated dicalcium silicate phase, in *Proceedings 9th ICCC, New Delhi*, Vol. 4, pp. 397–403.

Pritt, I.M., and Daugherty, K.E. (1976) The effect of stabilizing agents on the hydration rate of β-C_2S. *Cement and Concrete Research* **6**, 783–796.

Rajczyk, K. (1997) Hydraulic activity of low temperature dicalcium orthosilicate modified by BaO addition, in *Proceedings 10th ICCC, Göteborg*, paper 2ii045.

Rajczyk, K., and Nocun-Wczelik, W. (1992) Studies of belite cement from barium containing by-products, in *Proceedings 9th ICCC, New Delhi*, Vol. 2, pp. 250–254.

Rajczyk, K., Kurdowski, W., and Werynsky, B. (1992) Belite cement with barium oxide addition (in Polish). *Cement-Wapno-Gips* **45**, 9–13.

Roy, D.M., and Oyefesobi, S.O. (1977) Presentation of very reactive Ca_2SiO_4 powder. *Journal of the American Ceramic Society* **60**, 178–180.

Saraswat, I.O., Mathur, V.K., and Ahluwalia, S.C. (1986) Thermal composition and phase transformation studies of $CaCO_3$-SiO_2 (2:1) system in presence of sodium and potassium carbonate, in *Proceedings 8th ICCC, Rio de Janeiro*, Vol. 2, pp. 162–166.

Sasaki K. *et al.* (1993) Highly reactive β-dicalcium silicate. V. Influence of specific surface area on hydration. *Journal of the American Ceramic Society* **76**, 970–974.

Singh, V.K., Pyare, R., and Singh, V.K. (1997) Formation kinetics of β-dicalcium silicate, in *Proceedings 10th ICCC Göteborg*, paper 1i016.

Stark, J. *et al.* (1979, 1980) On active belite cements. Parts 1–3 (in German). *Silikattechnik (Berlin)* **30**, 357–362, **31**, 50–52, 168–171.

Stark, J. *et al.* (1986) Conditions of the existence of hydraulically active belite cement, in *Proceedings 8th ICCC, Rio de Janeiro*, Vol. 2, pp. 306–309.

Stark, J., Müller, A., and Milke, I. (1987) On belite clinkers containing SO_3 (in German). *Silikattechnik (Berlin)* **38**, 381–385.

Weisweiler, W., Osen, E., and Lemperle, M. (1986) Effect of Fe_2O_3 and Al_2O_3 admixtures on the reaction kinetics between CaO and SiO_2 powder compacts, in *Proceedings 8th ICCC, Rio de Janeiro*, Vol. 2, pp. 83–88.

Yang, N., and Zhong, B. (1982) A study on active β-dicalcium silicate. *Journal of the Chinese Silicate Society* **10**, 161–166.

Yang, N., Zhong, B., and Wang, Z. (1986) An active β-dicalcium silicate preparation and hydration, in *Proceedings 8th ICCC, Rio de Janeiro*, Vol. 3, pp. 22–27.

Yang, N., Zhang, H., and Zhong, B. (1992) Study on hydraulic reactivity and structural behavior of very active β-C_2S, in *Proceedings 9th ICCC, New Delhi*, Vol. 4, pp. 285–291.

Yang, S., and Song, H. (1992) The structure and properties of hydration of doped dicalcium silicate, in *Proceedings 9th ICCC, New Delhi*, Vol. 4, pp. 291–296.

4 Cements containing calcium sulfoaluminate

4.1 CALCIUM SULFOALUMINATE

Two ternary compounds exist in the system C-S-A-$\bar{S}$: calcium sulfoaluminate and calcium sulfosilicate.

Calcium sulfoaluminate [tetracalcium trialuminate sulfate ($4CaO.3Al_2O_3.SO_3$ or $Ca_4(Al_6O_{12})(SO_4)$, abbreviation $C_4A_3\bar{S}$], also called also "kleinite" or "Klein's compound," may be produced from a starting blend of an appropriate oxide composition by burning it at a temperature of about 1250–1300 °C (Santoro et al., 1986; Havlica and Sahu, 1992; Zhang et al., 1992; Sharp, 1999):

$$3CaCO_3 + 3Al_2O_3 + CaSO_4.2H_2O \rightarrow 4CaO.3Al_2O_3.SO_3 \\ + 3CO_2 + 2H_2O \quad (4.1)$$

The compound is stable up to about 1350–1400 °C.

The second compound of the system, calcium sulfosilicate [$5CaO.2SiO_2.SO_3$ or $Ca_5(SiO_4)_2.(SO_4)$, abbreviation $C_5S_2\bar{S}$, also sometimes called sulfospurrite] is stable only in a narrow temperature range, 1100–1180 °C, and is of little significance for cement chemistry.

The following sequence of reactions takes place in the system C-S-A-$\bar{S}$ at high temperatures:

1 800–900 °C: formation of C_2S and C_2AS;
2 above 1000 °C: disappearance of C_2AS and formation of both $C_4A_3\bar{S}$ and $C_5S_2\bar{S}$;
3 above 1180 °C: disappearance of $C_5S_2\bar{S}$ and the formation of a'-C_2S and $C\bar{S}$.

The crystalline structure of Klein's compound consists of a three-dimensional framework of AlO_4 tetrahedra sharing corners, with Ca^{2+} and SO_4^{2-} ions located in the existing cavities. It belongs to the tetragonal system (Zhang et al., 1992). Above about 1350 °C this phase becomes unstable and starts to decompose (Puertas et al., 1995).

The Al^{3+} within the structure of the $C_4A_3\bar{S}$ phase may be partially substituted by Fe^{3+} ions (Muntean *et al.*, 1998). The reactivity of this phase will decline with increasing iron content in the crystalline lattice.

The hydration of $C_4A_3\bar{S}$ depends on whether calcium sulfate and calcium hydroxide are also present, and progresses at temperatures up to 75 °C as follows (Ogawa and Roy, 1981; Santoro *et al.*, 1986; Hanic *et al.*, 1989; Havlica and Sahu, 1992):

1   In pure water $C_4A_3\bar{S}$ yields monosulfate ($C_4A\bar{S}H_{12}$) and aluminum hydroxide (AH_3) as products of hydration:

$$C_4A_3\bar{S} + 18H \rightarrow C_4A\bar{S}H_{12} + 2AH_3 \qquad (4.2)$$

2   Mixes of $C_4A_3\bar{S}$ and $C\bar{S}H_2$ (gypsum) yield ettringite alone, if mixed at a molar ratio of at least 1:2, and a combination of ettringite and monosulfate if the amount of gypsum is reduced. Simultaneously aluminum hydroxide is formed as a reaction product:

$$2C_4A_3\bar{S} + 2C\bar{S}H_2 + 52H \rightarrow C_6A\bar{S}_3H_{32} + C_4A\bar{S}H_{12} + 4AH_3 \quad (4.3)$$

$$C_4A_3\bar{S} + 2\ C\bar{S}H_2 + 36H \rightarrow C_6A\bar{S}_3H_{32} + 2AH_3 \qquad (4.4)$$

3   Mixes of $C_4A_3\bar{S}$ with calcium hydroxide, but without calcium sulfate, yield the hydrogarnet phase C_3AH_6 and an AFm phase of the approximate composition $C_3A.\frac{1}{2}C\bar{S}.\frac{1}{2}CH.xH$:

$$C_4A_3\bar{S} + 7CH + 2x\ H \rightarrow C_3AH6 + 2C_3A.\frac{1}{2}C\bar{S}.\frac{1}{2}CH.xH \qquad (4.5)$$

4   In the presence of sufficient amounts of both gypsum and calcium hydroxide ettringite is formed as the sole reaction product:

$$C_4A_3\bar{S} + 8C\bar{S}H_2 + 6CH + 74H \rightarrow 3C_6A\bar{S}_3H_{32} \qquad (4.6)$$

As to the phase relations in the system $C_4A_3\bar{S}$–$C\bar{S}H_2$–CH–H, it was found that it includes three quaternary phases, namely $C_3A.\frac{1}{2}C\bar{S}.\frac{1}{2}CH.H_x$, $C_4A\bar{S}.H_x$ and $C_6A\bar{S}_3.H_{32}$, which coexist with CH, $C\bar{S}H_2$, AH_3, and C_3AH_6 in eight four-phase assemblages.

Strontium and barium analogs of $C_4A_3\bar{S}$ in which Ca^{2+} has been partially replaced by Sr^{2+} and Ba^{2+} have also been synthesized and studied (Long *et al.*, 1992; Toreanu *et al.*, 1992; Yan *et al.*, 1992), but cements containing these phases are not in use.

4.2 SULFOBELITE CEMENTS

Sulfobelite cements, also called sulfoaluminate-belite cements, contain the phases belite (C_2S) and tetracalcium trialuminate sulfate ($C_4A_3\bar{S}$) as their main constituents. They do not contain alite or tricalcium alumi-nate, but may contain variable amounts of calcium aluminate ferrite. They also contain calcium sulfate in amounts higher than normal in Portland cement. Some minor phases that may also be present include CA, $C_{12}A_7$, M (periclase), C_2AS (gehlenite), and $C_5S_2\bar{S}$ (sulfospurrite) (Andac and Glasser, 1995). In clinkers made from raw meals with a high MgO content the phase C_3MS_2 (merwinite) may also be present (Ali *et al.*, 1997).

Table 4.1 shows the phase composition of sulfobelite cements as devel-oped by different investigators. The amounts of the individual phases in the clinker may vary in the following ranges:

- C_2S: 10–60%
- $C_4A_3\bar{S}$:   10–55%
- $C\bar{S}$:   0–25%
- C: 0–25%
- C_2(A,F): 0–40%
- CA: 0–10%
- $C_{12}A_7$: 0–10%

Table 4.1 Composition of sulfobelite cements developed by different investigators.

Source	Phases
Ikeda (1980)	a-C_2S 34–41%; $C_4A_3\bar{S}$ 29–54%; CA 6–20%; $C\bar{S}$ 4–8%; glass
Sudoh *et al.* (1980)	C_2S 32%; $C_4A_3\bar{S}$ 41–54%; C 0–9%; C_4AF 0–22%
Mehta (1980)	C_2S 25–65%; $C_4A_3\bar{S}$ 10–20%; C_4AF 15–40%; $C\bar{S}$ 1–20%
Wang *et al.* (1986)	C_2S 15–30%; $C_4A_3\bar{S}$ 35–60%; C_4AF 15–45%
Mudbhatkal *et al.* (1986)	β-C_2S, $C_4A_3\bar{S}$
Lang and Sauman (1988)	β-C_2S, $C_4A_3\bar{S}$, $C\bar{S}$, C
Wang *et al.* (1992a)	a-$C_2\bar{S}$ 10–17%; $C_4A_3\bar{S}$ 33–56%; $C\bar{S}$ 5–36%
Beretka *et al.* (1993)	β-C_2S 29%; $C_4A_3\bar{S}$ 42%; $C\bar{S}$ 29%
Sahu and Majling (1994)	C_2S 41–60%; $C_4A_3\bar{S}$ 13–18%; $C\bar{S}$ 9–17%; C 0–21%
Knöfel and Wang (1994)	C_2S 30%; $C_4A_3\bar{S}$ 60%; C_4AF 5%; $C_{11}A_7.CaF_2$ 5%
Kasselouri *et al.* (1995)	β-C_2S 47%; $C_4A_3\bar{S}$ 20%; $C\bar{S}$ 20%; C_4AF 13%
Andac and Glasser (1995)	C_2S; $C_4A_3\bar{S}$; C_4AF
Beretka *et al.* (1996)	C_2S 8–27%; $C_4A_3\bar{S}$ 15–55%; $C_5S_2\bar{S}$ 18–77%
Ali *et al.* (1997)	C_2S 40%; $C_4A_3\bar{S}$ 32%; C_4AF 20%; $C\bar{S}$ 8%
Su *et al.* (1997)	C_2S 10–25%; $C_4A_3\bar{S}$ 50–70%; C_4AF 3–10% C_2S 10–20%; $C_4A_3\bar{S}$ 40–60%; C_4AF 15–25%

Sulfobelite cements may perform either as non-expansive high early strength cements, or as expansive cements, depending on the proportion of the individual phases in the clinker and the amount of interground calcium sulfate.

The terminology of various sulfobelite cements is rather confused. Tentatively the following designations are being suggested:

- **Sulfoaluminate-belite cement (SAB)**: cement in which belite is the main constituent; the content of $C_4A_3\bar{S}$ is relatively low.
- **Sulfoferroaluminate-belite cement (SFAB)**: cement similar in composition to SAB, but with a significant replacement of Al by Fe.
- **Belite-sulfoaluminate cement (BSA)** or just **Belite-sulfoaluminate cement**: cement in which the sulfoaluminate phase predominates over the belite phase.
- **Belite-sulfoferroaluminate cement (BSFA)** or just **Belite-sulfoferroaluminate cement**: belite-sulfoaluminate cements with a partial replacement of Al by Fe in the sulfoaluminate phase.
- **Belite-sulfoaluminate-ferrite cement (BSAF)**: sulfobelite cement with a particularly high iron content that – in addition to the Fe-doped sulfoaluminate phase – also contains distinct amounts of the ferrite phase.

Belite (dicalcium silicate; C_2S): is usually the most abundant phase in sulfobelite cements. Its amount in sulfobelitic clinkers may vary over a wide range. It is usually present as β-C_2S, but may sometimes also be present in its α modification. Belite contributes only little to the strength of the cement after short hydration times, but is the phase mainly responsible for the ultimate strength.

Tetracalcium trialuminate sulfate (calcium sulfoaluminate, $C_4A_3\bar{S}$): is the phase that is mainly responsible for the early strength development of the cement. For best performance the $C_4A_3\bar{S}$ content in non-expansive sulfobelite cements should lie between 30% and 40% (Beretka *et al.*, 1996). In the presence of iron oxide in the raw meal, a small amount of this oxide may enter into solid solution with the sulfoaluminate phase, giving $C_4A(3-x)F_x\bar{S}$ with x around 0.15 (Sharp *et al.*, 1999).

The **ferrite** phase (calcium aluminate ferrite; $C_2(A,F)$) is formed in the presence of Fe_2O_3 in the raw meal. The ferrite phase present in sulfoaluminate cements possesses a higher reactivity than in ordinary Portland cement, presumably because of its formation at a lower temperature (Sharp *et al.*, 1999). It contributes to both the short-term strength and the ultimate strength of the cement.

The **calcium aluminate** phases CA and $C_{12}A_7$ may be present in sulfobelitic clinkers if the SO_3 content in the raw meal is insufficient to convert the whole amount of Al_2O_3 into $C_4A_3\bar{S}$. Both phases exhibit a fast hydration.

Calcium sulfate (anhydrite, $CaSO_4$; $C\bar{S}$) remains in the clinker in free form at SO_3 contents in the raw meal higher than those needed to convert the existing Al_2O_3 into $C_4A_3\bar{S}$.

Free lime (CaO; C) may also be present in some sulfobelitic clinkers. Owing to the low burning temperature employed in the manufacture of such clinkers, it is more reactive than free CaO present in Portland clinker/cement, and hydrates rather readily to $Ca(OH)_2$ upon mixing with water.

To produce a sulfobelitic clinker, a starting mix of appropriate oxide composition must be burnt at maximum temperatures of 1200–1300 °C. In addition to limestone, clay, bauxite and gypsum or anhydrite, a variety of industrial by-products, such as fly ash, blast furnace slag, high-SO_3 ashes, or phosphogypsum, may also be employed as starting materials (Beretka *et al.*, 1992, 1993; Santoro *et al.*, 1986; Sahu and Majling, 1994). It has been observed that the use of such materials, especially phosphogypsum, enables both the temperature and the time needed for completing the burning process to be reduced. The clinkering process and the texture of the clinkers that are formed may also be affected by the addition of some minor constituents to the raw meal, such as calcium phosphate, barium sulfate, or borax (Chae and Choi, 1996a, 1996b). On an industrial scale the burning process may be performed in conventional rotary kilns (Mudbhatkal *et al.*, 1986).

The clinker formation process is initiated by the dehydration of gypsum to anhydrite at around 100–120 °C, followed by a decomposition of clay minerals at around 300–600 °C. The decarbonization of the calcium carbonate that is present starts at about 700 °C, and is completed before the temperature reaches 900 °C.

By 800–900 °C gehlenite ($2CaO.Al_2O_3.SiO_2$, abbreviation C_2AS) starts to be formed as an intermediate phase; it decomposes at temperatures below 1200 °C.

At 1000–1100 °C the phase CA is formed, which reacts further with CaO to $C_{12}A_7$. As the burning temperature increases, the latter reacts with calcium sulfate, to yield $C_4A_3\bar{S}$.

The formation of $C_4A_3\bar{S}$ takes place between about 1000 and 1250 °C. This phase may be formed either directly from CaO, Al_2O_3, and $CaSO_4$, or in a solid-state reaction between intermediately formed $C_{12}A_7$ and anhydrite:

$$3(12CaO.7Al_2O_3) + 7CaSO_4 \rightarrow 7(4CaO.3Al_2O_3.SO_3) + 15CaO \quad (4.7)$$

or in a heterogeneous solid-gas reaction between primary formed tricalcium aluminate and sulfur oxide formed in the thermal dissociation of the calcium sulfate present (Strigac and Majling, 1997):

$$3(3CaO.Al_2O_3) + SO_2 + \tfrac{1}{2}O_2 \rightarrow 4CaO.3Al_2O_3.SO_3 + 6CaO \quad (4.8)$$

At temperatures above 1200 °C $C_4A_3\bar{S}$ starts to decompose: it reacts with free lime, if still present, to yield C_3A and $CaSO_4$.

Dicalcium silicate is formed in the temperature range 1000–1200 °C.

Owing to the presence of free calcium sulfate, sulfospurrite [$Ca_5(SiO_4)_2(SO_4)$, abbreviation $C_5S_2\bar{S}$] may also be formed as an intermediate phase, starting at about 900 °C (Beretka *et al.*, 1992, 1993; Takuma *et al.*, 1994). It decomposes to dicalcium silicate and calcium sulfate above 1200–1280 °C.

The amount of free $CaSO_4$ declines in the course of burning as it is consumed in the formation of $C_4A_3\bar{S}$ and sulfospurrite. Above 1200 °C its amount increases again, as sulfospurrite decomposes. The ferrite phase starts to be formed at about 1100 °C in the form of C_2F. At higher temperatures it takes up alumina, and eventually it attains a composition of about C_6AF_2.

To preserve the $C_4A_3\bar{S}$ that is formed, and to prevent the thermal dissociation of calcium sulfate at high temperatures, the burning temperature in the production of sulfobelitic clinkers should not exceed 1300 °C.

The kinetics of belite sulfoaluminate clinker formation may be altered by adding to the original raw mix some foreign ions (Chae and Choi, 1996a, 1996b). Calcium phosphate stabilizes the β-C_2S phase while having a negative effect on C_4AF formation. Barium sulfate lowers the formation temperature of C_2S and accelerates the formation of C_4AF. Borax stabilizes a'- and β-C_2S and lowers the formation temperature of C_4AF. All three additives retard the formation of $C_4A_3\bar{S}$.

The texture of sulfobelitic clinkers, as studied by optical microscopy, is characterized by the presence of well-developed crystals of the sulfoaluminate phase, 1–10 μm in size. The SO_3 content of this phase is usually higher than the corresponding theoretical value (Takuma *et al.*, 1994). $C_4A_3\bar{S}$ is embedded in a C_2S + C_4AF matrix. Here C_2S forms an agglomerated microstructure of fine particles several micrometers in size. Significant amounts of SO_3 are incorporated into the crystalline lattice of this phase (Takuma *et al.*, 1994). The ferrite phase consists of columnar crystals 2–8 μm in size. It is low in Al_2O_3, and its composition is close to that of C_6AF_2 (Takuma *et al.*, 1994). The clinker has a high porosity, and is easy to grind (Beretka *et al.*, 1992; Ivashchenko, 1992; Andac and Glasser, 1995; Takuma *et al.*, 1994; Kasselouri *et al.*, 1995).

The chemical enthalpy of the formation of sulfobelitic clinkers is lower than that of ordinary Portland clinkers and even of belitic clinkers. This is due mainly to a lower overall CaO content in the clinker to be produced, as well as to the fact that a fraction of the needed CaO is introduced in the form of calcium sulfate rather than calcium carbonate. The thermal losses are also reduced because of the lower burning temperature required. A potential problem in large-scale production may be the partial thermal dissociation of $CaSO_4$ present in the raw meal, which may require appropriate measures to control the emission of SO_3. On the other hand, the emission of both CO_2 and NO_x will be significantly reduced (Takuma *et al.*, 1994).

To obtain sulfobelite cement the clinker that is produced must be ground to a specific surface area similar to that of ordinary Portland cement. Gypsum or anhydrite must be added only if the amount of anhydrite in the clinker is insufficient. The grindability of sulfobelitic clinkers is better than that of typical Portland cement clinkers.

The initial stage of hydration of sulfobelite cement is dominated by a fast formation of ettringite. A significant fraction of the sulfoaluminate phase is consumed within a few hours of hydration (Mudbhatkal *et al.*, 1986; Wang *et al.*, 1992a; Kasselouri *et al.*, 1995; Ali *et al.*, 1997). About 60–70% of it is consumed within the first 24 hours, and most but not all within 28 days. Generally, the degree of hydration increases with increasing amount of calcium sulfate in cement (Wang *et al.*, 1992a). The initial hydration of the cement may be accelerated by the presence of sulfate ions in the liquid phase. Aluminate and calcium ions delay hydration by creating a retarding layer, which may be considered a co-precipitate of aluminum and calcium hydroxide (Palon and Majling, 1996).

The $C_2(A,F)$ phase also hydrates very rapidly, yielding the AFt phase as a product of reaction. A large fraction of the ferrite phase is consumed within the first day of hydration (Kasselouri *et al.*, 1995).

Initially, the hydration of $C_4A_3\bar{S}$ progresses mainly according to equation (4.4) (section 4.1). However, if the dissolution rate of the calcium sulfate cannot keep pace with that of the $C_4A_3\bar{S}$, the hydration may also progress partly according to reaction (4.3). Aluminum hydroxide is formed simultaneously.

If non-reacted calcium sulfate is still present, after all the $C_4A_3\bar{S}$ and $C_2(A,F)$ have been consumed, the formation of ettringite may continue, though at a slower rate, in a reaction between AH_3, formed as a by-product in the hydration of $C_4A_3\bar{S}$, and calcium sulfate and calcium hydroxide, formed in the hydration of C_2S:

$$AH_3 + 3C\bar{S}H_2 + 3CH + 20H \rightarrow C_6A\bar{S}_3H_{32} \tag{4.9}$$

In cements in which the calcium sulfate content has been too low for complete conversion of all the Al_2O_3 present in the original cement to ettringite, a formation of the AFm phase may get under way in a reaction between ettringite and AH_3, after all the calcium sulfate has been used up:

$$C_3A.3C\bar{S}.32H + 2AH_3 \rightarrow 3C_3A.C\bar{S}.12H + 2H \tag{4.10}$$

For optimum formation of ettringite, the amount of calcium sulfate that is dissolved in the liquid phase must match the dissolution rate of $C_4A_3\bar{S}$ and $C_2(A,F)$. This may be controlled by the amount of calcium sulfate present, its form, and its degree of dispersion. If the rate of calcium sulfate dissolution is too low, an AFm phase or even C_3AH_6 may be formed, in addition to, or instead of, AFt (ettringite), even in the presence of free calcium sulfate (Mudbhatkhal *et al.*, 1986). Conversely, if the dissolution of the anhydrite is too fast, it may convert into gypsum before reacting with $C_4A_3\bar{S}$ (Sahu *et al.*, 1992, 1993). Free gypsum may also be found in mature pastes made from cements with calcium sulfate contents higher than needed for complete conversion of the $C_4A_3\bar{S}$ and $C_2(A,F)$ phases to ettringite. As well as $C_4A_3\bar{S}$ and $C_2(A,F)$, the phases CA and $C_{12}A_7$ – if present in the clinker – may also react with calcium sulfate and calcium hydroxide to yield ettringite.

The hydration of the belite is much slower than that of $C_4A_3\bar{S}$, and becomes significant only after longer hydration times. Very little of it is hydrated within the first 7 days, and between about 5% and 40% hydrates within 28 days (Sudoh *et al.*, 1980; Wang *et al.*, 1992; Kasselouri *et al.*, 1995; Malami *et al.*, 1996). The rate of hydration generally increases with increasing SO_3 content in cement (Wang *et al.*, 1992). The C-S-H phase formed in the hydration is mainly responsible for the ultimate strength of the hardened cement paste. Most or all of the calcium hydroxide produced in the hydration of C_2S is consumed in the formation of ettringite, just like the free calcium oxide present in the original clinker.

In the absence of free lime and at low alkali contents the ettringite formed within the first hours and days of hydration appears in the form of a fine crystalline material (crystals about 1–2 μm long). It is formed in a through-solution process, and gradually fills the empty space between the original cement particles. In the later stages of hydration limited amounts of this phase may also be formed as larger needle-like crystals, to be precipitated in larger pore spaces still existing in the paste (Kasselouri *et al.*, 1995).

In mature sulfobelite cement pastes the C-S-H phase and AFt (ettringite) are the two main hydration products present. Variable amounts of AFm and residual gypsum may also be present, depending on the amount of calcium sulfate in the original cement. The material contains no, or very little, free calcium hydroxide, as it has been consumed in the formation of ettringite. The alkalinity of the pore liquid is relatively low, at around pH = 9.5–10 (Wang *et al.*, 1992a).

At equal starting water/cement ratios and degrees of hydration the porosity of pastes made from sulfobelite cements is distinctly lower than that of ordinary Portland cement pastes. This is due to the higher amount of water bound within the crystalline lattice of the solid phases formed in the hydration of the former cement, especially in that of the AFt phase.

Some sulfobelite cements exhibit distinct expansion in the course of hardening, and may qualify as expansive cements (see section 21), whereas others do not. Whether or not the paste will expand will depend on the way and time of formation of ettringite in the course of hydration. As long as most of the ettringite is formed before the development of a rigid skeletal structure in the paste, the paste will not expand. An expansion *will* take place if ettringite is formed in a topochemical reaction at a later stage of hydration after the paste has attained a distinct but limited rigidity.

It has been postulated that, to prevent expansion, the formation of ettringite must take place in the absence of free calcium hydroxide [according to equations (4.4) or (4.3), section 4.1]. Under these conditions $C_4A_3\bar{S}$ will hydrate rapidly, but will give rise to a dimensionally stable product (Kasselouri *et al.*, 1995; Lan and Glasser, 1996; Beretka *et al.*, 1997). However, according to other investigators (Sudoh *et al.*, 1980; Sahu and Majling, 1994), a limited amount of free lime (calcium hydroxide) is desirable, to achieve rapid strength development and a high early strength of the cement.

The development of strength and the extent of expansion will depend on the composition of the cement and the proportions of the phases involved. Within reasonable limits and at appropriate calcium sulfate contents the early strength will increase with increasing amounts of the sulfoaluminate phase in the cement and – to a lesser extent – with increasing content of the ferrite phase. However, the belite phase will affect mainly the final strength of the cement. Table 4.2 shows the effect of cement composition on the compressive strength development of a series of sulfobelite cements.

The strength achieved and the extent of expansion will also depend on the sulfate content in cement. It has been suggested that for optimum strength development the SO_3/Al_2O_3 molar ratio in the cement should range between 1.3 and 1.9 (Sudoh *et al.*, 1980). At this ratio most or all of the ettringite will be formed prior to setting, and the main calcium

Table 4.2 Effect of composition of sulfobelite cements on compressive strength development.

	Rapid hardening	Rapid hardening	Normal	Slow hardening	Slow hardening
Composition (%)					
C_2S	25	30	45	50	65
$C_4A_3\bar{S}$	20	20	20	10	10
C_4AF	40	30	15	30	15
$\bar{S}$	15	20	20	10	10
Compressive strength (MPa)					
8 h	n.d.	15.6	n.d.	n.d.	n.d.
1 d	34.8	28.3	9.5	5.6	5.2
3 d	36.9	33.8	19.3	7.6	8.9
7 d	37.4	37.7	27.4	11.7	12.4
28 d	n.d.	n.d.	49.8	14.1	14.5
90 d	n.d.	n.d.	n.d.	21.4	22.4
120 d	51.8	53.8	86.2	n.d.	n.d.

n.d. = not determined
Source: Mehta (1980)

aluminate sulfate hydrate phase formed in the later stages of hydration will be AFm, whose formation does not cause expansion. Non-reacted residual gypsum will be absent in the mature paste.

Figures 4.1 and 4.2 show the effect of the amount of gypsum added to a

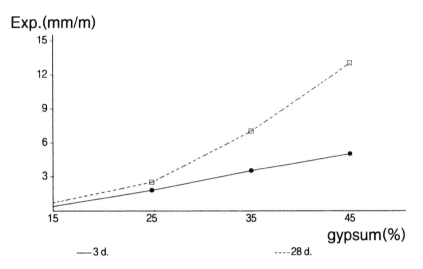

Figure 4.1 Effect of gypsum interground with a sulfobelitic clinker on the expansion of the resultant cement. SO_3 in clinker = 10.0 wt%.
Source: Wang *et al.* (1992a)

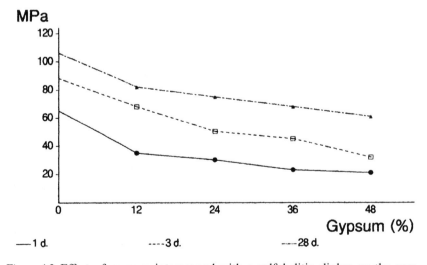

Figure 4.2 Effect of gypsum interground with a sulfobelitic clinker on the compressive strength of the resultant cement. SO_3 in clinker = 10.0 wt%.
Source: Wang *et al.* (1992a)

sulfobelitic clinker (with an SO_3 content of 10.0 wt%) on expansion and compressive strength development.

The extent of expansion increases because of the increasing amounts of ettringite formed after setting (Wang *et al.*, 1992). Along with it the strength of the cement declines as a consequence of excessive expansion. The available data indicate that both a non-expansive high early strength cement and an expansive cement may be produced from the same clinker just by varying the amount of interground calcium sulfate. To prevent expansion, the cement must not contain excessive amounts of aluminates, calcium sulfate, or free lime (Lan and Glasser, 1996; Beretka, 1997).

The composition of non-expansive sulfobelite cements is usually adjusted for them to perform as high early strength binders. Such binders are characterized by a short setting time (less than 1 hour) and a fast development of short-term strength, which is due mainly to fast formation of the ettringite phase. Measurable strengths may be attained within 3 hours of hydration. After most of the sulfoaluminate phase has been converted to ettringite, the strength development slows down significantly before an additional strength increase, associated with hydration of the belite phase, gets under way. The strength after 28 days is relatively low, as at this time a significant fraction of dicalcium silicate still remains non-hydrated, but a significant additional increase of strength takes place between 28 days and 1 year.

It has been claimed (Huo *et al.*, 1996) that the compressive strength of sulfobelite cement may be increased by up to 35% by exposing the mixing water to a magnetic field before mixing, but this claim has not so far been confirmed from other sources.

Non-expansive sulfobelite cements exhibit good chemical resistance in contact with chloride or sulfate solutions. Storage in such solutions (sodium chloride: 20 g/L; sodium sulfate: 3 g/L) appears to have little effect on the development of strength (Malami *et al.*, 1996). No chloroaluminate hydrates were formed in samples stored in the chloride solution for 90 days.

Heat liberation in the hydration of sulfobelite cements is characterized by two peaks within the first few hours, attributed to the formation of gypsum from anhydrite and the precipitation of ettringite (Sudoh *et al.*, 1980; Sahu *et al.*, 1993). In spite of rapid heat liberation at the beginning of hydration, due mainly to the formation of ettringite, the total amount of heat liberated in the hydration of sulfobelite cements is lower than that of Portland cement, and amounts to about 200 kJ/kg (Su *et al.*, 1997). This is due to the absence of C_3S and C_3A in this cement, which are phases whose hydration is highly exothermic. The overall heat of hydration of sulfobelite cement will decline with increasing belite content.

In high-iron sulfobelite cements (also called calcium sulfoaluminate-ferrite cements) the amount of Fe_2O_3 present is increased, and they may contain up to 45% of the ferrite phase (Mehta, 1980; Wang *et al.*, 1986, 1992b;

Palon and Majling, 1995; Su *et al.*, 1997). Other phases present include C_2S, $C_4A_3\bar{S}$ and $C\bar{S}$, out of which $C_4A_3\bar{S}$ may contain distinct amounts of Fe^{3+} in its crystalline lattice (Muntean *et al.*, 1998). The amount of free calcium sulfate in the cement may be adjusted by varying the amount of gypsum or anhydrite added to the raw meal and interground with the finished clinker. By altering the amount of this constituent the cement may be produced as either an expansive or a non-expansive binder (Wang *et al.*, 1986).

The following phases are formed in the hydration of high iron sulfobelite cement: C-S-H, AFt and smaller amounts of AFm, AH_3 and FH3 (Wang, 1986). The AFt phase formed in the hydration process contains variable amounts of Fe^{3+} ions substituting for Al^{3+} ions, the amount being dependent on the $C_4A_3\bar{S}/C_4AF$ ratio in the clinker. The AFt phase contributes significantly to both the short-term and the ultimate strength of the hardened cement paste. The strength development of these cements may be controlled by varying the proportions of the clinker phases present and the amount of interground calcium sulfate.

Wang *et al.* (1986) suggested replacing part of the gypsum to be interground with the clinker with limestone, to eliminate a retrogression of bending strength at later ages. Under these conditions the phases C_4ACH_{11} and C_2AH_8 will also be formed in the hydration.

Wang *et al.* (1992b) combined ferroaluminate cement with micro-silica and fly ash. Mortars made from such blends with an added superplasticizer attained higher strengths even at fly ash contents of 30 per cent. Remarkably, the flexural/compressive strength ratio of such mortars was exceptionally high, i.e. around 0.2. (Table 4.3).

Toreanu *et al.* (1986) produced a **white sulfobelite cement** by omitting Fe_2O_3 completely from the raw meal, and by increasing its MgO content to 10% to increase its burnability.

In high-$C_4A_3\bar{S}$ sulfobelite cements the amount of $C_4A_3\bar{S}$ may reach 70%, and the amount of C_2S is reduced to 10–25%. The amount of the ferrite

Table 4.3 Strength properties of blended high-iron sulfobelite cements

Micro-silica (%)	Fly ash (%)	Superplas-tisizer (%)	Compressive Strength (MPa)		Flexural Strength (MPa)	
			3 d.	28 d.	3 d.	28 d.
–	–	2	75	–	11	–
10	–	2	99	124	17	26
10	10	2	98	126	18	26
10	20	2	103	131	21	27
10	30	2	103	125	23	24

Note: clinker composition: $C_4A_3\bar{S}$=61%, C_2S=15%, C_4AF=24% mortar, w/c=0.20 amount and form of SO_3 in cement: not published
Source: Wang et al. (1992b)

phase may vary between 3% and 25% (Su *et al.*, 1997). To produce a rapid-hardening cement the amount of interground gypsum should range between 10% and 15%, whereas for expansive cements those amounts should be 20–25%. In addition to fast setting and short-term strength development the cement also exhibits excellent resistance to the alkali-silica reaction (Su *et al.*, 1997).

It has been suggested that sulfobelitic blended cements could be produced by combining sulfobelitic clinkers with granulated blast furnace slag or fly ash, in addition to calcium sulfate (Sudoh *et al.*, 1980; Beretka *et al.*, 1992, 1993). Such an approach makes little sense, however, because calcium hydroxide is virtually absent among the hydrates formed, and the pH of the pore solution is relatively low. Under these conditions the added fly ash cannot undergo a pozzolanic reaction, and the hydraulic activation of the slag is insufficient.

In assessing the durability of hydrated sulfobelite cements the following facts must be considered. Both ettringite and residual gypsum undergo dehydration at temperatures that are significantly lower than those at which C-S-H decomposes. Ettringite loses most of its combined water and turns amorphous below 120 °C, even though its basic structural framework remains intact up to about 400 °C. The amorphous calcium sulfoaluminate decomposes above 600 °C, forming $C_4A_3\bar{S}$, $CaSO_4$, and CaO (Park *et al.*, 1997). The thermal decomposition of the hardened cement is associated with a distinct loss of strength. In high-$C_4A_3\bar{S}$ sulfobelite cement the hardened cement paste loses about 15% of its original strength upon heating to 100–150 °C. At 300 °C the strength loss is about 45%, and at 500 °C about 70% (Su *et al.*, 1997).

Sulfobelite cement is more sensitive to carbonation than ordinary Portland cement. In addition to carbonation as it exists in ordinary Portland cement, which involves the portlandite and the C-S-H phases, the ettringite present in the hardened paste may also react with the CO_2 in the air, yielding calcite or aragonite, gypsum and alumina gel (Beretka *et al.*, 1992, 1997; Sherman *et al.*, 1995):

$$C_6A\bar{S}_3H_{32} + 3\bar{C} \rightarrow 3C\bar{C} + 3C\bar{S}H_2 + AH_3 + 23H \qquad (4.11)$$

The process may result in a decrease of strength due mainly to an increase of porosity, brought about by the conversion of part of the water bound within the crystalline lattice of the ettringite into its liquid form. The degree to which carbonation may be relevant for the durability of structures made with sulfobelite cement will also depend on the ratio of C_2S to $C_4A_3\bar{S}$ in the binder that is used. In general, the resistance to carbonation will increase with increasing C_2S content in the cement. In laboratory tests Beretka *et al.* (1992) observed a loss of strength of 30–70% in test specimens stored for 6–12 months in air. By contrast Mudbhatkal *et al.* (1986) reported no noticeable damage in a concrete structure four years after it was erected, and a steady increase of strength during this period. Su *et al.*

(1997) observed only a moderate carbonation of hardened sulfobelite cement in accelerated tests.

Data on the pH of the pore solution in hardened sulfobelite cements are not uniform. According to Wang and Su (1989), high-$C_4A_3\bar{S}$ sulfoaluminate cement possesses a low alkalinity, with the pH of the pore fluid below 11.5. Nevertheless it is claimed that concretes made with such cements are able to adequately protect steel reinforcement from corrosion. According to Zhang *et al.* (1999), ferroaluminate cements with a higher Fe_2O_3 content have a higher pH of their pore solution and a better capacity to protect steel from corrosion than sulfoaluminate cements low in Fe_2O_3. Janotka and Krajci (1999) found an insufficient degree of steel passivation in a sulfobelite cement mortar after 90 days. However, a suitable increase of pH and adequate corrosion protection of the steel reinforcement may be achieved by adding Portland cement to the mix in amounts of at least 15%. In separate work Andac and Glasser (1999) found in the pore solution of a commercial china-made calcium sulfoaluminate cement pH values of 12.8–13.8 between 1 and 60 days of hydration. The main components in the solutions were K and Al, with a maximum for potassium at about three days.

Sulfobelite cements do not meet the requirements of most specifications for Portland cement, because their SO_3 content is higher than permitted. However, special standards for sulfobelite cements were issued in China, and about a million tons of this type of cement are produced annually in that country (Zhang *et al.*, 1999).

4.3 SULFOALITE CEMENT

Sulfoalite cement is characterized by the simultaneous presence of the phases C_3S and $C_4A_3\bar{S}$ in its clinker. Other phases may include C_2S, $C_2(A,F)$ and C_3A. The possibility of producing such clinkers initially appeared uncertain, as the optimum temperatures for the synthesis of the two phases differ significantly: whereas $C_4A_3\bar{S}$ is formed at around 1250–1300 °C and decomposes above 1300 °C, a temperature of at least 1350 °C is required to synthesize C_3S at an acceptable rate. Moreover, it has been reported that the presence of greater amounts of SO_3 in the raw meal tends to reduce or prevent completely the formation of C_3S (Strunge *et al.*, 1985; Cheong *et al.*, 1992; Moranville-Regour and Boikova, 1992). However, it has been found that clinkers containing both phases may be produced at burning temperatures of 1230–1300 °C if small amounts of fluorspar (CaF_2) are added to the raw mix (Abdul-Maula and Odler, 1992; Odler and Zhang, 1996). At the same time the phase composition of the clinker that is produced may be varied over a wide range by varying the oxide composition of the raw meal. The energy consumption in the production of sulfoalitic clinkers is lower than in the production of ordinary Portland clinker, but higher than in the production of sulfobelitic clinkers, owing to the presence of tricalcium silicate. Just like sulfobelitic clinkers, sulfoalitic clinkers are easy to grind.

The strength development of sulfoalite cements may be controlled by varying the phase composition of the clinker and the amount of inter-ground calcium sulfate. This amount is generally higher than is common in ordinary Portland cement, but it must not exceed an optimum amount, to prevent unwanted expansion of the hardened paste. An advantage of sul-foalite cements compared with sulfobelite cements is their more favorable strength development, especially between 1 and 28 days. Table 4.3 shows the strength development of a typical sulfoalite and sulfobelite cement and that of a Portland and a belite cement, for comparison.

Unlike sulfobelite cements, the hardened paste of sulfoalite cements will contain distinct amounts of free calcium hydroxide, in addition to C-S-H and AFt. As with sulfobelite cements, hydrated sulfoalite cement pastes will be more susceptible to thermal degradation than pastes made with ordinary Portland cement. Their resistance to carbonation is significantly improved compared with that of sulfobelite cements. This is because CO_2 reacts preferentially with the free $Ca(OH)_2$ that is present, yielding $CaCO_3$, which results in an increase of the density and a decrease of the perme-ability of the hardened cement paste. Sulfoalite cements do not comply with specifications for ordinary Portland cement, because their SO_3 con-tent is too high.

Table 4.4 Compressive strength development of different types of cement.

	Portland cement	Belite cement	Sulfobelite cement	Sulfoalite cement
Phase composition of clinker (%)				
C_3S	80	0	0	80
C_2S	0	80	60	0
C_3A	10	10	0	0
$C_4A_3\bar{S}$	0	0	20	10
C_4AF	10	10	20	10
SO_3 opt.	3	3	7	5
Compressive strength (MPa)				
1 d	22	9	21	35
3 d	50	12	27	59
7 d	73	23	29	77
28 d	82	35	33	86
365 d	85	96	78	102

Note: Laboratory-made cements
　　SO_3 opt: Gypsum addition yielding optimum strength

Source: Abdul-Maula and Odler (1992)

4.4   CALCIUM SULFOALUMINATE MODIFIED PORTLAND CEMENT

Calcium sulfoaluminate modified Portland cements may be produced by combining ordinary Portland clinker with limited amounts of calcium sulfoaluminate, produced separately (see section 21.3), and calcium sulfate.

Taczuk *et al.* (1992) studied sulfoaluminate-modified cements with $C_4A_3\bar{S}$ contents between 5% and 20% and with a sulfoaluminate/calcium sulfate molar ratio of 1:8. Calcium sulfate was added in the form of anhydrite or gypsum. The form of calcium sulfate present determined its dissolution rate (the dissolution rate of gypsum is much faster than that of anhydrite) and thus the properties of the produced cement. It was found that the presence of anhydrite tended to favor monosulfate precipitation at the expense of ettringite. As a consequence, such cements exhibited significantly shorter setting times and higher short-term strengths.

While sulfoaluminate-modified Portland cements that contain limited amounts of $C_4A_3\bar{S}$ and calcium sulfate exhibit no expansion, those with higher amounts of these phases expand and are used as expansive cements (type K; see section 21.3).

4.5   SULFOALUMINATE CEMENT

Beretka *et al.* (1997) produced a series of cements that contained only the phases $C_4A_3\bar{S}$ (15–50%), $C_5S_2\bar{S}$ (sulfospurrite) (25–77%), and $C\bar{S}$ (anhydrite) (8–25%), but no calcium silicates. Such cements exhibit a very fast initial ettringite formation and strength development, but no additional strength gain at later ages. Mixes with water contents too low for complete conversion of $C_4A_3\bar{S}$ to ettringite exhibit expansion due to delayed ettringite formation, associated with an uptake of water from the environment.

Upon exposure to CO_2 the hardened paste loses strength, because of carbonation. Such strength loss may be prevented if the primary formed ettringite is subsequently converted to monosulfate. This may be achieved by introducing C_4AF to the original cement in an appropriate stoichiometric ratio.

REFERENCES

Abdul-Maula, S., and Odler, I. (1992) SO_3-rich Portland cements: synthesis and strength development. Materials Research Society Symposium Proceedings **245**, 315–320.

Ali, M.M., Bhargava, R., and Ahluvalia, S.C. (1997) Hydration characteristics of magnesia assimilated sulfoaluminate-belite cement, in *Proceedings 10th ICCC, Göteborg*, paper 2ii006.

Andac, M., and Glasser, F.P. (1995) Microstructure and microchemistry of calcium sulfoaluminate cement. *Materials Research Society Symposium Proceedings* **370**, 135–142.

Andac, M., and Glasser, F.P. (1999) Pore solution composition of calcium sulfoaluminate cement. *Advances in Cement Research* **11**, 15–26.

Beretka, J. *et al.* (1992) Synthesis and properties of low energy cement based on $C_4A_3\bar{S}$, in *Proceedings 9th ICCC, New Delhi*, Vol. 3, pp. 195–200.

Beretka, J. *et al.* (1993) Hydraulic behaviour of calcium sulfoaluminate-based cements derived from industrial process waste. *Cement and Concrete Research* **23**, 1205–1214.

Beretka, J. *et al.* (1996) The influence of $C_4A_3\bar{S}$ content and W/S ratio on performance of calcium sulfoaluminate based cements. *Cement and Concrete Research* **26**, 1673–1682.

Beretka, J. *et al.* (1997) Effect of composition on the hydration properties of rapid hardening sulfoaluminate cement, in *Proceedings 10th ICCC, Göteborg*, paper 2ii29.

Chae, W.-H., and Choi, S.-H. (1996a) The effect of borax on formation of modified belite cement minerals (in Korean). *Yoop Hakhoechi* **33**, 1163–1169 [ref. CA 126/161062].

Chae, W.-H., and Choi, S.-H. (1996b) The effect of $CaHPO_4.2H_2O$ and $BaSO_4$ on formation of modified belite cement minerals (in Korean). *Yoop Hakhoechi* **33**, 1024–1030 [ref. CA 125/337102].

Cheong, H.M. *et al.* (1992) Effect of sulphate on the reaction of C_3S formation, in *Proceedings 9th ICCC, New Delhi*, Vol. 2, pp. 335–341.

Hanic, F., Kapralik, I., and Gabrisova, A. (1989) Mechanism of hydration reactions in the system $C_4A_3\bar{S}$-$C\bar{S}$-CaO-H_2O referred to hydration of sulfoaluminate cements. *Cement and Concrete Research* **19**, 671–682

Havlica, J., and Sahu, S. (1992) Thermodynamics and kinetics in the system $Ca_4(Al_6O_{12})(SO_4)$-$CaSO_4.2H_2O$-H_2O up to 75 °C, in *Proceedings 9th ICCC, New Delhi*, Vol. 4, pp. 157–163.

Huo, G. *et al.* (1996) Magnetic treatment for improving the strength of pack material with high water content (in Chinese). *Dougbei Daxue Xuebao Ziran Kexueban* **12**, 163–166 [ref. CA 125/255283].

Ikeda, K. (1980) Cements along the join $C_4A_3\bar{S}$-C_2S, in *Proceedings 7th ICCC, Paris*, Vol. 2, pp. III 31–36.

Ivashchenko, S.I. (1992) Cements based on modified Portland and sulfoaluminate-belite clinkers, in *Proceedings 9th ICCC, New Delhi*, Vol. 3, pp. 222–225.

Jan, A. *et al.* (1980) Sulfoaluminate cement series, in *Proceedings 7th ICCC, Paris*, Vol. 4, pp. 381–386

Janotka, I., and Krajci, L. (1999) An experimental study on the upgrade of sulfoaluminate-belite cement systems by blending with Portland cement. *Advances in Cement Research* **11**, 35–41.

Kapralik, I., and Hanic, F. (1989) Phase relations in the subsystem $C_4A_3\bar{S}$-$C\bar{S}H_2$-CH-H_2O of the system CaO-Al_2O_3-$C\bar{S}$-H_2O referred to hydration of sulfoaluminate cement. *Cement and Concrete Research* **19**, 89–102.

Kasselouri, V. *et al.* (1995) A study on the hydration products of a non-expansive sulfoaluminate cement. *Cement and Concrete Research* **25**, 1726–1736.

Knöfel, D., and Wang, J.F. (1994) Properties of three newly developed quick cements. *Cement and Concrete Research* **24**, 801–812.

Lan, W., and Glasser, F.P. (1996) Hydration of calcium aluminosulfate cements. *Advances in Cement Research* **8**, 127–134.

Lang, K., and Sauman, Z. (1988) Study on the chemism of fast setting sulphoaluminate cements (in Czech). *Silikaty (Prague)* **32**, 79–84.

Lau, W., and Glasser, F.P. (1996) Hydration of calcium sulfoaluminate cements. *Advances in Cement Research* **8**, 127–134.

Long, S., Wu, Y., and Liao, G. (1992) Investigations on the structure and hydration of new mineral phase $3SrO.3Al_2O_3.CaSO_4$, in *Proceedings 9th ICCC, New Delhi*, Vol. 4, pp. 418–423.

Majling, J. *et al.* (1993) Relationship between raw mixture and mineralogical composition of sulfoaluminate belite clinkers in the system $CaO-SiO_2-Al_2O_3-Fe_2O_3-SO_3$. *Cement and Concrete Research* **23**, 1351–1356.

Malami, C. *et al.* (1996) A study on the behaviour of non-expansive sulfoaluminate cement in aggressive media. *World Cement* **27**, 129–133.

Mehta, P.K. (1980) Investigations on energy saving cements. *World Cement Technology* **11**, 166–177.

Moranville-Regour, M., and Boikova, A.I. (1992) Chemistry, structure, properties and quality of clinker, in *Proceedings 9th ICCC, New Delhi*, Vol. 1, pp. 3–45.

Mudbhatkal, G.A. *et al.* (1986) Non-alitic cement from calcium sulfoaluminate clinker – optimization for high strength and low temperature applications, in *Proceedings 8th ICCC, Rio de Janeiro*, Vol. 4, pp. 364–370.

Muntean, M. *et al.* (1998) Synthesis of calcium sulfoferroaluminates and their properties. *Materiale de Constructii (Bucharest)* **281**, 60–64 [ref. CA 129/57741].

Odler, I., and Zhang, H. (1996) Investigations on high SO_3 Portland clinkers and cements. *Cement and Concrete Research* **26**, 1307–1313, 1315–1324.

Ogawa, K., and Roy, D.M. (1981, 1982) $C_4A_3\bar{S}$ hydration, ettringite formation and its expansion mechanism. *Cement and Concrete Research* **11**, 741–750; **12**, 101–109.

Palon, M.T., and Majling, J. (1995) Preparation of the high iron sulfoaluminate belite cements from raw mixtures incorporating industrial waste. *Ceramics Silikaty (Prague)* **39**, 63–67.

Palon, M.T., and Majling, J. (1996) Effect of sulfate, calcium and aluminum ions upon the hydration of sulfoaluminate belite cement. *Journal of Thermal Analysis* **46**, 549–556.

Park, C.-K. *et al.* (1997) Microstructural changes of calcium sulfoaluminate cement paste due to temperature, in *Proceedings 10th ICCC, Göteborg*, paper 41v068.

Puertas, F., Varela, M.T.B., and Molina, S.G. (1995) Kinetics of thermal decomposition of $C_4A_3\bar{S}$ in air. *Cement and Concrete Research* **25**, 572–580.

Sahu, S., and Majling, J. (1994) Preparation of sulfoaluminate belite cement from fly ash. *Cement and Concrete Research* **24**, 1065–1072.

Sahu, S., Majling, J., and Havlica. J. (1992) Influence of anhydrite particle size on the hydration of sulfoaluminate belite cement, in *Proceedings 9th ICCC, New Delhi*, Vol. 4, pp. 443–448.

Sahu, S., Tomkova, V., and Majling, J. (1993) Influence of particle sizes of individual minerals on the hydration process in the system $C_2S-C_4A_3\bar{S}-C\bar{S}$. *Cement and Concrete Research* **23**, 693–699.

Santoro, L., Garofano, R., and Valenti, G.K. (1986) Calcium sulphoaluminate made from phosphogypsum and its hydration properties, in *Proceedings 8th ICCC, Rio de Janeiro*, Vol. 4, pp. 389–394.

Sharp, J.H., Lawrence, C.D., and Yang, R. (1999) Calcium sulfoaluminate cements

– low-energy cements, special cements or what. *Advances in Cement Research* **11**, 3–13.

Sherman, N. *et al.* (1995) Long term behaviour of hydraulic binders based on calcium sulfoaluminate and calcium sulfosilicate. *Cement and Concrete Research* **25**, 113–126.

Strigac, J., and Majling, J. (1997) Phase composition development of calcium sulfoaluminate belite cement in the SO_x atmosphere. *World Cement* **28**, 82–86.

Strunge, J., Knöfel, D., and Dreizler, I. (1985–1990) Effect of alkalis and sulfur on cement properties. Parts I-IV (in German). Zement-Kalk-Gips **38**, 150–158, 441–450, 451–456; **43**, 199–208.

Su, M. *et al.* (1997) Preliminary study on the durability of sulfo/ferro-aluminate cements, in *Proceedings 10th ICCC, Göteborg*, paper 4lv029.

Sudoh, G., Ohta, T., and Harada, H. (1980) High strength cement in the $CaO-Al_2O_3-SiO_2-SO_3$ system and its application, in *Proceedings 7th ICCC, Paris*, Vol. 3, pp. V 152–157.

Taczuk, L. *et al.* (1992) Understanding of the hydration mechanism of $C_4A_3\bar{S}$ – Portland clinker – $CaSO_4$ mixes, in *Proceedings 9th ICCC, New Delhi*, Vol. 4, pp. 278–284.

Takuma, Y. *et al.* (1994) Effect of reducing the emission of CO_2 and the characteristics of clinker in the system $C_2S-C_4A_3\bar{S}-C_4AF-C\bar{S}$. *Journal of the Ceramics Society of Japan* **102**, 1115–1121 [ref. CA 122/37325]; also *Chichibu Onoda Kenkyu Hokoku* **46**, 56–66 (1995) [ref. CA 123/348089].

Toreanu, I., Muntean, M., and Bedo, M. (1986) Belitic Portland cements with various contents of MgO and SO_3, in *Proceedings 8th ICCC, Rio de Janeiro*, Vol. 2, pp. 217–222.

Toreanu, I. *et al.* (1992) Mechanism and the formation of some unitary strontium and barium sulphoaluminates, in *Proceedings 9th ICCC, New Delhi*, Vol. 3, pp. 250–255.

Wang, Y., and Su, M. (1989) Recent development of new kind of cement in China, in *Second Beijing Conference Report*.

Wang, Y., Deng, J., and Su, M. (1986) An investigation into cement $CaO-SiO_2-Al_2O_3-Fe_2O_3-SO_3$ system, in *Proceedings 8th ICCC, Rio de Janeiro*, Vol. 2, pp. 300–305.

Wang, Y. *et al.* (1992a) A quantitative study of paste microstructure and hydration characteristics of sulfoaluminate cement, in *Proceedings 9th ICCC, New Delhi*, Vol. 4, pp. 454–460.

Wang, Y. *et al.* (1992b) Study of high-strength blended ferro-aluminate cement mortar and concrete, in *Proceedings 9th ICCC, New Delhi*, Vol. 5, pp. 621–627.

Yan, P. (1992) Investigation on the hydration of Sr- and Ba-bearing sulfoaluminates, in *Proceedings 9th ICCC, New Delhi*, Vol.4, pp. 411–417.

Zhang, L., Su, M., and Wang, Y. (1999) Development of the use of sulfo- and ferroaluminate cements in China. *Advances in Cement Research* **11**, 15–21.

Zhang, P. *et al.* (1992) The crystal structure of $C_4A_3\bar{S}$, in *Proceedings 9th ICCC, New Delhi*, Vol.3, pp. 201–208.

5 Cements containing the phases $C_{12}A_7$ or $C_{11}A_7.CaF_2$

5.1 THE PHASES $C_{12}A_7$ AND $C_{11}A_7.CAF_2$

The phases dodecacalcium heptaaluminate ($12CaO.7Al_2O_3$ or $Ca_{12}Al_{14}O_{33}$, abbreviation $C_{12}A_7$) and calcium fluoroaluminate ($11CaO.7Al_2O_3.CaF_2$ or $Ca_{12}Al_{14}O_{32}F_2$, abbreviation $C_{11}A_7.CaF_2$ or $C_{11}A_7F$) are closely related. The crystal structure of the former consists of an incomplete framework of corner-sharing (AlO_4) tetrahedra with the empirical composition $Al_7O_{16}^{11-}$ and Ca^{2+} ions. One additional O^{2-} ion per formula unit is distributed statistically between 12 sites. Upon heating in air of normal humidity the material tends to take up water to an amount corresponding to the formula $C_{12}A_7H$. Under these conditions the O^{2-} ions that statistically occupy one of 12 possible sites in a formula unit are replaced by two OH^- ions.

In the crystalline lattice of $11CaO.7Al_2O_3.CaF_2$ the statistically distributed O^{2-} ion is replaced by two F^- ions. Other halide analogs, namely $11CaO.Al_2O_3.CaCl_2$, $11CaO.7Al_2O_3.CaBr_2$ and $11CaO.7Al_2O_3.CaI_2$, and a sulfide analog $11CaO.7Al_2O_3.CaS$ are also known. In blends of CaO, Al_2O_3 and CaF_2, $C_{11}A_7.CaF_2$ is formed at 1100–1300 °C. Above 1300 °C it reacts with excessive CaO – if present – to produce C_3A. The chemical properties of $C_{11}A_7.CaF_2$ are rather similar to those of $C_{12}A_7$. Both compounds form a continuous range of solid solutions.

The initial hydration product in the hydration of $C_{12}A_7$ is C_2AH_8, which is gradually replaced by C_3AH_6. In addition to these phases, significant amounts of an amorphous calcium aluminate hydrate are also formed. An alumina gel that is formed in parallel to the calcium aluminate hydrates converts gradually to crystalline gibbsite (AH_3) (Edmonds and Majumdar, 1988). The aluminate gel is not formed, or its amount is significantly reduced, if the hydration takes place in the presence of free calcium hydroxide (Hannawayya, 1992). In the hydration of $C_{11}A_7.CaF_2$ the phase C_3AH_6, together with alumina gel, is formed as the main product of hydration; below 25 °C small amounts of C_2AH_8 and below 15 °C even CAH_{10} may also be formed. Again, the primary formed alumina gel converts gradually to gibbsite (Yu et al., 1997).

The hydration of both compounds is quite rapid, and – at ambient temperature – is virtually completed within the first 24 hours. Up to a temperature of 30 °C the reaction accelerates with increasing temperature, but it slows down if the temperature is even higher. This is due to an increase of the diffusion resistance of the hydrates formed under these conditions (Yu *et al.*, 1997). The hydration may also be retarded by Zn doping; at the same time the conversion of CAH_{10} and C_2AH_8 to C_3AH_6 may be prevented (Quian *et al.*, 1992).

With increasing water/cement ratio the rate of the hydration process increases significantly, but the hydration products remain the same.

In the presence of calcium sulfate and calcium hydroxide, ettringite rather than calcium aluminates is formed in the hydration of both $C_{12}A_7$ and $C_{11}A_7CaF_2$. Small amounts of monosulfate may also be formed. In pastes made of C_2S and $C_{12}A_7$ hydrated for 16 years gehlenite hydrate (strätlingite, C_2ASH_{18}) was also found among the products of hydration (Hannawayya, 1992).

5.2   ALITE-FLUOROALUMINATE CEMENT

Alite-fluoroaluminate cement, also called regulated set cement (in the USA) or jet-cement (in Japan), contains tricalcium silicate (50–75%) and calcium fluoroaluminate (20–30%) as its main constituents. Other constituents may include the ferrite phase, dicalcium silicate, and free CaO.

Alite-fluoroaluminate clinker can be produced from raw meals containing – in addition to CaO and SiO_2 – appropriate amounts of alumina and fluorspar (CaF_2), by burning them to temperatures of about 1300 °C (Chvatal, 1973; Odler and Abdul-Maula, 1980; Knöfel and Wang, 1994). The added CaF_2 acts as a mineralizer, and allows the formation of tricalcium silicate at temperatures lower than usual in the production of ordinary Portland cement clinker. At temperatures above about 1000 °C the added fluorspar reacts with lime and alumina to yield calcium fluoroaluminate. At the same time it prevents the formation of tricalcium aluminate, which would be readily formed in the absence of fluorine ions. An appropriate maximum burning temperature is essential, because if the temperature is too high, $C_{12}A_7.CaF_2$ loses its stability and converts to tricalcium aluminate. However, it has also been observed that in the range 1250–1450 °C the reactivity of the $C_{12}A_7.CaF_2$ phase is dependent on the clinkering temperature, and increases as this temperature increases. The burning of the clinker may be performed conveniently in a rotary kiln. The cooling rate in the production process must not be too high, as rapid cooling promotes the incorporation of SiO_2 into the structure of $C_{11}A_7.CaF_2$, which – in due course – reduces the reactivity of this phase (Kanaya *et al.*, 1998).

Unlike ordinary Portland clinker, alite-fluoroaluminate clinker is brownish in color, rather than dark gray. It typically possesses a very low porosity, and the alite and belite crystals that are present are distinctly smaller than those in ordinary Portland clinker. Only some of the fluorine is in the $C_{11}A_7.CaF_2$ phase; the rest is incorporated into alite (Odler and Abdul-Maula, 1980).

To obtain alite-fluoroaluminate cement, the alite-fluoroaluminate clinker must be interground with anhydrite. The amount of anhydrite must be adjusted to the amount of fluoroaluminate present in the clinker to enable its hydration to ettringite, and is usually appreciably higher than in ordinary Portland cement.

Upon mixing with water, alite-fluoroaluminate cement exhibits very rapid hydration. In paste hydration at ambient temperature about 40–50% of the $C_{11}A_7.CaF_2$ is hydrated within the first 2 hours, and about 70% within 7 days (Uchikawa and Tsukiyma, 1973; Knöfel and Wang, 1994). The hydration of alite is significantly slower: only about 10% of it hydrates within 2 hours, and about 40–65% within one day.

The main product of hydration at shorter hydration times is ettringite (AFt phase), which is formed in a reaction between calcium fluoroaluminate and anhydrite. At the same time aluminum hydroxide is also formed. The fluorine in the fluoroaluminate phase enters the structure of aluminum hydroxide, yielding aluminum hydroxide fluoride. The process can be expressed by the following equation:

$$11CaO.7Al_2O_3.CaF_2 + 12CaSO_4 + 137H_2O$$
$$\rightarrow 4(3CaO.Al_2O_3.3CaSO_4.32H_2O) + 4Al(OH)_3 + 2Al(OH)_2F \quad (5.1)$$

Tricalcium silicate hydrates to yield the C-S-H phase and calcium hydroxide, which in due course may participate in the hydration of the fluoroaluminate phase and the formation of ettringite:

$$11CaO.7Al_2O_3.CaF_2 + 6Ca(OH)_2 + 18CaSO_4 + 188 H_2O$$
$$\rightarrow 6(3CaO.Al_2O_3.3CaSO_4.32H_2O) + 2Al(OH)_2F \quad (5.2)$$

The formed C-S-H phase varies in its composition, and also contains distinct amounts of Al_2O_3 and SO_3 (Uchikawa *et al.*, 1978).

In the later stage of hydration, after the supply of calcium sulfate has been exhausted, some AFm may also be formed at the expense of AFt:

$$3CaO.Al_2O_3.3CaS_4.32H_2O + 4Al(OH)_3 + 6 Ca(OH)_2$$
$$\rightarrow 3(3CaO.Al_2O_3.CaSO_4.12H_2O) + 8H_2O \quad (5.3)$$

Compared with ordinary Portland cement pastes made with the same water/cement ratio, the porosity of mature alite-fluoroaluminate cement pastes is distinctly lower, mainly because of the high combined water content of the resulting ettringite phase.

At ambient temperature, pastes made from alite-fluoroaluminate cement exhibit a setting time between about 5 and 30 minutes, depending on their

$C_{12}A_7$ content, and a measurable strength within about 2 hours. If desired, the setting time may be extended by the use of suitable retarders, such as citric acid, oxalic acid, boric acid, calcium sulfate hemihydrate, or some polyphosphates (Chvatal, 1973; Uchikawa and Uchida, 1973; Knöfel and Wang, 1994). The short-term strength depends on the SO_3/Al_2O_3 ratio of the cement, and may be adjusted by the amount of gypsum interground with the clinker. Within the range 0.26–0.51 the strength increases with this ratio; however, further increase of the ratio results in a drastic strength reduction and uncontrolled expansion (Knöfel and Wang, 1993).

After a very rapid initial strength development, associated with the formation of the ettringite phase, the strength increases rather slowly between about 3 hours and one day, as the formation of additional ettringite is slowed down, and the amount of formed C-S-H is still low. Subsequently, the strength development accelerates again as more C-S-H is formed. The ultimate strength of alite-fluoroaluminate cements is comparable to those of ordinary Portland cements.

At low temperatures the setting time of alite-fluoroaluminate cements is extended up to several hours, and the development of strength is retarded (Knöfel and Wang, 1992, 1994; Jäger *et al.*, 1993).

Alite-fluoroaluminate cements do not exhibit expansive properties, unless the amount of interground calcium sulfate is excessively high. This absence of expansion is explained by the formation of the ettringite phase in a through-solution process, rather than in a topochemical reaction (Chvatal, 1993).

Just like ordinary Portland cement pastes, alite-fluoroaluminate cement pastes undergo carbonation in a reaction with the CO_2 in air. However, the rate of this process is slower than in similar ordinary Portland cement pastes, mainly because of lower porosity (Kim *et al.*, 1992). X-ray diffraction studies have revealed a gradual disappearance of peaks belonging to the AFt and AFm phases, and the formation of gypsum, calcite, and vaterite (Knöfel and Wang, 1994). The strength of the material is not adversely affected by carbonation (Knöfel and Wang, 1992).

In **blended alite-fluoroaluminate cements** the alite-fluoroaluminate clinker is interground with granulated blast furnace slag and/or limestone, in addition to calcium sulfate (Knöfel and Wang, 1994). Such binders have similarities with blended cements on the basis of ordinary Portland clinker, but still exhibit a short setting time and a distinct short-term strength.

Alite-fluoroaluminate cement is particularly suitable for emergency repair works, where very fast setting and short-term strength development are desired. Other possible applications include shotcrete engineering and precast concrete product production. Alite-fluoroaluminate cement must not be used for steam curing or any applications at high temperatures, because the ettringite phase is thermodynamically unstable at temperatures above about 70 °C.

5.3   BELITE-FLUOROALUMINATE CEMENT

Belite-fluoroaluminate cement is produced from a clinker that contains the phases $C_{11}A_7$.CaF_2 and belite, but no tricalcium silicate, as its main constituents. The clinker may be produced by burning a blend of limestone, bauxite, and fluorspar of appropriate composition in a rotary or shaft kiln at temperatures between 1250 and 1400 °C (Liu *et al.*, 1992). The content of the $C_{11}A_7$.CaF_2 phase in the clinker is usually higher than that in alite-fluoroaluminate cements, and typically ranges between about 40% and 80%. The belite content ranges between 10% and 55%. Limited amounts of the ferrite phase may also be present (Knöfel and Wang, 1992; Liu *et al.*, 1992). To produce belite-fluoroaluminate cement the clinker must be ground with anhydrite, to obtain an SO_3/Al_2O_3 ratio of about 0.5–0.6. Higher SO_3/Al_2O_3 ratios must be avoided to prevent a decline of strength associated with sulfate expansion and crack formation within the cement paste under these conditions (Knöfel and Wang, 1992, 1994; Liu *et al.*, 1992).

Just as in alite-fluoroaluminate cements, ettringite, together with aluminum hydroxide, is the main product of hydration at shorter hydration times, whereas the C-S-H phase, together with calcium hydroxide, is formed in the later stage of hydration. The hydration of dicalcium silicate and the formation of C-S-H in belite-fluoroaluminate cements is much more sluggish than the hydration of tricalcium silicate in alite-fluoroaluminate cements. Owing to existing differences in the compositions of the clinkers used, the hydrated paste made from belite-fluoroaluminate cement contains more ettringite and aluminum hydroxide and less C-S-H. The amount of calcium hydroxide in the paste is particularly low. The porosity of a hydrated belite-fluoroaluminate paste is significantly lower than that of a similar ordinary Portland cement paste. Such low porosity affects positively both the impermeability and the corrosion resistance of the material.

The setting time of belite-fluoroaluminate cements is similar to or even shorter than that of alite-fluoroaluminate cements, and the short-term strength may be somewhat higher, owing to the larger amount of ettringite that is formed. However, subsequent strength development may be more sluggish, owing to the slower hydration rate of dicalcium silicate. As for alite-fluoroaluminate cements, the setting time may be extended by the addition of suitable retarders.

The liberation of heat of hydration is faster and the overall heat of hydration is higher than in ordinary Portland cement. The following values were reported by Liu *et al.* (1992):

- 4 hours: 60 cal/g (251 k J/kg);
- 1 day: 66 cal/g (276 k J/kg);
- 7 day: 75 cal/g (314 k J/kg).

Belite-fluoroaluminate cement is more susceptible to carbonation than its alite-fluoroaluminate counterpart, owing to its higher ettringite content, and this process may be accompanied by a moderate decline of strength (Knöfel and Wang, 1994).

Belite-fluoroaluminate cement may be used in similar applications as alite-fluoroaluminate cement. It is especially suitable for emergency repair works in which extremely fast setting and initial development of strength are required.

5.4   IRON-RICH BELITE-FLUOROALUMINATE CEMENT

Iron-rich belite-fluoroaluminate cement is characterized by the presence of significant amounts of the ferrite phase in its clinker, in addition to calcium fluoroaluminate and belite. The composition of the clinker may vary in the following range: $C_{11}A_7.CaF_2$, 5–60%; C_2S, 10–50%; $C_2(A,F)$, 20–80% (Huang *et al.*, 1992b).

The clinker may be produced by burning a raw meal containing limestone, bauxite, iron ore, and fluorspar at temperatures between 950 and 1050 °C. In clinkers burnt at such low temperatures the ferrite phase is present as C_6AF_2, regardless of the A/F ratio in the original raw meal (Huang *et al.*, 1992a). A calcium aluminate ferrite phase of this composition exhibits a relatively high reactivity, especially if it is produced by high-heating-rate burning (Huang *et al.*, 1992b). To obtain an iron-rich fluoroaluminate cement, the clinker must be interground with 10–20% of gypsum.

Just as in low-iron belite-fluoroaluminate cements, the initial hydration of high-iron belite-fluoroaluminate cement is characterized by a very fast reaction of the $C_{11}A_7.CaF_2$ phase, and the formation of the AFt phase and aluminum hydroxide. The F^- ions that are present interact with aluminum hydroxide, yielding aluminum hydroxide fluoride. The hydration of the belite and ferrite phases is distinctly slower, though faster than is typical in ordinary Portland cements. C_2S hydrates to yield the C-S-H phase and limited amounts of calcium hydroxide, most of which is consumed in the formation of AFt. The hydration products of the ferrite phase include an AFt phase containing Fe^{3+}, together with C_4AH_{19} and C_2AH_8, which also incorporate limited amounts of Fe^{3+} in their structure. The excess of Fe^{3+} yields an iron hydroxide gel (Huang *et al.*, 1992b).

The setting time of high-iron fluoroaluminate cement is distinctly longer than that of its low-iron counterpart, but is still shorter than 1 hour. Generally, the setting time is extended with increasing content of the ferrite phase, and may also be extended by adding small amounts of ZnO to the raw meal. Thus in most instances the cement may be used without any retarders. The cement exhibits measurable strength after several hours of hydration, whose magnitude increases with increasing $C_{11}A_7.CaF_2$ content in the clinker and – within reasonable limits – with increasing amount of added gypsum. The final strength is comparable to that of ordinary Portland cement.

5.5   BELITE FLUORO-SULFO-ALUMINATE CEMENT

Belite fluoro-sulfo-aluminate cement contains simultaneously the phases calcium fluoroaluminate ($C_{11}A_7$.CaF_2) and calcium sulfoaluminate ($C_4A_3\bar{S}$), together with belite and calcium sulfate. It also may contain limited amounts of the ferrite phase.

Iron-rich belite fluoro-sulfo-aluminate cement is characterized by a higher iron content and the presence of significant amounts of the ferrite phase in the clinker. Its clinker composition may vary in the following range (Feng and Zhu, 1986):

- $C_2(A,F)$:   30–75 wt%;
- C_2S: 0–23 wt%;
- $C_{11}A_7$.CaF_2:   0–32 wt%;
- $C_4A_3\bar{S}$:   17–34 wt%.

It may be produced by firing pertinent raw meals at temperatures of around 1250 °C. To produce the cement, the clinker must be ground with gypsum in amounts that depend on the composition of the clinker phase. Owing to the presence of the highly reactive calcium fluoroaluminate and calcium sulfoaluminate phases, the cement exhibits very fast setting and a rapid initial strength development. The ferrite phase contributes both to the short-term and the ultimate strength, whereas the belite phase is responsible only for strength after longer hydration times.

5.6   CALCIUM ALUMINATE AND CALCIUM FLUOROALUMINATE MODIFIED PORTLAND CEMENTS

Calcium aluminate modified Portland cement is essentially a Portland cement with added dodecacalcium heptaaluminate ($C_{12}A_7$), produced separately, and with an elevated calcium sulfate content (Chvatal, 1973; Kondo *et al.*, 1997). The phase composition of the binder differs from that of a conventional alite-fluoroaluminate cement by the presence of tricalcium aluminate, and of $C_{12}A_7$ instead of $C_{11}A_7$.CaF_2. The cement exhibits a very short setting time and a rapid strength development. It has been produced and studied only under laboratory conditions. Potentially it could be employed in similar applications as an alite-fluoroaluminate cement.

Calcium fluoroaluminate modified Portland cement may be produced by joint grinding of an ordinary Portland clinker, a special high-$C_{11}A_7$.CaF_2 clinker, and calcium sulfate. Such a high-$C_{11}A_7$.CaF_2 clinker, with a fluoroaluminate content of 40–50 wt%, may be produced by burning a blend of kaolin, limestone and fluorspar at about 1300 °C (Chvatal, 1973). The resultant binder – unlike alite-fluoroaluminate cement – will contain the C_3A phase in addition to $C_{11}A_7$.CaF_2. It exhibits a short setting time and a fast strength development, which depends on the amount of the fluoroaluminate and calcium sulfate present. Excessive amounts of calcium sulfate must be avoided to prevent undesired expansion of the hardened paste. The cement may be used in similar applications to those of normal alite-fluoroaluminate cement.

REFERENCES

Chvatal, Th. (1973) New rapid setting cement (in German). *Zement-Kalk-Gips* **26**, 385–391.

Edmonds, R.N., and Majumdar, A.J. (1988) The hydration of $12CaO.7Al_2O_3$ at different temperatures. *Cement and Concrete Research* **18**, 473–478.

Feng, X., and Zhu, Y. (1986) Research on an early strength cement containing high content of iron, in *Proceedings 8th ICCC, Rio de Janeiro*, Vol. 2, pp. 286–292.

Hannawayya, F. (1992) Chemistry of hydration products formed in paste of β-C_2S with $C_{12}A_7$ aged 16 years, in *Proceedings 9th ICCC, New Delhi*, Vol. 4, pp. 257–264.

Huang, W. *et al.* (1992a) The research on $C_2A_xF_{(1-x)}$ iron-rich fluoroaluminate cements, in *Proceedings 9th ICCC, New Delhi*, Vol. 3, pp. 9–15.

Huang, W. *et al.* (1992b) New energy-saving cement – iron-rich fluoroaluminate cement, in *Proceedings 9th ICCC, New Delhi*, Vol. 3, pp. 277–282.

Jäger, R.G., Esser, G., and Knöfel, D. (1993) Development of compressive strength and porosity of some regulated set cements. *Cement and Concrete Research* **23**, 700–710.

Kanaya, M., Matsumi, M., and Ichikawa, M. (1998) Effect of cooling rate on fine texture and hydraulic activity of clinker in the C_3S-C_2S-$C_{11}A_7$-CaF_2-C_4AF system (in Japanese). *Muk. Materiaru* **5** (274), 215–219 [ref. CA 119/84891].

Kim, S., Tsurumi, T., and Daimon, M. (1992) Carbonation of jet cement pastes (in Japanese). *Semento Konkurito Ronbunshu* **46**, 586–591 [ref. CA 119/102017].

Knöfel, D., and Wang, J. (1992) Investigation on pore structure of quick cement, in *Proceedings 9th ICCC, New Delhi*, Vol. 4, pp. 370–376.

Knöfel, D., and Wang, J.-F. (1993) Early hydration of quick cements in the system $C_{12}A_7$.CaF_2-C_2S-$C_2(A,F)$. *Advances in Cement Based Materials* **1**, 77–82.

Knöfel, D., and Wang, J.-F. (1994) Properties of three newly developed quick cements. *Cement and Concrete Research* **24**, 801–812.

Kondo, N. *et al.* (1997) Relationship between ettringite formation and development of strength for rapid hardening cement, in *Proceedings 10th ICCC, Göteborg*, paper 2ii017.

Liu, K. *et al.* (1992) Study on cement in the $C_{11}A_7$.CaF_2 system, in *Proceedings 9th ICCC, New Delhi*, Vol. 3, pp.268–276.

Odler, I., and Abdul-Maula, S. (1980) Structure and properties of Portland cement clinker doped with CaF_2. *Journal of the American Ceramic Society* **63**, 654–659.

Quian, G. *et al.* (1992) Effect of ZnO on the hydration behaviour of $C_{11}A_7.CaF_2$, in *Proceedings 9th ICCC, New Delhi*, Vol. 4, pp.384–390.

Uchikawa, H., and Tsukiyma, K. (1973) The hydration of jet cement at 20 °C. *Cement and Concrete Research* **3**, 263–277.

Uchikawa, H., and Uchida, S. (1973) The influence of additives upon the hydration of mixture $11CaO.7Al_2O_3.CaF_2$, $3CaO.SiO_2$ and $CaSO_4$ at 20 °C. *Cement and Concrete Research* **3**, 607–624.

Uchikawa, H., Uchida, S., and Mihara, Y. (1978) Characterization of hydrated ultra rapid hardening cement. *Il Cemento* **73**, 59–70.

Yu, Q., Feng, X., and Sugita, S. (1997) Hydration characteristics of $11CaO.7Al_2O_3.CaF_2$, in *Proceedings 10th ICCC, Göteborg*, paper 2ii43.

6 Cements containing the alinite phase

6.1 ALINITE

Alinite is a calcium oxy-chloro-aluminosilicate that is related to alite, the main constituent of Portland cement. Whereas in alite (tricalcium silicate) Ca^{2+} cations are balanced by SiO_4^{4-} and O^{2-} anions, the structure of alinite consists of Ca^{2+} (and Mg^{2+}) cations balanced by SiO_4^{4-}, AlO_4^{5-}, O^{2-}, and Cl^- anions. Crystallographic studies have revealed that the main feature of alinite is the peculiar position of chlorine atoms surrounded by eight calcium atoms (Noudelman and Gadaev, 1986).

Table 6.1 summarizes the chemical formulas suggested for alinite by different investigators. Originally, the formula $Ca_{11}[3(SiO_4).(AlO_4).O_2.Cl]$ was suggested for this phase, but this oversimplifies the actual situation. In reality, alinite may be stable only if some Mg^{2+} is also present in the crystalline lattice, substituting for Ca^{2+}. The proportions of

Table 6.1 Composition of alinite as suggested by various investigators.

Source	Formula
Ilyukhin *et al.* (1977)	$Ca_{11}(Si_{0.75}Al_{0.25})_4O_{18}Cl$ or $\quad Ca_{11}[3(SiO_4)(AlO_4)]O_2Cl$ or $\quad 21CaO.6SiO_2.Al_2O_3.CaCl_2$
Noudelman *et al.* (1980)	$Ca_{11}(Si,Al)_4O_{18}Cl$
Massaza and Gilioli (1983)	$3[CaO_{0.875}Mg_{0.070})(CaCl_2)_{0.055}][(SiO_2)_{0.885}(Al_2O_3)_{0.115}]$
von Lampe *et al.* (1986)	$Ca_{9.9}Mg_{0.8\square0.3}[(SiO_4)_{3.4}(AlO_4)_{0.6}/O_{1.9}/Cl]$
Bikbaou (1988)	$Ca_{21}Mg[(Si_{0.75}Al_{0.25})O_4]_8O_4Cl_2$
Neubauer and Pöllmann (1994)	$Ca_{10}Mg_{1-x/2\square x/2}[(SiO_4)_{3+x}(AlO_4)_{1-x}/O_2Cl] \quad 0.35 < x < 0.45$

SiO_4/AlO_4 in alinite are not constant, and may vary between about 3.35/0.65 and 3.45/0.55 (Neubauer and Pöllmann, 1994). Thus the actual composition of alinite may be most accurately expressed by the formula

$$Ca_{10}Mg_{1-x/2} \square_{x/2}[(SiO_4)_{3+x}(AlO_4)_{1-x}/O_2/Cl]$$

with $0.35 < x < 0.45$.

The crystalline lattice of alinite can also accommodate limited amounts of additional ions, such as Fe^{3+}, P^{5+}, Ti^{4+}, Na^+, and K^+ (Noudelman et al., 1980; Boikova et al., 1986). At least four modifications of alinite have been identified, and designated α, α', β, and γ (Noudelman and Gadaev, 1986).

Bromine (Kurdowski and Moric, 1989) and fluorine (Bürger and Ludwig, 1986) analogues of alinite have been also synthesized.

Alinite is isostructural with jasmundite, a silicate sulfide mineral $[Ca_{22}(SiO_4)_8O_4S_2]$, which does not, however, exhibit any hydraulic reactivity (Bikbaou, 1986).

A chlorine-containing phase similar to belite and designated **belinite** has also been synthesized (von Lampe et al., 1986; Pradip and Kapur, 1990). It has the chemical formula $Ca_8Mg[(SiO_4)_4Cl_2]$. Whereas the hydration characteristics of alinite are superior to those of alite, belinite does not hydrate as easily as belite.

Another silicate phase containing chlorine that may also be present in alinite cements in limited amounts is calcium chloride orthosilicate, $Ca_3SiO_4Cl_2$ (Agarwal et al., 1986; Miskiewicz and Pyzalski, 1988).

Alinite may be synthesized from a starting mix containing $CaCO_3$, SiO_2, Al_2O_3, MgO, and $CaCl_2$ by burning it at a temperature of 1000–1200 °C. At this temperature the rate of alinite formation exceeds that of tricalcium silicate (at 1400–1500 °C) by about 7–8 times (Bikbaou, 1986). At even higher temperatures alinite converts to alite, and this process is accompanied by a release of chlorine (Agarwal et al., 1986).

Alinite exhibits a high hydraulic reactivity, which is distinctly greater than that of tricalcium silicate. The following degrees of hydration have been reported for both compounds (Boikova et al., 1986; Ji et al., 1997):

hydration time	Degree of hydration (%)		
	Tricalcium silicate (Boikova)	Alinite (Boikova)	Alinite (Ji)
15 min	–	–	9
1 h	10	22	–
6 h	15	64	36
12 h	20	70	–
1 d	25	80	43
3 d	40	85	53
28 d	–	–	62

The hydration of alinite takes place without any distinct induction period (Boikova *et al.*, 1986), and is associated with a single exothermic peak about 60–120 min after mixing (Neubauer and Pöllmann, 1994). The rate of hydration is not affected by the SiO_4/AlO_4 ratio in the crystalline lattice (Neubauer and Pöllmann, 1994). It has been observed, however, that alinite without impurities in the crystalline lattice is more reactive than that which constitutes alinite clinker, and contains distinct amounts of foreign ions (Boikova *et al.*, 1986; Ji *et al.*, 1997). The hydration of alinite is significantly accelerated by the presence of free chloride ions in the liquid phase (Neubauer and Pöllmann, 1994).

Upon hydration, alinite yields a C-S-H phase similar to that formed in the hydration of tricalcium silicate, together with $Ca(OH)_2$ (portlandite) and an AFm phase that is similar to, but not identical with, Friedel's salt ($C_3A.CaCl_2.10H_2O$). In it, up to 15 mol% of CaCl+ is replaced by $Ca(OH)_2$ and $CaCO_3$ (contamination from air) (Pöllmann, 1986; Neubauer and Pöllmann, 1994). The composition of this phase may be expressed by the general formula $(1-x-y)C_3A.xCaCl_2.yCaCO_3.Ca(OH)_2$. At the same time distinct amounts of chloride, aluminate and calcium ions also enter the liquid phase. The following concentrations were reported by Ji *et al.* (1997) after 15 min of hydration:

- Cl^-: 3.3 mg/ml;
- Al_2O_3: 22.5 µg/ml;
- CaO: 3.3 µg/ml.

A microprobe analysis did not indicate any incorporation of Cl^- ions into the C-S-H phase (Neubauer and Pöllmann, 1994).

6.2 ALINITE CEMENT

Alinite cement was developed in the former Soviet Union, and was patented by Noudelman in 1977. It is characterized by the presence of the alinite phase in clinker instead of alite.

Alinite clinker may be produced from a raw meal containing limestone, clay, MgO and $CaCl_2$ (6–18 wt%) by burning it to temperatures of 1000–1300 °C. A variety of industrial waste products, such as fly ash, magnesite dust, steel plant wastes, or municipal waste with a significant amount of chlorine-containing materials, may be also utilized as raw meal constituents (Pradip and Kapur, 1990).

In the course of burning, the calcium chloride that is present melts at around 780 °C, and the formed melt accelerates the overall rate of the clinkering process. Owing to the lower thermal losses associated with the lower burning temperature and the lower $CaCO_3$ content in the raw mix, the energy consumption in the burning process is reduced by about 1250 kJ/kg, corresponding to about 30%, as compared with ordinary Portland cement (Locher, 1986; Kostogloudis *et al.*, 1998). Additional energy savings may be achieved owing to the improved grindability of the clinker. However, significant chlorine corrosion of the production equipment has to be taken into account when producing alinite clinker.

Table 6.2 gives the chemical composition of alinite clinker/cement as published by different investigators. The key parameter in raw meal design of alinite cement is the "lime index," defined as $LI = C/(S + A + F)$. It should range between 1.5 and 1.8 (Pradip and Kapur, 1990). The optimum $CaCl_2$ addition to the raw meal has been found to be about 8% (Pradip and Kapur, 1990); however, some of the chlorine escapes in the course of clinker burning, thus reducing the chlorine content in the final product.

The phase composition of alinite clinkers varies typically in the following range:

- alinite   50–80%;
- belite   10–40%;
- calcium aluminochloride   5–10%;
- calcium aluminoferrite   2–10%.

The clinker may contain also limited amounts of belinite, calcium chloride orthosilicate, calcium ferrite (CF), periclas (MgO), alite (C_3S), and γ-C_2S. If the burning temperature is increased to 1370 °C and above, some of the alinite formed at lower temperatures converts to alite, and at the same time the content of chloroaluminate is increased (Tsuchida *et al.*, 1996; Uchikawa and Obana, 1996).

To produce alinite cement, the clinker does not need to be interground with calcium sulfate to control setting; however, both the short-term and long-term strengths increase with increasing additions of gypsum (Pradip and Kapur, 1990).

Alinite cement exhibits a fast initial hydration, which exceeds that of Portland cement. Table 6.3 summarizes the progress of hydration of this cement and its main constituents. The hydration is associated with a distinct heat evolution, with peaks immediately after mixing and after about 6 hours (Ji *et al.*, 1997). The products of hydration are (Ji *et al.*, 1997):

1 a C-S-H phase similar to that formed in the hydration of Portland cement and portlandite ($Ca(OH)_2$), both produced in the hydration of the alinite phase and belite.
2 Friedel's salt ($C_3A.CaCl_2.10H_2O$) or – more likely – a product in which part of the $CaCl_2$ is substituted by $Ca(OH)_2$ or $CaCO_3$. This phase is formed in the hydration of calcium aluminochloride and also the ferrite phases. In the latter case Al^{3+} may be partly substituted by Fe^{3+}.

Table 6.2 Chemical composition of alinite cement as published by various investigators (in wt. %).

	Agarwal et al. (1986)	Boikova et al. (1986)	Pradip and Kapur (1990)	Tsuchida et al. (1996)	Kostogloudis et al. (1998)
CaO	62.5–63.4	63.5	43–55	60.4–62.2	60.2
SiO$_2$	22.7–23.3	23.1	13–19	21.1–23.4	22.8
Al$_2$O$_3$	4.4–4.6	4.1	9–12	4.2–4.8	3.6
Fe$_2$O$_3$	3.6–4.8	0.1	4–10	2.5–2.6	3.5
MgO		3.9	1–10	2.8–3.1	3.0
Cl total	1.0–2.9	5.2			1.9
CaCl$_2$			6–18	2.8–4.0	
CaO free	<0.5				1.4
Cl free	<0.5				
L.O.I.	0.7–1.0	2.3			0.4

Table 6.3 Progress of hydration and strength development of alinite cement and its constituents.

	w/s	1 d	3 d	7 d	28 d
Degree of hyration (%)					
Alinite	0.25		53	54	62
	10	43	51	79	>95
Calcium aluminochloride	0.25		34	48	55
	10	45	>90		
Belite	10	35	47	64	82
Alinite cement	10	35	47	64	82
Compressive strength (MPa)					
Alinite	0.25	–	42.3	58.9	65.4
Calcium aluminochloride	0.36	–	48.2	53.2	64.8
Alinite cement	0.25	–	49.5	57.1	69.3

Source: Ji *et al.* (1997)

3   If calcium sulfate is interground with alinite clinker, an AFt phase may be also formed. It converts to monosulfate in the latter course of hydration.

4   Other phases that may be present in small amounts include CAH_{10} and C_3AH_6.

A significant part of the chlorine originally present in the alinite clinker becomes bound within the $C_3A.CaCl_2.10H_2O$ phase in the course of hydration. It has also been reported (Ji *et al.*, 1997) that the formed C-S-H phase may accommodate up to 3.5% Cl^-, probably by adsorption. Significant amounts of Cl^-, however, enter the liquid phase. Cl^- concentrations of around 1 mg/ml have been reported for hydration times between 1 and 28 days (Ji *et al.*, 1997). At the same time the aluminate concentration stayed at around 16–18 μg/ml and the CaO concentration at around 2 μg/ml. The following chlorine concentrations in the pore solution were reported by Kostogloudis *et al.* (1998):

Time	continuous storage in tap water (g/L)	interrupted storage in tap water (g/L)
0 h	12.0	12.0
4 d	6.5	6.5
7 d	8.5	8.5
13 d	6.0	6.0
20 d	5.5	–
38 d	4.5	–
2 m	3.5	4.5
4 m	3.5	4.0
8 m	4.5	5.0

The decrease of Cl concentration in the early hydration stages was attributed to chloride ion binding associated with the formation of Friedel's salt, and the subsequent increase, observed between 4 and 7 days, to hydration of the alinite phase, in which Cl⁻ ions are released into the pore solution. The observed decrease thereafter resulted from the diffusion of the Cl⁻ ions toward the surface of the test specimens.

The alkalinity of the pore solution in hardened alinite cement pastes lies at around pH = 11 (Kostogloudis *et al.*, 1998), and thus is distinctly lower than that common in ordinary Portland cement pastes (pH 12.5–13.5).

Data on the corrosion of steel reinforcement embedded in concrete made from alinite cement are not uniform. In most investigations an intensified corrosion of steel was observed under such conditions; however, in a recently published study (Kostogloudis *et al.*, 1998) no significant corrosion was observed in test specimens that were stored continuously or in interrupted fashion under water for up to 12 months.

The initial strength development of alinite cement is similar or even superior to that of ordinary Portland cement, but the 28–day strength is lower. It was found that the strength values increase with increasing lime index (between 1.5 and 1.8) and with increasing gypsum additions to the final cement (Pradip and Kapur, 1990). The strength of the binder may also be increased by blending it with Portland cement or ground granulated blast furnace slag (Tsuchida *et al.*, 1996).

Alinite cement may be produced from municipal waste incineration ashes, which usually contain significant amounts of chlorine and thus can rarely be utilized in other applications. A cement made from this and other waste materials with elevated chlorine contents is usually called ecocement (see section 6.4).

The use of alinite cement in steel-reinforced constructions must be discouraged because of possible corrosion of the reinforcement. The cement may serve well in applications in which reinforcement is not used.

6.3   FLUORALINITE CEMENT

Fluoralinite cement is a fluorine analogue of alinite cement, and is characterized by the presence of fluoralinite ($C_{19}S_7.CaF_2$) (Ludwig and Urbonas, 1993). Fluoralinite cement clinker may be produced by firing a suitable raw meal, with fluorspar (CaF_2) content between about 3.5 and 8 wt%, at temperatures of 950–1100 °C. The clinker contains between about 35 and 80 wt% of the fluoralinite phase. Other phases that may or may not be present include C_3S, C_2S, $C_{11}A_7.CaF_2$, and C_4AF. In contrast to alinite cement, fluoralinite cement does not cause corrosion of the reinforcement steel in the resulting concrete.

6.4   ECOCEMENT

The term "ecocement" is used to designate cements made from incineration ash produced in the incineration of municipal waste, from sewage sludge and from other waste materials, especially those that contain distinct amounts of chlorine. The main reason for producing this kind of cement is to avoid the necessity of disposing these wastes, and to utilize them as raw materials for a technically useful product instead.

The composition of the inorganic fraction of municipal waste, and along with it the composition of the ash produced from it, may vary depending on the area where the waste is collected. It may vary also seasonally, especially in rural regions. This has to be taken into consideration whenever such ash is employed as a raw material in the cement industry. The oxide composition of the ash may vary in the following range:

- CaO 25–40%;
- SiO_2 14–29%;
- Al_2O_3 16–30%;
- Fe_2O_3 4–7%;
- MgO 2–12%;
- Na_2O 2–4%;
- K_2O 2–3%;
- SO_3 1–3%.

The ash always also contains significant amounts of Cl^- (typically between 5 and 11 wt%) and variable amounts of heavy metals. As to the phase composition, the ash contains usually dicalcium silicate and significant amounts of free calcium oxide, which may be partially converted to hydroxide or carbonate in the course of storage. The chlorine may be present in the form of calcium, sodium or potassium chloride, and also as a constituent of belinite (see section 6.1). Other phases that may be present in variable amounts include periclase, silicon dioxide, and even metallic aluminum (Uchikawa and Hanehara, 1997).

In the production of ecocement a blend of the ash, limestone and possibly other constituents is burnt at temperatures between 1000 and 1200 °C. A potential problem in the production process is the possible deposition of low-melting-point chlorine compounds in the equipment.

Because of their distinct chlorine content, most but not all ecocements may be considered special forms of alinite cement. Their typical chemical composition lies in the following range (Moriya *et al.*, 1996; Tsuchida *et al.*, 1996; Uchikawa and Obana, 1996; Takuma *et al.*, 1997):

- CaO: 55–63%
- SiO_2:   16–24%
- Al_2O_3:   3–14%

- Fe_2O_3: 1–3%
- MgO: 1–3%
- Cl: 1–6%

The main phases present in ecocement clinkers are alinite, belite, calcium chloroaluminate ($11CaO.7Al_2O_3.CaCl_2$), and the ferrite phase (Moriya *et al.*, 1996; Tsuchida *et al.*, 1996; Takuma *et al.*, 1997). Some free lime may also be present. In cements with a relatively low chlorine content and burnt to a relatively high temperature (above 1300 °C) alite may be formed instead of alinite, and the total amount of chlorine is bound within the chloroaluminate phase (Takuma *et al.*, 1997). Virtually all the chlorine is incorporated into the alinite and chloroaluminate phases, while only negligible amounts are present in the C_3S and C_2S crystal lattices (Takuma *et al.*, 1997). Ecocement is obtained from eco-clinkers by grinding them with limited amounts of calcium sulfate. Na_2SO_4 or $CaCO_3$ may be also interground with the clinker (Moriya *et al.*, 1996, Tsuchida *et al.*, 1996).

The hydration products formed in the hydration of ecocement include the C-S-H phase, portlandite, Friedel's salt ($C_3A.CaCl_2.10H_2O$), ettringite, monosulfate, and $CaCl_2$ (Moriya *et al.*, 1996).

Ecocements typically exhibit a somewhat faster rate of hardening than Portland cement, but the ultimate strength tends to be lower (Takuma *et al.*, 1997). The strength may be increased by interblending the clinker with ground granulated blast furnace slag (Moriya *et al.*, 1996).

The setting time of the binder will depend on the amount of calcium chloroaluminate present, on the form of SO_3 that is present, and on the SO_3/Al_2O_3 ratio. Usually the cement sets faster than Portland cement. If required, a hydroxycarboxylic acid, such as citric acid, may be added to retard the setting process (Takuma *et al.*, 1997).

In the course of hydration the chlorine present in the original cement becomes partially bound by the formed hydrates, but also enters the pore solution. The released fraction amounts to about 10–40% of the total amount, depending on the composition of the original cement (Takuma *et al.*, 1997). Because of possible corrosion caused by free chlorine ions, ecocement must not be used in applications in which the hardened material is in contact with steel, such as in the production of steel-reinforced concrete, prestressed concrete, or metallic-fiber-reinforced concrete.

In contact with water, soluble chlorides may be eluated from the hardened paste/concrete, and may enter the environment. The elution rate and the eluted amount of Cl^- will increase with increasing SO_2/Al_2O_3 ratio in cement, and will decline with increasing amount of interblended ground granulated blast furnace slag (Moriya *et al.*, 1996).

The content of heavy metals in ecocements is determined by their amount in the original waste material. The amount of these cement constituents that becomes eluted in the course of elution with groundwater stays within acceptable limits in most instances (Moriya *et al.*, 1996).

Ecocement may be considered for soil stabilization and as a constituent of grouting materials (Moriya, 1966).

REFERENCES

Agarwal, R.K., Paralkar, S.V., and Chatterjee, A.K. (1986) Chloride salts as reaction medium for low temperature clinkerization – A probe into alinite technology, in *Proceedings 8th ICCC, Rio de Janeiro*, Vol. 2, pp. 327–333.

Bikbaou, M. (1980) Mineral formation processes and phase composition of alinite clinker, in *Proceedings 7th ICCC, Paris*, Vol. 4, pp. 371–376.

Bikbaou, M.Y. (1986) Formation, crystal chemistry and properties of alinite and jasmundite, in *Proceedings 8th ICCC, Rio de Janeiro*, Vol. 2, pp. 352–357.

Boikova, A.I., Grishchenko, L.V., and Domansky, A.I. (1986) Hydration activity of chlorine containing phases, in *Proceedings 8th ICCC, Rio de Janeiro*, Vol. 3, pp. 275–276.

Bürger, D., and Ludwig, U. (1986) Synthesis of calcium silicates at low temperatures and influences on their reactivity, in *Proceedings 8th ICCC, Rio de Janeiro*, Vol. 2, pp. 372–278.

Ftikos, Ch., and Kiatos, D. (1994) The effect of chlorides on the formation of belite and alinite phase. *Cement and Concrete Research* **24**, 49–54.

Ftikos, Ch., Georgiades, A., and Philipou, Th. (1991) Preparation and hydration study of alinite cement. *Cement and Concrete Research* **21**, 1129–1136.

Ftikos, Ch., Philippou, Th., and Marinos, J. (1993) A study of the effect of some factors influencing alinite clinker formation. *Cement and Concrete Research* **23**, 1268–1272.

Ilyukhin, V.V. *et al.* (1977) Crystal structure of alinite. *Nature* **269**, 397–398.

Ji, L., Ren, X., and Su, H. (1997) Effect of chloride on hydration properties of alinite cement, in *Proceedings 10th ICCC, Göteborg*, paper 2ii030.

Kostogloudis, G.C. *et al.* (1998) Comparative investigation of corrosion resistance of steel reinforcement of alinite and Portland cement mortars. *Cement and Concrete Research* **28**, 995–1010.

Kurdowski, W., and Moric, U. (1989) Once more about bromine alinite. *Cement and Concrete Research*, **19**, 657–661.

Locher, F.W. (1986) Low energy clinker, in *Proceedings 8th ICCC, Rio de Janeiro*, Vol. 1, pp. 57–67.

Ludwig, U., and Urbonas, L. (1993) Synthesis and reactivity of fluoralinites and fluoralinite cement (in German). *Zement-Kalk-Gips* **46**, 568–572.

Massazza, F., and Gilioli C. (1983) Contribution to the alinite knowledge. *Il Cemento* **80**, 101–106.

Miskiewicz, K., and Pyzalski, M. (1988) Polymorphism of the $Ca_3SiO_4Cl_2$ – DTA studies. *Cement and Concrete Research* **18**, 819–822.

Moriya, M. *et al.* (1996) A consideration on reducing the amount of chloride elution in the special cement made from municipal waste. *Journal of Research of the Chichibu Onoda Cement Corporation* **47** (130), 62–72.

Neubauer, J., and Pöllmann, H. (1994) Alinite – chemical composition, solid solution and hydration behaviour. *Cement and Concrete Research* **24**, 1413–1422.

Noudelman, B.I., and Gadaev, A.I. (1986) Physico-chemical aspects of the crystallization of chlorsilicates after clinkerization at low temperatures in melted salts (in French), in *Proceedings 8th ICCC, Rio de Janeiro*, Vol. 2, pp. 347–351.

Noudelman, B. *et al.* (1980) Structure and properties of alinite and alinite cement (in French), in *Proceedings 7th ICCC, Paris*, Vol. 3, pp. V-169–174.

Obana, H., Anzai, T., and Fukunada, T. (1994) Ecocement recycled from urban garbage and waste materials, in *International Symposium on Environmental Issues of Ceramics*, Sapporo, Japan, pp. 63–67.

Oberste-Padberg, R., and Neubauer, J. (1989) Laboratory experiments to produce alinite cement from municipal waste incineration plants (in German). *WLB Wasser Luft und Boden*, pp. 10/62–65

Oberste-Parberg R. *et al.* (1992) Alinite cement a hydraulic binder made from refuse incineration residues (in German). *Zement-Kalk-Gips* **45**, 451–455.

Pöllmann, H. (1986) Solid solutions of complex calcium aluminate hydrates containing Cl^-, OH^- and CO_3^{2-} anions, in *Proceedings 8th ICCC, Rio de Janeiro*, Vol. 3, pp. 300–306.

Pradip, A., and Kapur, P.C. (1990) Production and properties of alinite cements from steel plant waste. *Cement and Concrete Research* **20**, 15–24

Takuma, Y., Tsuchida, Y., and Uchida, S. (1997) Characteristics and hydration of cement produced from ash from incinerated urban garbage, in *Proceedings 10th ICCC, Göteborg*, paper 2ii118.

Tsuchida, Y. *et al.* (1996) Hydration and characterization of alinite rich clinker burnt at different temperatures. *Journal of Research of the Chichibu Onoda Cement Corporation* (In Japanese) **47** (130), 62–72.

Uchikawa, H., and Hanehara, S. (1997) Recycling of waste as an alternative raw material and fuel in cement manufacturing, in *Waste Materials Used in Concrete Manufacturing* (ed. S. Chandra), Noyed Publications, Westwood, NJ, pp. 430–553.

Uchikawa, H., and Obana, H. (1995) Ecocement – frontier of recycling of urban composite wastes. *World Cement* **26**, 33–40.

Uchikawa, H., and Obana, H. (1996) Ecocement – frontier of recycling of urban composite waste. *Journal of Research of the Chichibu Onoda Cement Corporation* (In Japanese) **47** (131), 153–162.

von Lampe, F. *et al.* (1986) Synthesis, structure and thermal decomposition of alinite. *Cement and Concrete Research* **16**, 505–510.

Yoshiaki *et al.* (1996) Hydration characteristics of alinite rich clinker burnt at different temperatures (in Japanese). *Journal of Research of the Chichibu Onoda Cement Corporation* **47**, 62–72.

7 Composite cements: general considerations

The term **composite cement** or **blended cement** is used to denote inorganic binders that contain a **mineral addition** in combination with Portland clinker and usually also with calcium sulfate.

A variety of industrial by-products and natural starting materials qualify as mineral additions, and may be used as constituents of inorganic binders or cementitious systems. These materials may possess cementitious properties by themselves, or they may be latent hydraulic, pozzolanic, or even non-reactive.

Hydraulically reactive materials exhibit cementing properties if mixed just with plain water, without the presence of other constituents to act as "activators." Some fluidized-bed ashes or industrial slags (steelwork slag, for example) are examples of such materials.

Latent hydraulic materials are also able to react hydraulically, but only in the presence of at least small amounts of a suitable activator. They are usually glassy or amorphous, and contain CaO, SiO_2, and Al_2O_3 as their main oxide constituents. Granulated blast furnace slag is the most widely used latent hydraulic material.

Pozzolanic materials or **pozzolanas** do not exhibit cementing properties if mixed with plain water. However, they possess the capacity to react at ambient temperatures with calcium hydroxide, in the presence of water, to yield strength-developing calcium silicate/aluminate hydrates. They include a variety of materials of natural and artificial origin, such as fly ash, microsilica, burnt clays, and diatomaceous earths.

Non-reactive constituents of inorganic cements do not react chemically in the course of hydration to any significant extent, even in the presence of calcium or alkali hydroxides. They may, however, alter the rheology of the fresh paste and some properties of the hardened material mainly by their physico-mechanical action.

Table 7.1 summarizes the typical composition and properties of a representative series of mineral additions used in blended cements. Binders based on granulated blast furnace slag and related materials are discussed in more detail in Chapter 8, and those based on natural and artificial pozzolanas in Chapter 9. Cements containing calcium carbonate are discussed in section 2.2.14.

Table 7.1 Properties of selected mineral additions for blended cements.

	Granulated blast furnace slag	Fly ash (low lime)	Fly ash (high lime)	Microsilica	Rice husk	Limestone
Chemical composition (wt %)						
SiO_2	27–40	34–60	25–40	85–98	80–97	—
Al_2O_3	5–33	17–30	8–17	0–3	—	—
Fe_2O_3	<1	2–25	5–10	0–8	—	—
CaO	30–50	1–10	10–38	<1	—	>50
MgO	1–21	1–3	1–3	0–3	—	<5
SO_3	—	1–3	1–5	—	—	—
$Na_2O + K_2O$	<1	<1	0–3	1–5	<1	—
C	—	<5	<5	0–4	2–20	CO_2 <40
Phase composition	G	G, M. Ma, H	G, M, Ma. H (L), (An)	A	A	C
Crystalline phases total	<10	10–50	10–50	<5	<5	100
Particle shape	Sharp, irregular	Mostly spherical	Mostly spherical	Spherical (very small)	Fiber-like	Irregular

A = amorphous; An = anhydrite; C = calcite; G = glass; H = hematite; L = lime; M = mullite; Ma = magnetite; Q = quartz

Table 7.2 Nomenclature and composition of composite cements according to ENV 197-1.

Cement type	Designation	Notation	Clinker	Mineral addition	
I	Portland–slag cement	II/A–S	80–94	Granulated blast furnace slag	6–20
		II/B–S	65–79		21–35
II	Portland–silica fume cement	II/A–D	90–94	Silica fume	6–10
	Portland–pozzolana cement	II/A–P	80–94	Natural pozzolana	6–20
		II/B–P	65–79		21–35
		II/A–Q	80–94	Industrial pozzolana	6–20
		II/B–Q	65–79		21–35
	Portland–fly ash cement	II/A–V	80–94	Fly ash (siliceous)	6–20
		II/B–V	65–79		21–35
		II/A–W	80–94	Fly ash (calcerous)	6–20
		II/A–W	65–79		21–35
	Portland–burnt shale cement	II/A–T	80–94	Burnt shale	6–20
		II/B–T	65–79		21–35
	Portland–limestone cement	II/A–L	80–94	Limestone	6–20
		II/B–L	65–79		21–35
	Portland–composite cement	II/A–M	80–94	Any mineral addition	6–20
		II/B–M	65–79		21–35
III	Blast-furnace cement	III/A	35–64	Granulated blast furnace slag	
		III/B	20–34		
		III/C	5–19		
IV	Pozzolanic cement	IV/A	65–89	Silica fume	
		IV/B	45–64	Pozzolana or Fly ash (siliceous)	
V	Composite cement	V/A	50–64	Pozzolana or Fly ash (siliceous)	
		V/B	20–39		

All proportions in mass %.

The values in the table refer to the cement nucleus, excluding calcium sulfate and any additives.

Minor additional constituents in amounts not exceeding 5% may also be added. They may be one or more of the main constituents unless these are included as main constituents in the cement.

The proportion of silica fume is limited to 10%.

Mineral additions are used in composite cements in amounts between a few and several tens of a per cent.

The term "Portland cement" is considered in cement chemistry as describing a cement that contains no other constituents than Portland clinker and calcium sulfate. Small amounts (below 1%) of some chemical agents may also be tolerated. However, some specifications use the term "Portland cement" for binders that contain limited amounts (usually less than 5%) of selected mineral additions. Table 7.2 summarizes the nomenclature and composition of composite cement as specified in the European standard ENV197–1.

Composite cements may be produced by grinding the mineral addition (or additions) together with Portland clinker and calcium sulfate, or by blending them with Portland cement produced separately.

Mineral additions must be distinguished from **chemical admixtures**, which are added to the binder only in small amounts (usually below 1%), and which influence the properties of the fresh mix or the hardened paste, but do not alter substantially the product of the hydration reaction.

Composite cements may have properties that are particularly suitable for some applications, and may be superior to Portland cement in this respect. The use of composite cements is also motivated by efforts to utilize waste materials, and to reduce the overall energy consumption in the cement industry.

An extension of the concept of blended cements is cements/blends that consist solely of mineral additions without the presence of Portland cement or Portland clinker. The hydraulic properties of such systems tend to be poor, and heat curing may be required in some of them to attain measurable strengths in a reasonable time.

7.1 MULTICOMPONENT CEMENTS

Blended cements usually contain a single blending component in addition to Portland clinker and calcium sulfate. Those that contain two or even more blending components side by side are called "multicomponent cements." Some multicomponent cements may also be produced by combining two or more blending components without adding Portland clinker to the system. The individual constituents of multicomponent cements may be either ground together, or blended together after being ground individually. Multicomponent cements that contain the following combinations of constituents have been studied:

- Portland cement + granulated blast furnace slag + fly ash (Entin *et al.*, 1992; Härdtl, 1992, 1997; Roszcynialski, 1992; Tan and Pu, 1998);
- Portland cement + granulated blast furnace slag + silica fume

(Shizawa *et al.*, 1992; Nagataki and Wu, 1995; Wiens *et al.*, 1995; Xi *et al.*, 1997; Bagel, 1998);

- Portland cement + fly ash + silica fume (Giergiczny, 1995; Wiens *et al.*, 1995; Lam *et al.*, 1998);
- Portland cement + fly ash + gypsum (Singh and Garg, 1992; Yan and You, 1998);
- Portland cement + fly ash + limestone (Härdtl, 1997);
- Portland cement + metakaolin + silica fume (Chung, 1996);
- Portland cement + silica fume + gypsum (Kovler, 1998);
- kiln dust + granulated blast furnace slag + fly ash (Bhatty, 1985).
- granulated blast furnace slag + high calcium fly ash + microsilica (Pavlenko and Oreshkin, 1992; Bazenov and Pavlenko, 1997);
- granulated blast furnace slag + fly ash + lime (Runowa *et al.*, 1996);
- gypsum binder + microsilica + carbide sludge (Ajrapetov *et al.*, 1996);
- metakaolin + hydrated lime + β-anhydrite (or Na_2SO_4) (de Silva and Glasser, 1992).

REFERENCES

Ajrapetov, G.A. *et al.* (1996) Multicomponent clinkerless gypsum (in Russian). *Stroitel'nye Materialy (Moscow)*, **1996**(1), 28–29.

Bagel, L. (1998) Strength and pore structure of ternary blended cement mortars containing blast furnace slag and silica fume. *Cement and Concrete Research* **28**, 1011–1020.

Bazenov, Y., and Pavlenko, S. (1997) Physicomechanical and deformation properties of cementless fine-grained ash-slag concrete over a 5 year period, in *Proceedings 10th ICCC, Göteborg*, paper 4iv058.

Bhatty, M.S.Y. (1985) Kiln dust cements evaluated. *Rock Products* **88**, 47–52.

Chung, M.C. (1996) Studies on the properties of high performance and high strength cement mortar using metakaolin and silica fume (in Korean). *Yoop Hakhoechi* **33**, 519–523.

de Silva, P.S., and Glasser, F.P. (1992) The hydration behaviour of metakaolin-Ca(OH)$_2$-sulfate binders, in *Proceedings 9th ICCC, New Delhi*, Vol. 4, pp. 671–677.

Entin, Z.B., Krivoborodov, Yu.R., and Shubin, V.I. (1992) Structure formation in the hardening stone of multicomponent cements, in *Proceedings 9th ICCC, New Delhi*, Vol. 4, pp. 324–330.

Giergiczny, Z. (1995) Fly ash cements with addition of silica fume (in Polish). *Cement-Wapno-Gips (Warsaw)*, **1995**(1), pp. 17–19, 22.

Härdtl, R. (1992) Chemical bonding of water during the hydration of Portland cements and blastfurnace slag cements blended with fly ash, in *Proceedings 9th ICCC, New Delhi*, Vol. 4, pp. 678–690.

Härdtl, R. (1997) The pozzolanic reaction of fly ash in connection with different types of cement, in *Proceedings 10th ICCC, Göteborg*, paper 3ii082.

Kovler, K. (1998) Setting and hardening of gypsum – Portland cement – silica fume blends. *Cement and Concrete Research* **28**, 423–438.

Lam, L., Wong, Y.L., and Poon, C.S. (1998) Effect of fly ash and silica fume on compressive and fracture behaviour of concrete. *Cement and Concrete Research* **28**, 271–284.

Nagataki, S., and Wu, C. (1995) *A study of the properties of Portland cement incorporating silica fume and blast furnace slag*. American Concrete Institute SP-153, pp. 1051–1068.

Pavlenko, S.I., and Oreshkin, A.B. (1992) Structure formation of cementless concrete made with slag sand and high-calcium ash from thermal power plants, in *Proceedings 9th ICCC, New Delhi*, Vol. 5, pp. 647–652.

Roszcynialski, W. (1992) The influence of some fly ashes and blast-furnace slags on blended cement hydration and properties, in *Proceedings 9th ICCC, New Delhi*, Vol. 4, pp. 698–704.

Runowa, R.F., Koczewych, M.A., and Rudenko, I.I. (1996) Use of ash-slag-lime mixes for manufacture of building materials (in Polish). *Cement-Wapno-Beton (Warsaw)*, **1996**(5), 190–193.

Shizawa, M. *et al.* (1992) Study on hydration properties of slag and silica fume blended cement for ultra-high strength concrete, in *Proceedings 9th ICCC, New Delhi*, Vol. 4, pp. 658–664.

Singh, M., and Garg, M. (1992) Development of cementitious properties in phosphogypsum-fly ash system, in *Proceedings 9th ICCC, New Delhi*, Vol. 4, pp. 489–494.

Tan, K., and Pu, X. (1998) Strengthening effect of finely ground fly ash, granulated blast furnace slag, and their combination. *Cement and Concrete Research* **28**, 1819–1825.

Wiens, U., Breier, W., and Schiessel, P. (1995) *Influence of high silica fume and high fly ash contents on alkalinity of pore solution and protection of steel against corrosion*. American Concrete Institute SP-153 (Vol. 2), pp. 749–761.

Xi, Y., Siemer, D.D., and Scheetz, B.E. (1997) Strength development, hydration reaction and pore structure of autoclaved slag cement with added silica fume. *Cement and Concrete Research* **27**, 75–82.

Yan, P., and You, Y. (1998) Study on the binder of fly ash – fluorogypsum – cement. *Cement and Concrete Research* **28**, 135–140.

8 Cements containing ground granulated blast furnace slag

8.1 BLAST FURNACE SLAG

Blast furnace slag is a by-product in the production of iron. In the blast furnace limestone reacts with SiO_2- and Al_2O_3-rich constituents of the iron ore and coke, yielding melted blast furnace slag, which leaves the furnace at a temperature of about 1350–1550 °C.

If allowed to cool down slowly in air, the melt crystallizes, yielding a polycrystalline material. Its main constituent is mellilite, which is a solid solution of gehlenite (C_2AS) and akermanite (C_2MS_2) of variable composition. Other crystalline phases that may be present include merwinite (C_3MS_2), larnite (β-C_2S), rankinite (C_3S_2), wollastonite (CS), pyroxene (CMS_2), forsterite (M_2S), and monticelite (CMS). Such a material has virtually no cementing properties, and cannot be used as a constituent of inorganic binders.

If cooled rapidly to below 800 °C the melted slag converts to a glassy material suitable for use in inorganic binders. Such cooling is usually done by pouring the melt into a large excess of water or spraying it with water under high pressure. The product is a wet sandy material called **granulated blast furnace slag**. If dried and ground, it is called **ground granulated blast furnace slag** (abbreviation GGBFS).

The oxide composition of blast furnace slags may differ considerably between individual steel mills, but usually varies little in a given plant, as long as the source of the ore stays the same. Typically, the content of the individual oxides may vary in the following ranges:

- CaO: 30–50%
- SiO_2: 27–42%
- Al_2O_3: 5–33%
- MgO: 1–21%
- Fe_2O_3: less than 1%
- S^{2-}: less than 3%
- $Na_2O + K_2O$: 1–3%

The glass content in granulated blast furnace slags may be determined as the difference between the total and the content of crystalline phases as determined by quantitative X-ray diffraction. Its amount may vary in different slags, and depends on their oxide composition, the initiation temperature of cooling, and the cooling method employed. Generally, the tendency to crystallize increases with increasing CaO and MgO contents of the slag and decreasing rate of cooling. High-quality slags may contain up to 95% of glass. The composition of the glass phase is usually fairly uniform.

SiO_2 is present in the glass phase in the form of SiO_4 tetrahedra, the monomer and dimer being the predominant silicate species.

Aluminum, magnesium, and titanium may each be present in tetrahedral and octahedral coordinations, and the proportion of the two forms depends on the oxide composition of the melt. Generally, the fraction of sixfold-coordinated species increases with increasing CaO and MgO contents and increasing C/S and A/S ratios.

Calcium and alkalis are present in the glass in a sixfold coordination.

The glass phase tends to contain structural defects, whose abundance increases with increasing rate of cooling. The crystalline constituents of the granulated slag are finely distributed within the glass phase.

If ground granulated blast furnace slag is mixed with water, a leaching process gets under way, resulting in the formation of a thin SiO_2-enriched and Ca^{2+}-deficient hydrated layer at the surface. At the same time the alkalinity of the liquid phase may reach pH = 10 in systems with a low enough water/solid ratio. The presence of the hydrate layer inhibits further migration of water toward the slag surface and ion dissolution, thus resulting in a discontinuation of the hydration reaction.

To bring about continuous hydration a suitable **activator** has to be added to the mix. **Alkaline activators** are substances that increase the alkalinity of the liquid phase, bringing it to above pH = 12. Under these conditions the formation of the impermeable layer at the slag surface is prevented, and the dissolution of the glass phase may continue indefinitely. Within minutes the concentration of Ca^{2+}, Al^{3+}, and Si^{4+} ions in the liquid phase may reach 8–10, 1–2, and 3–4 ppm respectively, followed by a gradual decline, as a precipitation of hydration products gets under way. The role of the alkaline activator is to break down the existing bonds within the structure of the glass phase of the slag, and thus to promote the dissolution of ionic species. Calcium hydroxide, Portland cement or clinker, NaOH, KOH, Na_2CO_3, and sodium silicate are examples of suitable alkaline activators.

Another activator is calcium sulfate, which acts as a sink for Ca^{2+} and $Al(OH)^{4-}$ ions released from the slag, yielding ettringite as the product of reaction. This **sulfatic activator** is not very effective at lower pH, but its effectiveness increases significantly in combination with an alkaline activator.

Ground granulated blast furnace slag, a material that exhibits a distinct hydraulic reactivity only in the presence of suitable activators, is considered **latently hydraulic**, unlike **hydraulic** materials, which exhibit hydraulic properties even when mixed with water alone. Among a variety of materials known to possess latent hydraulicity, ground granulated blast furnace slag is the one that is produced and used in the largest volumes.

There are wide variations in the reactivity of granulated blast furnace slags, characterized by their different readiness to undergo hydration.

- First, the reactivity depends on the glass content of the slag. Generally, the reactivity increases with increasing glass content; however, a small proportion of crystalline materials embedded in the glass phase may influence the reactivity positively. This is probably due to the presence of mechanical stresses in the glass, produced by the presence of the crystalline material.
- Second, the reactivity depends on the oxide composition of the slag. Generally, the reactivity increases with increasing alkalinity of the slag, and in particular with increasing CaO content. MgO, in amounts of up to about 10%, may be considered equivalent to CaO in this respect. Increased contents of Al_2O_3 tend to increase early strength, mainly because the amount of ettringite formed is increased under these conditions.

  Efforts have been made to assess the reactivity of slags on the basis of various "hydraulic" moduli. Out of these the modulus HM = (CaO + MgO + Al_2O_3)/ SiO_2 is the one most widely used. The minimum value for this ratio should be not less than 1.0. However, an assessment of slag reactivity purely on the basis of chemical composition is associated with a significant degree of uncertainty.
- Third, the reactivity depends on the rate of cooling. Generally, an increased abundance of structural defects produced by rapid cooling increases the readiness of the slag to react.
- Fourth, the reactivity of the slag increases with increasing fineness. Freshly ground slag is especially reactive, and its reactivity gradually declines with increasing storage time.

If alkali hydroxides or their salts are used as activators, a C-S-H phase with a low C/S ratio is produced as the main hydration product. It contains larger amounts of Al_2O_3 and MgO than the C-S-H phase formed in the hydration of Portland cement, and may also contain small amounts of alkalis. Other hydrates formed under these conditions include tetracalcium aluminate hydrate (C_4AH_x) and gehlenite hydrate (C_2ASH_8). In slags with high MgO contents a C_4AH_x-C_4MH_x solid solution and a hydrotalcite-type phase may also be formed.

If calcium hydroxide is used as activator, it is gradually consumed in the hydration process as it enters the structure of the C-S-H phase formed. Under these conditions the C/S ratio of this phase increases proportionately to the amount of $Ca(OH)_2$ present.

In the presence of calcium sulfate, ettringite – in addition to the C-S-H phase – is the main hydration product formed. At lower $CaSO_4$ additions the primary formed ettringite may partially or completely convert to monosulfate in the later course of hydration.

8.2 SLAG-LIME BINDER

Slag-lime binder, called also "slag cement," is a combination of ground granulated blast furnace slag and lime hydrate, which serves both as an alkaline activator and a reactant. The amount of lime hydrate in the binder may vary between about 5 and 30 wt%. Alternatively, finely ground lime may also be used instead of lime hydrate. The main product of the hydration reaction is a C-S-H phase whose C/S ratio will depend on the composition of the slag and the amount of available lime hydrate. Also formed in the hydration is the phase C_4AH_{13} (Tango and Vaidergorian, 1992). Owing to the limited reactivity of the slag and a relatively low alkalinity of the mix (pH = 12.0–12.5), the hydration, and along with it the strength development of the binder, progresses rather slowly, and a significant fraction of the slag remains non-reacted even after long hydration times (Hanafi, 1991). The hydration process may be accelerated by fine grinding of the slag and an elevated curing temperature. An increase of strength may also be achieved by adding to the mix small amounts of gypsum (Tango and Vaidergorian, 1992) or Na_2SO_4 or $CaCl_2$ (Shi and Day, 1995a). With added gypsum or Na_2SO_4 the strength improvement is due to the formation of ettringite, whereas a solid solution of $C_3A.Ca(OH)_2.12H_2O$-$C_3A.CaCl_2.1OH_2O$ is formed in the presence of $CaCl_2$. The binder has an excellent resistance to sulfates, and thus may be considered for underground foundations, where exposure to sulfates may be expected.

8.3 PORTLAND-SLAG CEMENT

Portland-slag cement, also called Portland blast furnace cement or Portland blast furnace slag cement, consists of Portland clinker, granulated blast furnace slag, and calcium sulfate in variable proportions. The amount of slag in different types of Portland-slag cements may vary greatly. The types of cements that contain granulated blast furnace slag are listed in the European Specification ENV197–1 (see top of next page).

The ASTM Standard C595M-95 distinguishes the types of cement that may contain granulated blast furnace slag in the amounts listed in the second table on the next page.

Designation	Notation	Slag (wt %)
Portland-slag cement	II/AS	6–20
	II/BS	21–35
Blast furnace cement	III/A	36–65
	III/B	66–80
	III/C	81–95

In Germany the designations *Eisenportlandzement* (maximum 40% of slag) and *Hochofenzement* (slag content 41–85%) are commonly used (DIN 1164).

Cement type	Slag (wt %)
Slag-modified cement	< 25
Portland blast furnace slag cement	25–70
Slag cement	> 70

Calcium sulfate in the form of gypsum or anhydrite is added to the cement in amounts of a few per cent, to act as a set regulator. The rest of the material constitutes ordinary Portland clinker.

In most instances the constituents of Portland-slag cement are ground together in an industrial ball mill to a specific surface area of > 300 m^2/kg. In a cement produced in this way the average size of the slag particles is usually greater than that of the clinker, owing to the poorer grindability of the slag. Such a particle size distribution must be considered unfavorable, as the reactivity of the slag is generally lower than that of the clinker, even at equal fineness. A more favorable but technically more demanding solution is separate grinding of the slag and clinker (together with gypsum) and subsequent blending together of the two ground materials. Such an approach makes it possible to produce cements in which the less reactive slag is ground as finely as or even more finely than the clinker.

In the hydration of Portland-slag cement the different cement constituents undergo hydration at different rates, reflecting the existing differences in their reactivity. Generally, the existing tricalcium silicate, which is the main constituent of the clinker, hydrates significantly faster than the slag (Hinrichs and Odler, 1989). This difference in the rate of hydration is less in Portland-slag cements in which the slag has been ground to a higher fineness in a separate grinding process.

After mixing with water the tricalcium silicate of the clinker undergoes fast hydration in which calcium hydroxide is released into the liquid phase. In parallel, alkali sulfates – if present in the clinker – are dissolved, and react with calcium hydroxide to yield alkali hydroxides. Both factors increase the pH of the liquid phase to the values needed for alkaline activation of the slag. Calcium sulfate formed in the latter reaction, together with gypsum/anhydrite added to the original cement, is responsible for a sulfatic activation of the slag. Table 8.1 summarizes the progress of slag and tricalcium silicate formation in cements containing different amounts of granulated blast furnace slag and ground to different finenesses. Significant amounts of the slag remain non-reacted even after one year of hydration at ambient temperature, regardless of the slag/clinker ratio and fineness.

Investigations on a variety of slags of different origin have revealed considerable differences in their reactivity, the main factors being the hydraulic modulus and fineness. Typically 35–55% of the slag becomes hydrated after 28 days and 45–75% after 1–2 years (Hinrichs and Odler, 1989; Battagin, 1992; Roy and Malek, 1993; Lumley *et al.*, 1996).

The calcium hydroxide formed in the hydration of tricalcium silicate, and to a lesser extent also of the dicalcium silicate of the clinker, is partially consumed in the hydration of the slag. As a consequence, the amount of calcium hydroxide present increases in the initial stage of hydration, when the hydration of tricalcium silicate is extensive, and only a small amount of the slag has been hydrated. In the later course of hydration, however, the amount of calcium hydroxide in the paste may decline, as the hydration of tricalcium silicate gradually slows down, whereas the slag continues to hydrate. Thus the amount of calcium hydroxide at any stage of hydration is lower than that which would be present in the absence of a reaction of this phase with the slag. Table 8.2 shows the amount of free lime present in a series of pastes made from cements with different slag contents and hydrated for different periods of time. In Table 8.3 the amount of free lime actually found is compared with that calculated under the assumption of no consumption of free lime by the hydrating slag.

The main product of hydration of Portland-slag cement is a C-S-H phase whose average C/S ratio is lower than the value found in ordinary Portland cement pastes. The ratio generally declines with increasing slag fraction in the cement towards the value existing in the anhydrous slag (Richardson and Groves, 1992). By X-ray microanalysis C/S ratios of 1.55–1.79 were found in the vicinity of the clinker grains, whereas values as low as C/S = 0.9–1.3 were determined in the hydrated material formed *in situ* from the slag. As the C/S value of the formed C-S-H phase decreases, its A/C value increases. Aluminum substitutes for silicon in the bridging tetrahedra of the dreierketten chains constituting the C-S-H phase (Richardson and Groves, 1992; Richardson, 1997). In cements made from slags with high Mg contents a hydrotalcite phase may also be present, which is usually intermixed with the C-S-H phase formed as the inner product in the hydration of the slag. Just as in pastes of ordinary Portland cement, in mature Portland-slag cement pastes the C-S-H phase may be intimately mixed with the AFm phase, which is also formed in the hydration of this cement. There are indications that polysulfides may be formed

Table 8.1 Hydration of alite and slag in Portland-slag cements with different contents of granulated blast-furnace slag.

Fineness (m²/kg)	Slag in cement (%)	Alite					Slag				
Hydration time (d)		2	7	28	90	360	2	7	28	90	360
300	0	51	65	73	85	100	–	–	–	–	–
	25	50	75	84	94	100	16	24	37	43	51
	50	54	82	93	100	100	26	36	39	48	53
	70	44	75	88	100	100	24	39	59	n.d.	63
500	0	67	73	85	92	100	–	–	–	–	–
	25	64	77	82	93	100	31	44	57	57	60
	50	68	79	82	92	100	n.d.	57	67	69	72
	70	63	80	n.d.	100	100	39	56	58	59	64

All values indicate the amount of reacted alite/slag in per cent
s(alite) = ±16   s(slag) = ±10
Source : Hinrichs and Older (1989)

Table 8.2 Free lime content in Portland-slag cements.

Slag in cement (%)	0	25	50	70
1 d	8.0	8.5	5.5	1.9
2 d	15.8	12.0	7.8	2.8
7 d	18.4	12.0	7.3	2.5
28 d	18.8	11.3	6.0	2.0
365 d	20.1	11.1	5.5	1.3

All data indicate the amount of $Ca(OH)_2$ in wt%.
Specific surface area of cements: 500 m^2/kg.

Source: Hinrichs and Odler (1989)

Table 8.3 Calculated and found free lime content in pastes made from cements with different blast-furnace slag contents.

SSA (Blaine) (m^2/kg)	Slag content (%)	$Ca(OH)_2$ calculated	$Ca(OH)_2$ found
300	0	25.6	26.5
	25	19.1	15.5
	50	16.6	6.0
	70	7.0	2.5
500	0	32.0	32.0
	25	23.6	17.0
	50	15.2	9.5
	70	8.4	4.0

Hydration time 28 d
All data in g per 100 g of original cement

Source: Hinrichs and Odler (1989)

in the hydration of pastes that contain S^{2-} in the original slag, causing a greenish coloration of the hardened material.

The microstructure of Portland-slag cement pastes is not too different from that of plain Portland cement. The main difference consists in a reduced amount of free $Ca(OH)_2$. Just as with clinker grains, layers of reaction products formed *in situ* are also present at the boundaries of the slag grains, in addition to the hydrated material formed within the space originally filled with water.

At equal water/cement ratios and hydration times the amount of combined water declines and the porosity of the paste increases with increasing amount of slag in the original cement, reflecting the existing differences in the hydration kinetics of slag and the clinker phases (Hinrichs and Odler, 1989). At equal porosities, however, the pore size distribution and the specific surface area of the pastes appear to be only slightly affected by the slag/clinker ratio in the original cement (Hinrichs and Odler, 1989).

The strength development of Portland-slag cements is generally more sluggish than that of Portland cement. Thus – at shorter hydration times – the strength declines with increasing content of slag in the cement; however, if hydration times are long enough and slag contents are not too high the strengths achieved may not be too different from those of plain Portland cement. At 180 days and beyond, highest values were found for 50% slag replacements (Battagin, 1992).

The strength development of Portland-slag cements is highly influenced by the curing conditions. Curing under water yields the highest strengths (Nakamoto and Togawa, 1995; Nakamoto *et al.*, 1995). As expected, the strength – especially after short curing times – also increases with increasing curing temperature. Portland-slag cements are more sensitive to temperature variations than plain Portland cement, especially at higher slag contents (Nakamoto and Togawa, 1995; Nakamoto *et al.*, 1995).

A distinct increase of strength may be attained by separate grinding of the slag to extremely high finenesses: that is, up to 620 m^2/kg (Blaine) (Clarke and Helal, 1990; Nieminen and Pylkkänen, 1992). Grinding to high finenesses may be especially indicated in Portland-slag cements with very high slag contents, to compensate for the limited reactivity of the slag (Tomisawa and Fuji, 1995). It has also been reported that the strength after 1 and 3 days may be distinctly increased by adding small amounts of sodium silicate (a solution of water glass) to the fresh concrete mix (Ghosh *et al.*, 1997; Roy *et al.*, 1998). This additive appears particularly effective at early ages. At the same time sodium silicate reduces the setting time of the cement and is also effective as a grinding aid. An addition of small amounts of sodium sulfate to the fresh mix also has a positive effect on strength (Marciano and Battagin, 1997).

The rate at which the heat of hydration is released is slowed down and the maximum adiabatic temperature rise decreases with increasing amounts of slag in cement (Nakamoto and Togawa, 1995; Nakamoto *et al.*, 1995; Tomisawa and Fuji, 1995; De Schutter, 1999).

Cement compositions that contain granulated blast furnace slag are more disposed to carbonation than are plain Portland cement pastes (Matala, 1997). The rate of carbonation is influenced by the curing conditions, and generally decreases with increasing degree of hydration and compressive strength. Just as in ordinary cement, the depth of carbonation will be proportional to the square of the concrete age (Nakamoto and Togawa, 1995; Nakamoto *et al.*, 1995). Carbonation causes an increase in and coarsening of the porosity of the C-S-H phase (Matala, 1997).

The response of concrete mixes made from Portland-slag cement to corrosive agents is qualitatively similar to, but generally better than, that of mixes made with plain Portland cement. This is due mainly to the reduced free calcium hydroxide content in the former ones. The mechanism of corrosion protection of steel embedded in Portland-slag cement pastes differs from that existing in pastes made from plain Portland cement.

A layer of iron (II), rather than of iron (III), is formed in the Portland-slag cement pastes, especially at high slag contents, owing to the reducing environment (Cao *et al.*, 1992). Nevertheless, no significant differences were found in the corrosion behavior of steel under the influence of chlorides between cements containing blast furnace slag and plain Portland cement.

Under hydrothermal conditions (175 °C) α-C_2SH is formed as the main hydration product in cements with high slag contents, resulting in a gradual decline of strength under prolonged autoclaving (Xi *et al.*, 1997). This effect may be eliminated by adding a few per cent of microsilica to the starting mix. Under these conditions an amorphous C-S-H or semi-crystalline C-S-H phase is formed as the main hydration product. The hydration of the existing slag is slowed down, owing to the reduced pH of the pore solution (Xi *et al.*, 1997).

In gypsum-free slag cement, gypsum – a regular constituent of Portland-slag cement – is replaced by a combination of a modified lignine derivative (or sulfonated polyphenolate) and an alkali compound (hydroxide or silicate) (Skvara, 1986). Free-flowing pastes may be obtained at *w/c* as low as 0.24–0.26. The attained strength may exceed 100 MPa after 28 days. The hydrated material consists of non-crystalline low-lime calcium hydrosilicates/hydroaluminates. It is also possible to replace gypsum with limestone (Salem *et al.*, 1995).

Portland-slag cements may be employed in most applications in which plain Portland cement is also used, as long as a high short-term strength is not required. The main areas of their use are applications in which a high corrosion resistance or a slow hydration heat liberation of the binder are required.

8.4 SUPERSULFATED CEMENT

Supersulfated cement is produced by intergrinding or intimately blending granulated blast furnace slag with calcium sulfate and Portland clinker. Usually the cement is more finely ground than Portland cement, to enhance its reactivity.

The main constituent of the cement is the granulated blast furnace slag present typically in amounts of 80–85 wt%. A high Al_2O_3 content of the slag (above 13 wt%) is preferred, to enhance the formation of ettringite in the hydration process.

The content of calcium sulfate in the cement is typically 10–15 wt%. It may be introduced either as anhydrite ($CaSO_4$) or as gypsum ($CaSO_4.2H_2O$).

Portland clinker is present in the binder in amounts of a few per cent. It is added to the cement blend to ensure a high pH value of the liquid phase, since this is needed for the hydration reaction to take place. It may be replaced by Portland cement or even hydrated lime, but this has to be added in smaller amounts than Portland clinker.

The main product of hydration is a C-S-H phase with an average C/S ratio lower than typical in Portland cement pastes. It may also contain distinct amounts of Al_2O_3 in its structure. It is the hydration product that is mainly responsible for the strength development of the cement (Mohan and Ghosh, 1992). The second main product of the hydration process is ettringite ($C_3A.3C\bar{S}.32H$). Most of this phase is formed within the first 3–7 days, by which time virtually all the anhydrite (or gypsum) has reacted. Ettringite is present in the hydrated material in the form of relatively coarse crystals, and its presence contributes to the strength development within the first few days of hydration. Unlike many other cementitious systems the formation of ettringite is not associated with an expansion. Small amounts of portlandite (CH) may also be present after short hydration times, but this phase is absent in mature pastes. The slag that is present does not hydrate completely, and a significant fraction of it may be found even in mature pastes. The combined water content of the hydrated material is relatively high, mainly because of the presence of high amounts of ettringite, a phase with an exceptionally high fraction of crystalline water in its crystalline lattice.

The strength development of supersulfated cement is rather sluggish and its final strength is relatively low. The rate of strength development increases with temperature up to about 40 °C, but at higher temperatures the strength declines significantly, mainly because of thermal decomposition of the ettringite. For the same reason supersulfated cement is not suitable for heat curing.

If used above ground, proper curing of concrete made from supersulfated cement is essential. The surface must be kept damp, as otherwise a friable surface layer, which tends to dusting, may develop. The tendency to dusting may be reduced by increasing the Portland clinker content in the cement.

The heat of hydration of supersulfated cement is very low, typical values being 40–45 cal/g (167–188 k J/kg) at 7 days and 45–50 cal/g (188–209 k J/kg) at 28 days.

Supersulfated cement is highly resistant to a variety of aggressive agents, especially sulfate solutions. It also exhibits a fair resistance to diluted solutions of inorganic and organic acids down to a pH of about 3.5.

Supersulfated cement is particularly suitable for underground applications, especially where a contact of the finished construction with water containing significant amounts of corrosive constituents is to be expected. It has also been successfully used in harbor and breakwater constructions, where its resistance to seawater is advantageous.

8.5   ALKALI-ACTIVATED SLAG CEMENT

Alkali-activated slag cements (called also AAS cements) consist of ground granulated blast furnace slag and an alkaline activator whose role is to activate the hydration of the latent hydraulic slag.

Generally, granulated blast furnace slags with C/S ratios between 0.5 and 2.0 and A/S ratios between 0.1 and 0.6 may be used in AAS cements. The fineness of the slag may range between 250 and 700 m^2/kg. Alternatively, some other industrial slags (steelwork slag, and most non-ferrous slags) may also be employed (Bian *et al.*, 1992).

Usually, NaOH, Na$_2$CO$_3$, or sodium silicate are employed as activators. Because of their higher costs the corresponding potassium compounds are seldom used, unless they are constituents of an industrial waste. They are also indicated if the cement is to be used in high-temperature applications. The use of potassium-based rather than sodium-based activators also significantly reduces the occurrence of efflorescence, which may be of importance in some applications. If an alkali silicate solution is employed as activator, a gradual polycondensation of the silicate ions, associated with changes in rheology, may take place upon prolonged storage. In the resultant cement paste, alkali carbonate and silicate will convert to alkali hydroxide in a reaction with calcium hydroxide, which is formed in the hydration process as an intermediate product. In this way a pH of the liquid phase of about 13.5 is attained.

Some less common activators recommended by some investigators include a blend of NaOH and CaSO$_4$.2H$_2$O (Tango and Vaidergorian, 1992), calcium hydroxide alone or in combination with gypsum (Tango and Vaidergorian, 1992), sodium sulfate (Chen *et al.*, 1992), sodium aluminate (Chen *et al.*, 1992), and potassium aluminum sulfate (Chen *et al.*, 1992). However, neutral sodium or potassium salts exhibit only a very weak activating effect, unless the pH of the mix is kept high by the simultaneous presence of alkali or calcium hydroxide.

The activator is introduced in amounts of a few per cent, typically about 2–7% per weight of the slag. It may be interblended with the dry slag prior to mixing with water, or – more commonly – dissolved in the mixing water prior to mixing with the ground slag. For some special applications, the addition of some finely powdered materials to the AAS mix may be indicated. The addition of some fly ash, ground limestone, or granite may improve the workability of the mix; the addition of microsilica tends to suppress efflorescence; and finely powdered metallic iron improves watertightness and chemical resistance.

For producing AAS concrete, concrete aggregates common in combination with Portland cement may be employed. Non-ground industrial slags may be particularly suitable. For mixing, equipment used for producing ordinary concrete may be used. In producing and handling AAS cement based concrete mixes the high causticity of the activators used and the very high pH of the material must be taken into consideration.

The water requirement of alkali-activated slag cements is relatively low, owing to the plasticizing effect of the alkali compounds that are present, resulting in a lower total porosity of the hardened material, as compared

with Portland cement mixes. The produced fresh AAS cement based concrete mixes exhibit a distinct thixotropy and require continuous mixing, to
prevent a quick slump loss and setting. Owing to their thixotropic properties even stiff mixes may be compacted if vibration is applied.

The setting time of AAS pastes depends on the reactivity of the slag
employed, as well as on the quality and amount of the activator. It may be
shortened or extended by the addition of suitable accelerators or retarders
to the mix. Effective accelerators include ground Portland clinker or
Portland cement and some aluminum compounds. If alkali silicate is
employed as activator, the setting time shortens with increasing silicate
modulus (SiO_2/Me_2O), and at values above 1.5 setting may take place
within minutes (Shi *et al.*, 1992). Effective retarders of setting are borates,
phosphates ($Na_4P_2O_7$), or copper sulfate.

The alkaline activator causes the breaking of bonds between interlinked
SiO_4 and AlO_4 tetrahedra constituting the three-dimensional network of
the glass phase existing in the slag:

$$-\overset{|}{\underset{|}{Si}}-O-\overset{|}{\underset{|}{Si}}- + \ H-O-H \rightarrow 2\ -\overset{|}{\underset{|}{Si}}-OH \tag{8.1}$$

$$-\overset{|}{\underset{|}{Si}}-O-\overset{|}{\underset{|}{Al}}- + \ H-O-H \rightarrow -\overset{|}{\underset{|}{Si}}-OH + -\overset{|}{\underset{|}{Al}}-OH \tag{8.2}$$

The silicate and aluminate ions enter the liquid phase as hydrosilicates
and hydroaluminates. They are present mostly in a condensed form. At
the same time the free OH^- groups are neutralized in the highly alkaline
environment:

$$-\overset{|}{\underset{|}{Si}}-OH + NaOH \rightarrow -\overset{|}{\underset{|}{Si}}-ONa + HOH \tag{8.3}$$

The hydration in AAS cement pastes progresses quite rapidly. At ambient temperature about a third of the slag is consumed within the first hour,
and about two thirds within 24 hours after mixing (Schilling *et al.*, 1994;
Fernandez-Jimenez and Puertas, 1997a, 1997b).

As the hydration progresses, three layers are formed at the slag grain
surface:

- an inner layer consisting of a glass phase depleted in Ca^{2+} and enriched
  in Al^{3+} and Mg^{2+};
- an intermediate layer of a constant thickness (about $0.3\,\mu m$) of a
  hydrotalcite-type composition;
- an outer C-S-H-rich layer.

The overall phase composition of the hydrated material will depend on the composition of the slag, the selection of the activator, and the hydration temperature. At ambient temperature a C-S-H phase is always the main reaction product (Chen *et al.*, 1992; Kutti, 1992; Shi *et al.*, 1992; Su, 1992; Tango and Vaidergorian, 1992; Roy and Malek, 1993; Talling and Brandstetr, 1993; Song and Lee, 1994; Richardson, 1997; Song and Jennings, 1999). It differs from that formed in the hydration of Portland cement by a lower C/S ratio and by the presence of distinct amounts of aluminum and alkali ions in its structure. The C/S ratio of the C-S-H phase is also influenced by the pH of the pore solution, and decreases as this value goes up (Song and Jennings, 1999). ^{29}Si and ^{27}Al NMR studies (Schilling *et al.*, 1994; Richardson, 1997) have revealed that in the tobermorite-like hydration product aluminum substitutes for silicon in the central tetrahedron of the pentameric silicate chains. Roughly one third of the bridging sites are occupied by aluminum in tetrahedral coordination. Some octahedrally coordinated Al was also found, and was attributed to the presence of the phase $(C,M)_4AH_x$.

It has been suggested, but is not widely accepted (Kutti, 1992), that – in addition to this C-S-H phase – another amorphous phase that is silica-rich and which exhibits properties similar to those of silica gel is formed in the initial stage of hydration. It contains significant amounts of loosely bound water, which escapes upon drying and is responsible for the shrinkage of the hardened cement paste and the formation of microcracks.

In addition to non-crystalline phases some crystalline phases may also be formed in the hydration process. According to Shi (1997), in mixes made with NaOH as activator the phases C_4AH_{13} and C_2ASH_8 (strätlingite) may be formed; in the presence of Na_2CO_3 the carboaluminate $C_4A\bar{C}H_x$ is formed, and C_4AH_{13} may be formed in sodium silicate activated pastes.

In mixes made from slags with high Mg contents a hydrotalcite phase may also be present (Jiang *et al.*, 1997). If a high-basicity slag is used, gehlenite hydrate (C_2SAH_8) may be formed. In mixes activated with a soluble sulfate, ettringite may be produced. Free calcium hydroxide (portlandite) is not present in hardened alkali-activated slag cement pastes. The amount of crystalline phases increases at high temperatures. Above 100 °C the C-S-H phase tends to be converted into tobermorite and gyrolite. Small amounts of hydrated calcium-sodium silicate may also be formed under these conditions.

The electrolyte concentration in the pore solution of AAS cement pastes is rather high; at low water/solid ratios it may reach values as high as 25–30%. In the course of hydration the amount of free alkali hydroxide in the mix declines, as sodium (potassium) ions are incorporated into the C-S-H phase and become adsorbed on the surface of the hydration products. After 28 days the amount of non-incorporated alkalis drops to about 70% and after 1 year to about 10–20% of the original value, and it may decline even further (Roy and Malek, 1993; Talling and Brandstetr, 1993). Nevertheless, the final pH of the pore solution stays high enough (about 12.5) to prevent corrosion of the steel reinforcement.

At comparable consistencies of the starting mix, alkali-activated slag cement pastes exhibit lower porosity than comparable Portland cement pastes, owing to the lower initial water/solid ratio. The proportion of pores with $r < 10$ nm is usually higher in hardened AAS pastes (Shi *et al.*, 1992); however, the actual pore size distribution also depends on the activator used. In a comparative study, mixes produced with sodium silicate exhibited the finest and those made with NaOH the coarsest pore structure (Shi, 1996, 1997). The specific surface area of AAS cement pastes is higher (by about 35–55%) than that of comparable ordinary Portland cement pastes (Talling and Brandstetr, 1993).

Unlike concrete based on Portland cement, the microstructure at the aggregate/paste interface in hardened AAS concrete differs only slightly from that of the bulk paste. The main difference between the interfaces of concretes made with Portland and AAS cements is the absence of portlandite and ettringite in the AAS concretes. Owing to the high bond strength between the aggregate and the paste, the fracture cracks in AAS concrete usually propagate through the aggregate.

The strength development of alkali-activated slag cements is greatly influenced by the nature and dosage of the activator (Shi *et al.*, 1992; Shi, 1996; Collins and Sanjayan, 1998). There are indications of a distinct selectivity: the effect of an activator on strength development may be different for slags of different origin, or – conversely – different activators may produce different strength developments if combined with the same slag. In most instances the early strength development is quite fast, and compressive strengths of around 20–30 MPa after 1 day and 40–60 MPa after 28 days at 20 °C may be considered fairly typical. An acceleration of strength development may be achieved by a very fine grinding of the slag (Clarke and Helal, 1990) or by an increase of the curing temperature. The strength development continues even after 28 days, reaching in some mixes a 2.5–fold increase after 15 years as compared with the 28–day strength (Su *et al.*, 1992). With steam curing, concrete strengths exceeding 100 MPa may be attained. The flexural strength of the material depends greatly on the curing conditions, and declines distinctly upon drying, as a consequence of a formation of microcracks caused by drying shrinkage (Kutti, 1992).

The liberation of heat in the hydration of alkali-activated slag cements is characterized by one or two peaks within the first few minutes of hydration and an additional peak within a few hours after mixing. The overall heat evolution within the first 24 hours may or may not surpass that of Portland cement (Shi and Day, 1995a, 1995b, 1996).

AAS cement pastes exhibit a distinct, irreversible chemical shrinkage (Kutti, 1992), which may cause the formation of microcracks (Chen *et al.*, 1992; Kutti, 1992), especially in rapid-hardening systems. The extent of this shrinkage may be reduced by curing at elevated temperature (Talling and Brandstetr, 1993). Upon curing in air the material also exhibits a distinct drying shrinkage, which may result in the formation of additional cracks, and may cause a decline of flexural strength (Kutti, 1992). The extent of shrinkage may be reduced by adding bivalent iron compounds to the AAS concrete mix, which expand upon their subsequent oxidation by the oxygen of air in the existing alkaline environment.

Dry-air storage of young AAS concrete may result in the formation of efflorescence caused by alkali compounds not yet incorporated into the hydrates to be formed and present in the pore solution in high concentrations (Wang *et al.*, 1995).

The chemical corrosion resistance of AAS concrete is very high. The hardened paste is completely resistant to sodium sulfate, and has a very high resistance to magnesium chloride and nitrate attack (Talling and Brandstetr, 1993). AAS concrete also protects the steel reinforcement effectively against corrosion by chloride solutions, mainly because of a very low diffusion rate of chloride ions in the hardened paste (Dhir *et al.*, 1996). However, the carbonation of AAS concrete surfaces by the CO_2 of the air progresses faster than in comparable mixes produced with Portland cement.

Upon heating, the hardened paste exhibits a gradual dehydration and formation of crystalline phases, such as mellilite, gehlenite, wollastonite, or sodium and sodium-calcium aluminosilicates. Even so, acceptable strength and good performance are maintained up to about 800–1000 °C. This temperature may be extended to even higher values by adding to the mix suitable additives, and by using aggregates that are resistant to high temperatures. Potassium-based activators are preferred for high-temperature applications.

Owing to the high electrolyte concentration in the liquid phase, AAS concrete is well suited to concreting at low ambient temperatures, especially when produced with a low water/cement ratio. The fresh concrete mix sets and hardens even at temperatures below 0 °C; however, the development of strength is very sluggish at these temperatures. The hardened concrete exhibits good freeze-thaw resistance, which is due both to the fine pore structure and to the low freezing temperature of the highly concentrated pore solution.

The high alkali hydroxide concentration in the pore solution may cause a reaction with an alkali-sensitive aggregate, resulting in expansion and cracking. It has been reported that AAS concrete is more vulnerable to alkali-carbonate reaction and less susceptible to expansion due to an existing alkali-silica reaction than OPC concrete (Wang *et al.*, 1995; Gifford and Gillot, 1996).

F-cement is a product developed and marketed in Finland. It consists of a finely ground granulated blast furnace slag, an alkaline activator, and a superplastcizer. It is used in the precast industry.

Possible applications of AAS cements include precast concrete, concrete for corrosive environments, heat-resistant concrete, wood composites, radioactive waste encapsulation, and grouting.

8.6   BINDERS BASED ON OTHER INDUSTRIAL SLAGS

Some industrial slags other than granulated blast furnace slags may also be used as constituents of inorganic binders. They are, however, produced in significantly lower volumes, and are far less important than blast furnace slag.

Steel slag is a by-product in the production of steel, which involves the removal of impurities from pig iron. The main oxides present in this slag are CaO, SiO_2, FeO, Fe_2O_3, MgO, MnO, and P_2O_5. The mineralogical composition may vary, and depends on the chemical composition of the slag, which in turn may vary with the steelmaking process. Being rich in CaO, the slag typically contains significant amounts of dicalcium and tricalcium silicate, and also some free lime. The slag may also contain limited amounts of iron, which is predominantly present in the form of hematite and magnetite. Other phases present may include olivine (MS), dicalcium ferrite, and calcium aluminoferrite. The glass phase content in the slag is low, and usually does not exceed 30–40%. The slag exhibits weak but distinct hydraulic properties, which increase with increasing alkalinity. The hydraulicity of the slag may be enhanced by granulation, resulting in an increased glass content of the material (Sharma *et al.*, 1997).

Arc furnace slag is also usually lime rich. It may be obtained in granulated (glassy) form, and may be used in a similar way as granulated blast furnace slag.

Non-ferrous slags are by-products in the smelting of non-ferrous metals. Vitrified non-ferrous slags, such as copper and nickel slags, exhibit weak pozzolanic properties, and their blends with Portland cements may be used as binders of inferior cementing value.

REFERENCES

Bakharev, T., Sanjayan, J.G., and Chen, Y.-B. (1999) Alkali activation of Australian slag cements. *Cement and Concrete Research* **29**, 113–120.

Battagin, A.F. (1992) Influence of degree of hydration of slag cement, in *Proceedings 9th ICCC, New Delhi*, Vol. 3, pp. 166–172.

Bian, Q., Wu, X., and Tang, M. (1992) High strength alkali steel-iron slag binder, in *Proceedings 9th ICCC, New Delhi*, Vol. 3, pp. 291–297.

Cao, H. *et al.* (1992) Corrosion characteristics of steel in blast furnace slag – Portland cement, in *Proceedings 9th ICCC, New Delhi*, Vol. 5, pp. 309–315.

Chen, Q.H., Tagnit-Hamon, A., and Darkar, S.L. (1992) Strength and microstructural properties of water glass activated slag. *Materials Research Society Symposium Proceedings* **245**, 49–61.

Clarke, W.J., and Helal, M. (1990) Activated slag and Portland/slag ultrafine cement. *Materials Research Society Symposium Proceedings* **179**, 219–232.

Collins, F., and Sanjayan, J.G. (1998) Early age strength and workability of slag pastes activated by NaOH and Na2CO$_2$. *Cement and Concrete Research* **28**, 655–664.

Collins, F., and Sanjayan, J.G. (1999) Strength and shrinkage properties of alkaliactivated slag concrete placed into a large column. *Cement and Concrete Research* **29**, 659–666.

De Schutter, G. (1999) Hydration and temperature development of concrete made with blast-furnace slag cement. *Cement and Concrete Research* **29**, 143–149.

Dhir, R.K., El-Mohr, M.A.K., and Dyer, T.D. (1996) Chloride binding in GGBS concrete. *Cement and Concrete Research* **26**, 1767–1773.

Fernandez-Jimenez, A., and Puertas, F. (1997a) Alkali-activated slag cements: kinetic studies. *Cement and Concrete Research* **27**, 359–368.

Fernandez-Jimenez, A., and Puertas, F. (1997b) Kinetic study of alkali-activated blast furnace slag, in *Proceedings 10th ICCC, Göteborg*, paper 3ii098.

Ghosh, S.N. *et al.* (1997) Studies on activation of slag cements, in *Proceedings 10th ICCC, Göteborg*, paper 3ii092.

Gifford, P.M., and Gillot, J. (1996) Alkali-silicate reaction (ASR) and alkali-carbonate reaction (ACR) in activated blast furnace cement (ABFSC). *Cement and Concrete Research* **26**, 21–26.

Hanafi, S. (1991) Hydration kinetics and microstructure of low-porosity slag-lime pastes, in *Proceedings of the 13th International Conference on Cement Microscopy, Tampa*, pp. 294–303.

Hinrichs, W., and Odler, I. (1989) Investigations on the hydration of Portland blast furnace slag cement. *Advances in Cement Research* **2**, 9–13, 15–20.

Jiang, W., Silsbee, M.R., and Roy, D.M. (1997) Alkali activation reaction mechanism and its influence on microstructure of slag cement, in *Proceedings 10th ICCC, Göteborg*, paper 3ii1000.

Kutti, T. (1992) Hydration products of alkali activated slag, in *Proceedings 9th ICCC, New Delhi*, Vol. 4, pp. 468–474.

Lumley, J.S. *et al.* (1996) Degrees of reaction of the slag in some blends with Portland cement. *Cement and Concrete Research* **26**, 139–151.

Marciano, E. Jr., and Battagin, A.F. (1997) The influence of alkali activator on the early hydration and performance of Portland blast furnace slag cement, in *Proceedings 10th ICCC, Göteborg*, paper 3ii103.

Matala, S. (1997) Carbonation mechanism in the granulated blast furnace slag concrete, in *Proceedings 10th ICCC, Göteborg*, paper 4iv005.

Mohan, K., and Ghosh, S.P. (1992) Evaluation of physico-chemical and hydration characteristics of supersulfated cement, in *Proceedings 9th ICCC, New Delhi*, Vol. 3, pp. 331–337.

Nakamoto, J., and Togawa, K. (1995) *A study of strength development and carbonation of concrete incorporating big volume blast furnace slag.* American Concrete Institute SP-153, pp. 1121–1139.

Nakamoto, J., Togawa, K., and Fujii, M. (1995) Influence of curing conditions on the strength development and carbonation of high blast-furnace slag content concrete (in Japanese). *Semento Konkurito Ronbunshu* **49**, 231–239.

Nishokawa T. *et al.* (1992) Microstructure and mechanical properties of hardened cement paste incorporating ground granulated blast-furnace slag, in *Proceedings 9th ICCC, New Delhi*, Vol. 4, pp. 665–670.

Nieminen, P.J., and Pylkkänen, K.P.K. (1992) Properties of very fine blast furnace slag manufactured by jet mill, in *Proceedings 9th ICCC, New Delhi*, Vol. 3, pp. 108–114.

Richardson, I.G., and Groves, G.W. (1992) The composition and structure of C-S-H gel in cement pastes containing blast furnace slags, in *Proceedings 9th ICCC, New Delhi*, Vol. 4, pp. 350–356.

Richardson, I.G. (1997) The structure of C-S-H in hardened slag cement pastes, in *Proceedings 10th ICCC, Göteborg*, paper 2ii068.

Roy, D.M., and Silsbee, M.R. (1996) Alkali-activated cementitious materials, an overview. *Materials Research Society Symposium Proceedings* **245**, 153–164.

Roy, D.M., and Malek, R.I. (1993) Hydration in slag cement, in Mineral Admixtures in Cement and Concrete (ed. S.N. Ghosh), ABI Books, New Delhi, pp. 84–117.

Roy, D.M., Silsbee, M.R., and Wolfe-Confer, D. (1990) New rapid setting alkali-activated cement composition. *Materials Research Society Symposium Proceedings* **179**, 203–218.

Roy, S. *et al.* (1998) Investigation of Portland slag cement activated by water glass. *Cement and Concrete Research* **28**, 1049–1056.

Salem, Th.M., El-Didamony, H., and Mohamed, T. A. (1995) Studies on Portland blast furnace slag cement with limestone as a retarder. *Indian Journal of Engineering Materials Science* **21**, 32–135.

Schilling, P.J. *et al.* (1994) 29Si and 27Al MAS-NMR of Na-activated blast-furnace slag. *Journal of the American Ceramic Society* **77**, 2363–2368.

Sharma, K.M. *et al.* (1997) Hydration characteristics of steel slag in ordinary Portland cement blends, in *Proceedings 10th ICCC, Göteborg*, paper 3ii094.

Shi, C. (1996) Strength, pore structure and permeability of alkali-activated slag mortars. *Cement and Concrete Research* **26**, 1789–1799.

Shi, C. (1997) Early hydration and microstructure development of alkali-activated slag cement, in *Proceedings 10th ICCC, Göteborg*, paper 3ii099.

Shi, C., and Day, R.L. (1995a) *Chemical activation of lime-slag blends*. American Concrete Institute SP-153, pp. 165–177.

Shi, C., and Day, R.L. (1995b) A calorimetric study of early hydration of alkali-slag cement. *Cement and Concrete Research* **25**, 1333–1346.

Shi, C., and Day, R.L. (1996) Some factors affecting early hydration of alkali-slag cement. *Cement and Concrete Research* **26**, 439–447.

Shi, C. *et al.* (1992a) Composition of the microstructure and performance of alkali-slag and Portland cement paste, in *Proceedings 9th ICCC, New Delhi*, Vol. 3, pp. 298–304.

Shi, C. *et al.* (1992b) Comparison of the microstructure and performance of alkali-slag and Portland cement pastes, in *Proceedings 9th ICCC, New Delhi*, Vol. 3, pp. 298–304.

Skvara, F. (1986) Microstructure of hardened paste of gypsum-free Portland and slag cement, in *Proceedings 8th ICCC, Rio de Janeiro*, Vol. 3, pp. 356–362.

Song, J.T., and Lee, J.M. (1994) Hydration of granulated blast furnace slag in the presence of sodium silicate (in Korean). *Yoop Hakhoechi* **31**, 538–542 [ref. CA 121/211460].

Song, S., and Jennings, H.M. (1999) Pore solution chemistry of alkali-activated ground granulated blast-furnace slag. *Cement and Concrete Research* **29**, 159–170.

Su, M., Kurdowsky, W., and Sorrentino, F. (1992) Development in non-Portland cement, in *Proceedings 9th ICCC, New Delhi*. Vol. 1, pp. 317–354.

Talling, B., and Brandstetr, J. (1993) Clinker-free concrete based on alkali-activated slag. In *Mineral Admixtures in Cement and Concrete* (ed. S.N. Ghosh), ABI Books, New Delhi, pp. 296–341.

Tango, C.E.S., and Vaidergorian, E.Y.Y. (1992) Some studies on the activation of blastfurnace slag in cements without clinker, in *Proceedings 9th ICCC, New Delhi*, Vol. 3, pp. 101–107.

Tomisawa, T., and Fuji, M. (1995) *Effects of high fineness and large amounts of ground granulated blast furnace slag on properties and microstructure of slag cements*. American Concrete Institute SP-153, Vol. 2, pp. 951–973.

Tomkova, V., Sahu, S., and Majling, J.M. (1993) Alkali activation of granulated blast furnace slag. *Ceramic-Silikaty (Prague)* **37**, 61–66.

Wang, S.D. *et al.* (1995) Alkali-activated slag cement and concrete. A review of properties and problems. *Advances in Cement Research* **7**, 93–102.

Xi, Y., Siemer, D.D., and Scheetz, B.E. (1997) Strength development, hydration reaction and pore structure of autoclaved slag cement with added silica fume. *Cement and Concrete Research* **27**, 75–82.

9 Cements based on natural and artificial pozzolanas

Pozzolanas, or pozzolanic materials, are defined as siliceous and aluminous materials that on their own possess little or no cementitious value, but which will – if present in finely divided form and in the presence of moisture – react chemically with calcium hydroxide at ordinary temperature to form compounds possessing cementitious properties (ASTM C 619–89). The required calcium hydroxide may also be introduced in the form of materials that liberate this compound in the course of their hydration, such as Portland cement.

A variety of natural and artificial materials that differ in their chemical composition, mineralogical nature, and geological origin exhibit pozzolanic properties. Chemically they are rich in SiO_2 and – to a lesser extent – in Al_2O_3, both oxides being present as constituents of a reactive (glassy or amorphous) phase. The CaO content of pozzolanic materials is low. Sometimes they may also contain limited amounts of chemically bound water. Some pozzolanic materials may also contain non-pozzolanic constituents side by side with those that are reactive.

The pozzolanicity – that is, the readiness to react with calcium hydroxide – may vary greatly in different pozzolanic materials. In general, it increases with increasing glass phase content and increasing specific surface area; it declines with increasing degree of SiO_4 polymerization.

The main product of the pozzolanic reaction is an amorphous or nearly amorphous calcium silicate/aluminate hydrate phase similar to that formed in the hydration of calcium silicates constituting Portland cement.

9.1 FLY ASH BASED CEMENTS

9.1.1 Composition and properties of fly ashes

Fly ash (abbreviation FA), also known as pulverized fuel ash (abbreviation PFA), is a product formed – together with bottom ash – in the incineration of pulverized coal in suspension-fired furnace chambers of thermal power plants. It escapes as a suspension in the flue gases, and is collected by mechanical or electrical precipitators. It appears as a fine gray powder, and consists of mostly glassy spherical particles of complex chemical and mineralogical composition. Fly ash makes up about 75–85% of the total ash formed in the combustion process. The bottom ash is much coarser than fly ash, and is collected at the bottom of the furnace; it is not used as a constituent of inorganic binders.

The properties of fly ashes may vary greatly, and depend on the composition of the inorganic fraction of the coal, the degree of pulverization, the thermal history, and the oxidation conditions. Based on the chemical composition we can distinguish between low-calcium (ordinary) and high-calcium fly ash.

Low-calcium (ordinary) fly ash corresponds to class F ash, as defined in ASTM C 618–89. It is derived from anthracite or bituminous coal. It is poor in CaO and MgO and relatively rich in SiO_2 and Al_2O_3.

High-calcium fly ash corresponds to class C ash as defined in ASTM C 618–89. It is derived from sub-bituminous coal or lignite. It contains less SiO_2 and Al_2O_3 than class F ash, but higher amounts of CaO, of which a part is present in the form of free lime. Table 9.1 shows the ranges of chemical composition of fly ashes produced in different countries.

The main constituent of both class F and class C ashes is a glass phase, which typically constitutes 60% and more of the ash. It is formed in the cooling of a melt that has been created from the inorganic constituents of the original coal at temperatures existing in the combustion chamber. This phase is basically a SiO_2-Al_2O_3-(CaO)-(MgO) glass with a low degree of SiO_4 polymerization. In general, the CaO content in the glass phase of class C ashes is higher, which causes a lower degree of SiO_4 polymerization and an increased reactivity (Mehta, 1985). Owing to its readiness to react with calcium hydroxide, the glass phase is the constituent of fly ash that is mainly responsible for its pozzolanicity.

The main crystalline constituents of fly ashes are quartz (SiO_2), mullite ($3Al_2O_3.2SiO_2$), magnetite (Fe_3O_4), and hematite (Fe_2O_3). High-calcium ashes may also contain free lime (CaO), periclase (MgO), tricalcium aluminate ($3CaO.Al_2O_3$), dicalcium silicate ($2CaO.SiO_2$), and anhydrite ($CaSO_4$). Other crystalline phases that may be (rarely) present in some ashes include merwinite [$Ca_3Mg(SiO_4)_2$] and mellilite (a solid solution of gehlenite, $2CaO.Al_2O_3.SiO_2$, and akermanite, $2CaO.MgO.2SiO_2$). Fly ashes regularly also contain some residual carbon, the amount of which usually does not exceed 3%, but in rare instances may reach 10%. If present in larger amounts it forms porous particles with a high specific surface area.

Fly ash particles are typically spherical, as they are formed by solidification of droplets of a partially melted material, suspended in the flue gas, upon cooling. Particles that are formed at lower combustion temperatures may be irregular, owing to the reduced amount of melt present. Occasionally, tiny grains of volatile salts may be deposited on the ash particle surface. About 20% of fly ash particles are hollow, owing to an

Table 9.1 Chemical composition of fly ashes produced in different countries.

Country	SiO$_2$	Al$_2$O$_3$	Fe$_2$O$_3$	CaO	MgO	SO$_3$	K$_2$O	Na$_2$O	L.O.I	Source
Australia	9–63	4–33	1–30	0–33	0–24	0–15	0.1–2.2	0.1–5.6	0.1–15.2	Samarin *et al.* (1983)
Belgium	47–54	25–29	6–10	1–4	1–2	<1	2.2–3.3	0.7–1.1	1.3–1.9	Tenouta
Brazil	52–78	9–33	2–14	1–2	<1.4	0–2	0.2–2.8	0.2–1.4	0.1–3.4	CIENTEC
Bulgaria	40–60	12–32	5–16	4–12	1–5	1–10	–	–	0.5–20	Kovacs (1982)
France	14–53	6–33	4–7		1–59	1–5	0.7–6.0	0.1–0.9	0.3–15.2	Aiticin
Germany	2–77	2–30	1–16	1–41	2–23	0–27	0.1–4.7	0.2–10	0–20	Ullmann
Hungary	41–60	16–34	5–17	1–11	1–7	1–7	0–2.2	0.2–2.5	1–5	Kovacs
India	37–67	18–29	3–22	1–11	1–5	0–3	–	–	0.3–17	Rehsi 1
Japan	53–63	25–28	2–6	1–7	1–2	<0.8	1.8–3.2	0.8–2.4	0.1–1.2	Kovacs
Poland	35–50	6–36	5–12	2–35	1–4	0–8	0.1–2.7	0.1–2.0	1–10	Kovacs
Romania	39–53	18–29	7–16	3–13	1–4	1–6	0.3–2.2	0.1–1.8	0.2–4.5	Kovacs
Former USSR	36–63	11–40	4–17	1–32	0–5	0–3	1.1–3.6	0.5–1.2	0.5–23	Kovacs
UK	43–55	22–34	6–13	1–8	1–2	0–2	1.0–3.8	0.1–4.0	1.2–13	Hubbart
USA	23–58	13–25	4–17	1–29	1–8	0–8	0.4–3.2	0.4–7.3	0.4–4.9	Gebler

entrapment of gases by the molten phase in the course of burning. They are called cenospheres. Some of them, called plerospheres, may contain smaller particles inside.

The size of the individual ash particles may range between less than 1 μm and several hundreds of μm, and the specific surface of the ash typically varies between about 200 and 400 m^2/kg (Blaine) or 0.4 and 1.0 m^2/g (BET) (Iyer and Stanmore, 1995). Upon grinding, the ash particles tend to break down, leaving shell-shaped fragments from the original cenospheres and solid fragments that may or may not partially retain their original spherical shape (Paya *et al.*, 1995, 1996).

At ambient temperature, in the presence of calcium hydroxide and water, the glass phase of the fly ash undergoes a pozzolanic reaction, yielding an amorphous C-S-(A)-H phase as the reaction product. The formation of this phase is responsible for setting and hardening, if the reaction takes place in paste form. As a first step in the pozzolanic reaction, the SiO_2–Al_2O_3 framework of the glass is attacked and broken down by OH⁻ ions. After a sufficient number of Si–O–Si or Si–O–Al bonds have been broken, free silicate and aluminate anions are detached from the network, and react with calcium hydroxide and water to yield an amorphous calcium silicate aluminate phase.

In some high-lime class C ashes sufficient amounts of calcium hydroxide may be formed by hydration of the calcium oxide that is present. Such ashes do not require a separate source of calcium hydroxide. Moreover, if these ashes contain tricalcium aluminate and dicalcium silicate, the hydration of these constituents may also contribute to setting. Such ashes possess cementing properties, and may set and harden even without being combined with calcium hydroxide; however, the obtained strengths are usually rather low. Quartz, mullite, magnetite, and hematite – the main crystalline constituents of the ash – do not react noticeably with calcium hydroxide at ambient temperatures.

9.1.2 Lime-activated fly ash binder

Lime-activated fly ash binders consist of fly ash blended with hydrated or non-hydrated lime in amounts corresponding to 15–25 wt% of $Ca(OH)_2$.

If mixed with water, the existing calcium hydroxide, or that formed in the hydration of added lime, dissolves in the liquid phase and reacts with the glass phase of the ash. An amorphous C-S-(A)-H phase is formed as the product of reaction, which – at appropriate water/solid ratios of the starting mix – causes setting and hardening of the system. Limited amounts of ettringite (AFt) and monosulfate (AFm) may also be formed in the reaction, if sulfate ions are present in the ash.

At ambient temperature the reaction progresses very slowly, and setting occurs only after several days of hydration. A measurable strength development may be observed only after more than one month of hydration, and the final strength of fly ash–lime pastes remains low. The reaction progresses in three stages: a relatively rapid initial reaction; a period of slow reaction, which determines the overall rate of the process; and a stage of accelerated reaction (Wang *et al.*, 1996; Jalali, 1997).

The setting and hardening of lime-fly ash pastes may be accelerated by adding appropriate activators to the system (Shi, 1996): an addition of Na_2SO_4 raises the alkalinity of the liquid phase, which accelerates the formation of the C-S-(A)-H phase. In addition, ettringite may also be formed. As a result of both effects the strength development is accelerated and the ultimate strength is increased. In the presence of $CaCl_2$ a solid solution of calcium aluminate monosulfate hydrate ($3CaO.Al_2O_3.CaSO_4.12H_2O$) and calcium aluminate dichloride hydrate ($3CaO. Al_2O_3. CaCl_2. 10H_2O$), an AFm phase, is formed, which contributes to higher strengths, especially after longer hydration times.

9.1.3   Portland–fly ash cement and fly ash concrete

Portland–fly ash cement is produced by intergrinding Portland clinker, fly ash, and calcium sulfate, or by intimate blending of Portland cement with pulverized fly ash. Only fine ashes may be interblended, whereas a joint comminution is necessary to utilize ashes with a coarse particle distribution. Under most specifications the amount of fly ash in these cements is limited to 30–40 wt%, but an addition of about 15–25 wt% may be considered more typical.

The Portland clinker used should contain a high amount of tricalcium silicate, preferably more than 45%. This is necessary as the hydration of this phase produces the calcium hydroxide needed for a pozzolanic reaction of the ash. The hydration of the clinker minerals is mainly responsible for the setting and initial strength development of the cement, as the reaction rate of the fly ash is rather slow. The hydration of the ash contributes to strength only at longer hydration times, but also affects other properties of the hardened material. The calcium sulfate added in the form of gypsum or anhydrite serves to control the setting of the fresh paste in a similar way as in plain Portland cement.

Alternatively, fly ash may be introduced directly into the fresh concrete mix as a separate component, rather than as a constituent of the cement. To do this, Portland cement, fly ash, and mixing water, together with the aggregate, are mixed in the concrete mixer to produce fly ash concrete.

The chemical and physical requirements for fly ashes to be used as cement/concrete constituents, as required by the specifications of various countries, are summarized in Tables 9.2 and 9.3. The following physical parameters are commonly specified:

- particle size and specific surface area;
- volume stability (maximum permissible expansion);

Table 9.2 Chemical requirements for fly ashes to be used as constituents of concrete mixes.

Country/ specification	SiO_2 ($\geq wt\%$)	$SiO_2 + Al_2O_3$ $+Fe_2O_3$ ($\geq wt\%$)	MgO ($\leq wt\%$)	SO_3 ($\leq wt\%$)	Na_2O equiv ($\leq wt\%$)	$L.O.I$ ($\leq wt\%$)	H_2O ($\leq wt\%$)
Australia/AS 1129	–	–	–	2.5	–	8.0	1.5
Canada C	–	–	–	5.0	–	6.0	3.0
CAN3–A23.5M82 F	–	–	–	3.0	–	5.0	1.0
China/GB 1596–79	–	–	–	3.0	–	5.0	1.0
Germany/DIN 1045	–	–	–	4.0	4.0	5.0	–
India/IS 3812	35.0	70.0	5.0	3.0	1.5	12.0	–
Japan/JIS A6201	45.0	70.0	5.0	5.0	–	10.0	3.0
Former USSR/GOST 6269	40.0	–	–	3.0	–	10.0	–
UK/BS 3892	–	–	4.0	2.5	–	5.0	0.5
USA/ASTM C 618 C	–	50.0	5.0	5.0	1.5	6.0	3.0
F	–	70.0	5.0	5.0	1.5	12.0	3.0

Germany: $Cl \leq 0.1$ wt%; CaO total ≤ 8.0 wt%; CaO free ≤ 1.5 wt%
Japan: CaO-total ≤ 6.0 wt%

Table 9.3 Physical requirements for fly ashes to be used as constituents of concrete mixes.

Country/ specification	Particles < 0.04 mm (≥wt%)	Specific surface area (≥ cm²/g)	Expansion ASTM C–151 (≤ mm/m)	Relative compressive 7 d (≥%)	Relative compressive 28 d (≤%)	Relative water requirement (≤%)	FA-lime mortar strength 7 d (≥ MPA)
Australia/AS 1129	50	–	–	–	–	–	–
Canada/CAN3–A23.5M82	34	–	8.0	68	75	–	–
China/GB 1596–79	95 (80 μm)	–	–	–	–	95	–
Germany/DIN 1045	50	2000	8.0	70	70	–	–
India/IS 3812	–	3200	8.0	–	–	–	4.0
Japan/JIS A6201	–	2700	8.0	–	60	100	–
Former USSR/GOST 6269	–	–	–	–	85	105	5.5
UK/BS 3892	12.5	–	–	–	85	95	–
USA/ASTM C 618	34	–	8.0	–	75	105	5.5

Canada + USA: drying shrinkage ≤ 0.03%
Japan: drying shrinkage: ≤ 0.15%

- relative compressive strength of a standard mortar in which Portland cement has been partially replaced with fly ash, as compared with a control mortar of the same composition made with pure Portland cement;
- water requirement of a cement paste made with a partial replacement of Portland cement with fly ash, compared with that of pastes made with Portland cement alone (at equal consistency);
- absolute compressive strength of mortar of standard composition made from fly ash and hydrated lime;
- drying shrinkage of mortar of standard composition.

Fly ash plays a number of different roles in concrete mixes. First, it affects the rheology of the fresh concrete mix. This is basically a physical effect, related to the spherical shape of the ash particles. After setting, the ash participates in the hydration process, mainly in a reaction with the calcium hydroxide produced by hydration of the calcium silicate phases of the clinker. In class C ashes the hydraulically reactive phases may also be involved in the hydration process. Usually, the ash does not undergo complete hydration, and the residual particles act as a filler. The fraction of the total amount of ash that undergoes hydration is especially low in mixes with high amounts of added ash. Here the pozzolanic reaction cannot proceed to its end, as insufficient calcium hydroxide is liberated in the hydration of the calcium silicate phases present in the original fresh mix.

As a component of concrete, fly ash may be used to partially replace Portland cement (by equal weight). In principle this is also true if Portland cement is substituted by an equal amount of Portland–fly ash cement, as under these conditions the amount of Portland clinker present is also reduced. As the specific gravity of the ash (typically about 2.3 g/cm^3) is lower than that of Portland clinker (about 3.15 g/cm^3) a cement–ash replacement on an equal weight basis is always associated with an increase of the volume of the cement paste in the concrete mix, and along with it a reduction of the amount of aggregate. Because of the lower reactivity of the ash, the initial strength development is reduced if Portland clinker or cement is replaced by fly ash. However, the long-term strength may be higher if the degree of replacement does not exceed about 20–30%. Partial replacement of cement/clinker with fly ash makes it possible to reduce the amount of Portland cement/clinker per unit volume of concrete, without reducing the volume of the paste. In this way the amount of hydration heat released during early hydration may be reduced, thus minimizing the occurrence of cracks induced by thermal expansion in massive concrete constructions.

If the amount of Portland cement/clinker is kept constant, while fly ash is added to the mix as an aggregate replacement, the early strength development may be preserved, and the final strength may be increased even further. Obviously, it is also possible to combine both these approaches, and to reduce the amount of Portland cement/clinker in the mix while replacing it with an increased rather than equal amount of fly ash.

In the system Portland cement–fly ash–water the initial hydration of the clinker phases, especially those of tricalcium silicate and tricalcium aluminate, progresses much faster than that of the ash. Just as in the hydration of pure Portland cement, a C-S-H phase, ettringite, and calcium hydroxide are formed as the first hydration products. Any alkali sulfate salts, commonly present in the clinker, also dissolve and convert to alkali hydroxides and calcium sulfate in a reaction with calcium hydroxide. The alkalinity of the liquid phase increases, and may exceed pH $= 13$.

The hydration of the clinker constituents is accelerated in the presence of fly ash, and it has been suggested that the fly ash particles act as nucleation sites for the C-S-H and CH precipitation (Uchikawa, 1986; Masazza, 1998).

The fly ash undergoes hydration much more slowly, and the first signs of surface etching may be observed by SEM only after days or weeks of hydration. In class F fly ashes the only phase that undergoes hydration is the glass phase. A significant breakdown of the glass network and a dissolution of silicate and aluminate species gets under way only at pH values exceeding pH $= 13.2$ (Fraay *et al.*, 1989). It has been recommended that the reactivity of the ash should be increased by pretreatment with calcium hydroxide and sodium silicate (Fan *et al.*, 1999). Under these conditions sodium hydroxide is formed, which etches the glass at the surface of the ash grains and partially breaks it down. The silicate ions entering the liquid phase after mixing with the binder react with the Ca^{2+} ions that already exist in the liquid, and a C-S-H phase is precipitated at, or close to, the ash particle surface. If the alkalinity significantly exceeds pH $= 13.2$ because of a high alkali content of the clinker and thus also a high alkali hydroxide concentration in the liquid phase, the concentration of both $Ca(OH)_2$ and $CaSO_4$ declines to a very low level, owing to the common ion effect. Under these conditions the C-S-H precipitation takes place further away from the surface, because of the highly suppressed Ca^{2+} concentration in the liquid phase (Fraay *et al.*, 1989).

The formed C-S-H phase tends to have a C/S ratio similar to and an A/S ratio higher than those formed in the hydration of pure Portland cement. ^{29}Si solid-state MAS-NMR results indicate an increase in the amount of silicate middle groups (Q^2) at the expense of end groups (Q^1) compared with pure Portland cement pastes, which indicates the presence of longer C-S-H chains (Pietersen and Bijen, 1992). The aluminate ions entering the liquid phase yield calcium aluminate hydrates, preferentially C_4AH_{13}, and in the presence of sulfate ions also ettringite. Additional hydrate phases that may also be formed include strätlingite (gehlenite hydrate, C_2ASH_8) and silicon hydrogarnets (Pietersen and Bijen, 1992). Out of the two ionic species entering the liquid phase in the dissolution of glass, the aluminate ions tend to be dissolved more rapidly and therefore migrate further away than the silicate ions.

In mixes made with class C fly ash, the reactive constituents of this ash, other than the glass phase, such as C_3A, C_2S, C, and $C\bar{S}$, also participate in the hydration process.

The amount of calcium hydroxide in Portland–fly ash cement pastes undergoing hydration increases in the initial stage of the process, as the rate of the liberation of this phase in the hydration of the C_3S and C_2S is higher than the rate of its consumption in a reaction with the glass phase of the ash. However, after reaching a maximum, the amount of free calcium hydroxide starts to decline, as the hydration of C_3S slows down and a more intensive hydration of the glass phase gets under way. Table 9.4 shows the free calcium hydroxide in a series of pastes made with cements that contained 0%, 20%, and 50% of fly ashes with different CaO contents. Cements in which quartz was used to replace clinker, instead of fly ash, served as controls.

For concrete mixes that contain fly ash, high humidity during curing is vital, as the pozzolanic reaction starts rather late and may become severely restricted at relative humidities below 80%. Also, at shorter hydration times the pore structure of the paste is rather coarse, and this facilitates the evaporation of free water.

As expected, the rate of hydration accelerates with increasing temperature of curing. Under hydrothermal conditions crystalline compounds rich in lime are formed at the expense of $Ca(OH)_2$. They include α-C_2SH ($C_2SH_{0.3-1.0}$), hydroxyl ellestadite [$CaO_{10}(SiO_4)_3(SO_4)_3(OH)_2$], and hydrogarnets ($C_3AS_xH_{6-2x}$) (Kropp *et al.*, 1986).

The microstructure of the hardened paste of Portland–fly ash cements does not differ significantly from that of pure Portland cement. Amorphous or poorly crystalline phases formed in the hydration of the clinker phases and the pozzolanic reaction account for the main mass of the hardened cementitious matrix. The amount of portlandite is reduced at later stages of hydration, especially in mixes with high ash contents. The ash particles present become more and more etched as the hydration progresses, and as the exterior glass hull undergoes reaction, crystals of mullite and other non-reactive ash constituents emerge at their surface. In the mature paste non-reacted residua of the ash particles are embedded in the formed cementitious matrix. The amount of non-hydrated residual ash is especially high in high-volume fly ash mixes, in which the amount of calcium hydroxide formed in the hydration of clinker phases has been too low to allow extensive hydration of the ash.

The volume of pores of Portland–fly ash cements pastes tends to be greater than that of similar pastes made from pure Portland cement, but the pore structure tends to be finer as long as the cement/fly ash replacement ratio is not excessive (above about 30%) (Li and Roy, 1986; Berry *et al.*, 1989).

Table 9.4 Free Ca(OH)$_2$ content of Portland–fly ash cement pastes after different hydration times..

Hydration time	Pure PC	Quartz	Ash A	Ash B	Ash C	Ash D
CaO total	–	–	1.02	4.52	20.8	30.4
CaO free	0.40	–	0.07	0.73	1.53	4.50
Ash content in cement 20%						
1 d	10.0	9.1	8.4	9.7	9.8	11.0
7 d	13.3	11.6	12.1	10.4	12.5	11.1
90 d	15.0	14.9	14.0	12.6	12.8	11.6
180 d	15.6	15.1	11.2	11.5	13.2	12.9
360 d	15.9	15.3	11.3	10.8	13.2	13.2
Ash content in cement 50%						
1 d	10.0	5.7	5.3	5.8	7.2	10.1
7 d	13.3	5.9	7.2	6.8	8.6	10.0
28 d	14.5	8.5	5.6	8.0	8.5	9.0
90 d	15.0	10.5	4.7	4.9	6.4	8.5
180 d	15.6	10.0	2.3	2.9	4.4	7.4
360 d	15.9	10.5	1.9	3.4	4.8	7.0

Source: Piazza (1994)

In fresh concrete mixes the presence of fly ash often reduces the water requirement for a given workability (Rattanussorn *et al.*, 1987; Dhir *et al.*, 1988). This has been attributed to the spherical geometry and smooth surface of the fly ash particles (ball bearing effect). Another factor that also improves workability is the increased dispersion of the clinker grains in the presence of fly ash. Because of the overall reduced initial formation of hydrate phases the setting of concrete mixes containing fly ash tends to be extended, and thus the mix keeps its workability preserved for longer.

The development of strength is generally slowed down in Portland–fly ash cements as compared with pure Portland cement (Babu and Rao, 1994; Piazza, 1994), owing to the slow initial hydration of the glass phase of the ash. However, in mixes in which the cement/ash replacement ratio is not too high (below 20–30%) the final strength may reach or even exceed the strength of a similar mix made with pure Portland cement. In general, mixes made with class C ashes gain strength faster than those made with class F ashes (Mehta, 1985). Figures 9.1 and 9. 2 show the strength development of a series of cements containing increasing amounts of fly ash.

Along with the retarded strength development the liberation of the heat of hydration is also slowed down. The total amount of heat of hydration is reduced (Wei *et al.*, 1985).

The strength development of Portland–fly ash cement may be accelerated distinctly by curing the mix at elevated temperature. Elevated temperature accelerates the strength development of these cements more effectively than for similar mixes made with pure Portland cement (Carette

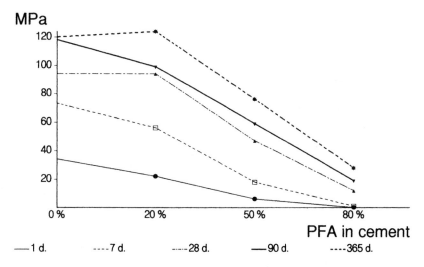

Figure 9.1 Compressive strength of Portland–fly ash cements made with different additions of class F fly ash. Composition of ash: SiO_2, 52.3%; Al_2O_3, 24.2%; Fe_2O_3, 10.0%; CaO, 4.4%.

Source: Piazza (1994)

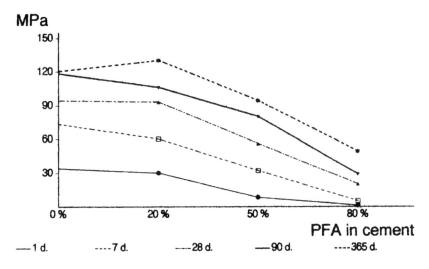

Figure 9.2 Compressive strength of Portland–fly ash cements made with different
additions of class C fly ash. Composition of ash: SiO$_2$, 40.4%; Al$_2$O$_3$,
20.0%; Fe$_2$O$_3$, 9.4%; CaO, 29.8%.
Source: Piazza (1994)

et al., 1986). It is also possible to increase the rate of strength development
by using a very fine ash, or by pregrinding the existing ash to a very high
fineness (Monzo *et al.*, 1995; Paya *et al.*, 1996; Haque and Kayali, 1998).
However, under these conditions the rheology of the fresh mix may be
degraded, especially at high ash-replacement ratios. It has also been suggested that the strength development of Portland–fly ash cement can be
accelerated by adding to the binder small amounts (about 5 wt%) of the
compounds C$_4$A$_3\bar{S}$ or C$_{11}$A$_7$.CaF$_2$ and gypsum (Yu and Zhang, 1991). The
accelerated setting and hardening under such conditions is brought about
by a fast formation of the ettringite phase.

The resistance to chemical attack, including that to sulfates, is improved
by the presence of fly ash in concrete mixes (Day and Ward, 1988;
McCarthy *et al.*, 1989). This is due mainly to the reduced amount of calcium hydroxide and the increased amount of the C-S-H phase formed,
which reduces the permeability of the paste for foreign ions such as SO$_4^{2-}$
or Cl$^-$. Moreover, in the case of sulfate corrosion the amount of ettringite
that may be formed is lower, because of reduced amounts of available tricalcium aluminate and free calcium hydroxide.

Reduced concrete deterioration due to alkali-silica reaction in mixes in
which Portland cement has been partially replaced by fly ash has been
widely reported (Hobbs, 1986, 1989; Meland, 1986; Shayan *et al.*, 1996).
Fly ash seems to act mainly as an alkali diluter, lowering the amount of
available alkalis in the system. The capability to reduce the alkali-aggregate

expansion may vary in different ashes, and depends on their own alkali content and fineness.

At low fly ash additions – that is, below 15% – the extent of carbonation in mature fly ash concrete tends to be equal to or lower than that in similar concrete mixes with no ash, in spite of the lower calcium hydroxide content of the formed hydrated cement paste (Buttler *et al.*, 1983; Hobbs, 1988; Goni *et al.*, 1997). This is due mainly to the reduced permeability of the paste to CO_2. However, at higher fly ash contents the resistance to carbonation is significantly reduced (Goni *et al.*, 1997).

The use of Portland–fly ash cement is indicated in applications where a reduced evolution of the heat of hydration and/or a high chemical resistance is required, and where slower strength development is acceptable.

9.1.4   High-volume fly ash concrete

High-volume fly ash concrete is characterized by a high fly ash content and a reduced amount of Portland cement in the mix. The cement/ash replacement ratio is high, and may exceed 50%. Typically, high-volume fly ash concrete mixes contain 100–200 kg/m^3 of cement and 150–300 kg/m^3 of fly ash.

Under these conditions only a small fraction of the ash present participates in the hardening process, as the amount of calcium hydroxide liberated in the hydration of the clinker phases and needed for a pozzolanic reaction is very limited. The products of the pozzolanic reaction concentrate preferentially in the vicinity of the residual ash particles (Zhang, 1995). The ash present in the fresh concrete mix affects its rheology positively, and acts as a filler in the hardened concrete.

The strengths attained in high-volume fly ash cement paste and concrete are rather low (Jiang and Guan, 1999). They may be improved somewhat by the use of chemical activators such as NaCl (Piazza, 1994; Xu, 1997). They may also be increased by fine grinding of the fly ash, especially if this is done together with the clinker (Bouzoubaa *et al.*, 1998). Owing to the limited amount of hydrates formed, the permeability of the hardened cement paste is relatively high, and as a consequence the concrete exhibits a high rate of carbonation and low corrosion resistance.

The use of high-volume fly ash concrete may be recommended in applications where the requirements for strength are low and where the release of very little heat of hydration is essential. It is unsuitable for steel-reinforced concrete constructions, owing to the insufficient protection of the steel from corrosion.

9.1.5   Alkali-activated fly ash binder

In alkali-activated fly ash binders a fly ash is combined with an alkali metal compound, preferably sodium silicate, which will secure a high alkalinity (pH = 13–14) of the liquid phase. The alkaline activator may be blended with the ash prior to mixing with water, or – preferably – dissolved in the mixing water before mixing with the ash.

In the hydration process the Si–O–Si and Si–O–Al bonds between the SiO_4 and AlO_4 tetrahedra constituting the glass phase of the ash are broken down in the highly alkaline environment, and become available for a reaction with the alkaline species of the liquid phase, yielding zeolite-type reaction products. Thus the alkaline activator not only produces the necessary high-pH environment, but also reacts with the original ash.

In type F ashes with a low CaO content a hydration product similar to natural analcide is formed as the reaction product:

$$Al_2O_3 + 4SiO_2 + 2NaOH + H_2O \rightarrow Na_2O.Al_2O_3.4SiO_2.2H_2O \quad (9.1)$$

In type C ashes that also contain distinct amounts of CaO, a sodium-calcium silicoaluminate similar to natural thomsonite

$$5Al_2O_3 + 10SiO_2 + 2NaOH + 4Ca(OH)_2 + 7H_2O$$
$$\rightarrow Na_2O.4CaO.5Al_2O_3.10SiO_2.12H_2O \quad (9.2)$$

or a calcium silicoaluminate similar to natural scolecite

$$Al_2O_3 + 3SiO_2 + Ca(OH)_2 + 2H_2O \rightarrow CaO.Al_2O_3.3SiO_2.3H_2O \quad (9.3)$$

may be formed (Lu, 1992).

La Rosa *et al.* (1991) studied pastes made with class F fly ash and Portland cement mixed with high-concentration NaOH and KOH solutions and cured at different temperatures. In many of these mixes zeolitic phases, as well as C-S-H and portlandite, were formed. The obtained strengths were higher in pastes mixed with alkali hydroxides and containing zeolites than in those mixed with water alone, in which no zeolites were formed. In similar experiments (Brough *et al.*, 1995) a significant formation of zeolites (sodalite, gismondine), together with amorphous and crystalline calcium silicate hydrates, was observed in a reaction of fly ash with a simulated alkali waste solution, at elevated temperatures.

According to Ikeda (1997) the solidification of fly ash–sodium silicate blends takes place by a polycondensation reaction in which metallic ions dissolved from fillers and hardeners play a crucial role:

$$(9.4)$$

Alkali-activated fly ash binders exhibit fast setting and strength development: compressive strengths of about 10 MPa after one day and 50 MPa after 50 days have been reported (Lu, 1986). In general, the strength of the material increases with increasing concentration of the alkaline activator in the mixing water and with curing temperature (Katz, 1998).

The hardened material has also been reported to possess good frost resistance, a thermal stability up to 1200 °C, and a high resistance to chemical attack, including attack by free sulfuric acid (Lu, 1986).

The binder may be suitable for applications in which high chemical corrosion, including acid attack, is likely. It has also been suggested that alkali-activated fly ash binders should be used for fixation of hazardous waste, including radioactive wastes (Brough *et al.*, 1995; Roy *et al.*, 1995).

9.2   CEMENTS MADE WITH FLUIDIZED BED ASHES

Fluidized bed combustion of coal is used to reduce the emission of sulfur dioxide and NO_x into the environment. In this technology the pulverized fuel is burnt in a fluidized bed combustor after being blended with limestone or dolomite, which act as sorbents for SO_2. The burning temperature is significantly lower than is common in the production of fly ashes in suspension-fired furnace chambers, at around 850 °C. The process may take place at atmospheric pressure (atmospheric fluidized bed combustion, AFBC) or elevated pressure (pressurized fluidized bed combustion, PFBC).

In the burning process, sulfur – originally present in the fuel in the form of organic compounds – is oxidized first to sulfur dioxide, which reacts further with the carbonate that is present and additional oxygen of the air to yield calcium sulfate (anhydrite, $CaSO_4$):

$$SO_2 + CaCO_3 + \tfrac{1}{2}O_2 \rightarrow CaSO_4 + CO_2 \tag{9.5}$$

The calcium sulfate becomes a constituent of the ash, and the SO_2 content of the flue gas is reduced accordingly.

The composition of the ash may vary greatly, depending on the composition of the fuel, on the sorbent added, and on the technology employed. It contains neither a glass phase nor mullite, as the burning temperature is too low for the formation of these phases. Instead, it contains an X-ray amorphous reactive material formed in the thermal decomposition of clay minerals present in the original coal. The amount of anhydrite present may vary greatly, depending on the sulfur content of the fuel. AFBC ashes regularly also contain distinct amounts of free lime formed in the thermal decomposition of excessive limestone added to the fuel. Such excess of the sorbent is needed to attain a sufficient absorption of sulfur oxide. The calcium oxide formed in this way is highly reactive, owing to the low temperature at which it has been formed. By contrast, PFBC ashes contain barely any free lime (calcium oxide), but calcium carbonate instead, owing to the high partial pressure of CO_2 in the combustion chamber. Other phases that may be present in the ashes include quartz, hematite, magnetite, and residual carbon. If dolomite has been employed as the sorbent, magnesium oxide and magnesium sulfate may also be present.

AFBC ashes exhibit hydraulic properties. The main reaction products are a C-S-H phase and ettringite formed in a reaction of the thermally activated clay minerals with the present free lime, calcium sulfate, and mixing water. PFBC ashes are similarly reactive; however, calcium oxide or calcium hydroxide must be added to the system to compensate for the lack of sufficient amounts of free lime in the ash.

Fluidized bed ashes may be employed in a similar way to fly ashes for producing Portland–ash cements. The amount of ash that may be introduced is limited by its anhydrite content, as the overall SO_3 content in the final binder must not exceed a critical value above which an undesired expansion of the hardened paste may get under way. An addition of additional calcium sulfate to the system is not needed in most instances. The properties of the cement produced are similar to those of conventional Portland–fly ash cement; however, its strength development is distinctly faster, owing to the higher hydraulic/pozzolanic reactivity of the fluidized bed ash.

Jiang and Roy (1995) developed an expansive cement by combining AFBC ash and an alkali sulfate.

9.3   BINDERS CONTAINING NATURAL POZZOLANAS AND RELATED PRODUCTS

A variety of materials of volcanic or sedimentary origin exhibit pozzolanic properties. They include tuffs, pozzolanas (from Italy), Santorin earth (from Greece), trass (from Germany), some rhyolites, and diatomaceous earth (consisting of siliceous skeletons of microorganisms). The active constituent of these materials is a glass phase rich in silica and/or a zeolite (most commonly analcite and phillipsite). Most of them also contain variable amounts of inactive minerals such as quartz, felspars, mica, pyroxenes, or magnetite. They usually exhibit a high specific surface area, and may also contain some chemically bound water. Such materials are called natural pozzolanas.

Unlike natural pozzolanas, artificial pozzolanas may be produced from inactive starting materials, or those with a limited reactivity, by thermal treatment. The required temperature varies in different materials, and ranges between about 400 and 900 °C.

Clay minerals exhibit a limited pozzolanicity in their original state, but this property may be enhanced by thermal treatment at an appropriate temperature. The exact selection of the temperature of calcination is essential, as the reactivity of the calcined product increases with increasing calcination temperature to an optimum, and declines again as the temperature increases even further. In general, the thermal degradation of clay minerals starts with a loss of adsorbed and interlayer water at 100–250 °C. Dehydroxylation may begin at 300–400 °C, and becomes rapid at 500–600 °C. This process is associated with a structural breakdown of the material and the formation of a highly distorted crystalline lattice. This results in a significant increase of the pozzolanicity of the material (Zhang and Malhotra, 1995). At even higher temperatures the structure starts to consolidate, and new crystalline phases are formed. These reactions result in a gradual decline of the pozzolanicity of the material.

Among different clay products **kaolin**, which contains as its main or sole mineral constituent **kaolinite** $(Al_4[(OH)_8/Si_4O_{10}]$ or $Al_2O_3.2SiO_2.2H_2O$, abbreviation AS_2H_2), has been most widely studied. Upon heating it converts first to metakaolinite (AS_2), a material with a highly distorted crystalline lattice, and eventually to mullite and quartz. Different optimum calcination temperatures have been reported by various investigators to convert natural kaolin to **metakaolin**. They range between about 600 and 850 °C. A product with a particularly high pozzolanicity may be obtained by flash calcination of the original kaolin (Salvador, 1995). Metakaolin products may differ distinctly in their pozzolanic reactivity, and the activity of some of them may be comparable to that of microsilica, if they are used in high-performance concrete (Curcio *et al.*, 1998).

The optimum calcination temperature varies in different clay minerals. For sodium and calcium montmorrilonite a calcination temperature of 830 °C has been found to be optimal (He *et al.*, 1996). Galal *et al.* (1990) reported the same optimum temperature range – 600–800 °C – for the calcination of kaolinite, montmorrilonite, and mixed type clays. He *et al.* (1995) found, for a series of different clay minerals, optimum calcination temperatures ranging between 650 and 960 °C (see Table 9.5).

Pastes obtained by blending metakaolin with lime hydrate and water exhibit setting and hardening. The reaction process may be activated and the obtained strength may be increased by using NaOH or $CaSO_4$ or a combination of both as activators (de Silva and Glasser, 1992b).

Studies on pure metakaolin–Portland cement pastes with a metakaolin content of 15% revealed that at ambient temperature the pozzolanic reaction reaches a maximum rate at about 14 days, followed by a period of significant retardation attributable to the formation of an inhibiting layer of reaction products on the metakaolin particles (Wild and Khatib, 1997). Consequently, the amount of free calcium hydroxide exhibits a minimum at 14 days, as at this stage more CH is being removed by the pozzolanic reaction than is being generated by hydration of the clinker minerals. In mature cement pastes the addition of metakaolin to the original cement results in a decline of the amount of calcium hydroxide present and an increase in the amounts of C-S-H and C_4AH_{13} formed (Morsy *et al.*, 1998). The presence of metakaolin causes a reduction of the pH value of the pore solution; however, this value stays within the range needed to provide

Table 9.5 Compressive strengths of mortars containing untreated and optimally calcined clay minerals.

	Calc. temp. (°C)	Compr. str. umtr. (MPa)	Compr. str. treat. (MPa)	Compr. str. untr. (%)
Kaolin	650	37.5	84.6	54
Illite	930	38.5	54.6	56
Calcium montmorrilonite	830	53.6	86.7	77
Sodium montmorrilonite	830	29.8	78.1	43
Mixed layer	960	25.2	77.9	46
Sepiolite	830	14.2	58.3	20
Pure Portland cement	–	69.2	–	–

Cement:clay = 70:30   Curing: 28 d at 40°C
Note: data expressed in per cent are compared with results found in a pure Portland cement paste

Source: He *et al.* (1995)

adequate corrosion protection of the embedded steel (Coleman and Page, 1997). The capacity of the hydrated material to bind chloride ions is enhanced (Coleman and Page, 1997).

Reactions similar to those taking place in the calcination of pure clay minerals also take place in the production of ceramic bricks and tiles in industrial kilns, and these products also exhibit pozzolanic properties if crushed to a fine powder (Ranogajec and Fisang, 1992). In experiments performed with Lower Oxford clays it was found that samples fired to 600–800 °C still contained substantial amounts of residual illite, whereas in those burnt at 900–1100 °C an amorphous glass phase was present that exhibited a distinct pozzolanic activity (Wild *et al.*, 1996b, 1997).

One group of natural pozzolanas that exhibits a strong pozzolanic activity is the **zeolites**. These are aluminosilicates that have a three-dimensional network structure created by corner-linked SiO_4 and AlO_4 tetrahedra. The structure of zeolites is very open, and contains channels and cavities that are normally occupied by alkali and alkali earth cations and loosely held water molecules. Their structure may be expressed by the formula

$$(Na,K)_x.(Ca,Mg)_y[(SiO_4)_z(Al_2O_3)_{x+2y}]mH_2O$$

Natural zeolites exhibit a distinct pozzolanicity in their original state, which may be enhanced further by thermal treatment.

Investigations performed on different samples of some types of natural pozzolana (volcanic tuffs and "suevite") revealed significant variations in their pozzolanicity and response to calcination, depending on their origin (Liebig and Althaus, 1998). However, the reactivity of most natural and synthetic glasses and zeolites is not too different. In the first weeks of hydration they typically exhibit a higher reactivity than most fly ashes, mainly because they have a higher specific surface area.

Pastes made from natural and artificial pozzolanas in combination with calcium hydroxide or Portland cement exhibit setting and hardening similar to that of plain Portland cement. In the latter case it is the calcium hydroxide liberated in the hydration of the clinker minerals that reacts with the pozzolanic constituent. As a first step Ca^{2+} ions are taken up by the existing zeolites, which can act as ion exchangers. Ultimately, however, the structure of the reactive components of the pozzolanic material is broken down in the highly alkaline environment, and new reaction products responsible for setting and hardening are precipitated.

The main product of the reaction between natural or artificial pozzolanas and calcium hydroxide is a C-S-H phase. It has a rather high C/S ratio at the beginning of hydration, which declines later (Galal, 1990). The average C/S ratio in mature pastes is estimated to be about 1.5–1.7 (Liu *et al.*, 1997). The Al content of the C-S-H phase formed from metakaolin and similar starting materials tends to be higher than that of mixes made with fly ash (Pietersen and Bijen, 1992). Other phases that may also be formed include strätlingite (gehlenite hydrate, C_2SAH_8), hydrogarnet ($C_3AS_{(3-x)}H_{2x}$), calcium aluminate hydrate (C_4AH_{13}), and calcium carboaluminate hydrate (C_4ACH_{11}) (Galal, 1990; Zampieri, 1990; Pietersen and Bijen, 1992; Ranogajec and Fisang, 1992; de Silva and Glasser, 1992b; He, 1995; Morsy *et al.*, 1998). The hydration process may be accompanied by a release of distinct amounts of alkali into the liquid phase, made free in ion exchange reaction with calcium hydroxide (de Silva and Glasser, 1992b). As a consequence, the alkalinity of the pore solution will gradually increase, and may reach a maximum of pH = 12.8. In the later stage of hydration the alkali content of the liquid phase may decline again, as alkali ions are adsorbed by the C-S-H phase formed. In cases in which a zeolite-based material has been used as a pozzolanic additive, the amount of alkalis in the pore solution may also be reduced significantly by an ion-exchange mechanism. Thus the addition of such materials may prevent expansion due to alkali-silica reaction (Feng, 1993).

The presence of natural and artificial pozzolanic materials contributes to the strength development of mixes made with either calcium hydroxide or Portland cement, especially after longer hydration times. This is due mainly to the formation of increased amounts of C-S-H. In general, the strengths are significantly higher in mixes with cement than in those with lime, owing to the simultaneous contribution of the cement to strength. It has been suggested (Wild *et al.*, 1996a) that there are three factors by which the artificial pozzolanas, in particular metakaolin, contribute to strength:

- a filler effect;
- an acceleration of the hydration of the Portland cement constituting this composite cement;
- the pozzolanic reaction of the pozzolanic constituent itself.

It has also been reported that the strength increases with increasing amounts of iron oxide in the pozzolanic material (Pera *et al.*, 1986).

The early strength development of pozzolana–lime mixes may be significantly improved by adding sodium sulfate (about 4 wt%) or gypsum to the system (de Silva and Glasser, 1992a; Shi and Day, 1993; Liu *et al.*, 1997). This is due to the formation of ettringite under these conditions, which later converts to monosulfate (de Silva and Glasser, 1990, 1992a). Also, sodium hydroxide alone or in combination with sulfates may act as an effective secondary activator (de Silva and Glasser, 1992a). Calcium chloride is not helpful in obtaining a higher early strength, but can substantially increase the late strength.

In addition to strength, the addition of pozzolanic materials, and in particular metakaolin, leads to a refinement of the pore structure (Khatib and Wild, 1996) and significantly reduces the permeability of the resulting concrete (Caldarone and Gruber, 1995). This, in due course, will improve the overall durability and resistance against chemical attack, including sulfates (Wild *et al.*, 1997).

As in the pozzolanic reaction free calcium hydroxide is consumed, and is replaced by phases of extremely low water solubility: lime mortars combined with natural or artificial pozzolanas attain a high degree of durability and water resistance if allowed to be precured for a sufficiently long time. Thus, not surprisingly, many structures built in ancient times – especially by the Romans – using these binders have been preserved until the present day, even when constructed to be used as aqueducts.

In pastes produced with an addition of metakaolin a gradual decomposition of the C-S-H and C_4AH_{13} phases occurs, if exposed to elevated temperatures. Nevertheless, a moderate increase of compressive strength of samples heated to temperatures between about 200 °C and 500 °C may be observed, and this increase may be attributed to the additional hydration of non-hydrated clinker still present in the paste and to an additional pozzolanic reaction of the non-reacted metakaolin with residual calcium hydroxide (Morsy *et al.*, 1998).

Under autoclave conditions (180 °C), in mixes of metakaolin with quartz and lime or Portland cement the hydrogarnet phase is consistently among the first hydration products formed, and appears before the appearance of 11A tobermorite. At low Al/(Si + Al) ratios the amount of hydrogarnet subsequently starts to decrease at prolonged autoclaving, whereas at higher Al/(Si + Al) ratios hydrogarnet coexists indefinitely together with the 11A tobermorite phase (Klimesch and Ray, 1998a, 1998b; Klimesch *et al.*, 1998). The formation of 11A tobermorite and hydrogarnet also depends on the fineness of the quartz present. The initial lime–kaolin reaction is retarded only when fine quartz is used, whereas a slower release of silicate anions from coarse quartz particles prolongs the stability of the hydrogarnet phase and retards the formation of tobermorite. The maximum amount of metakaolin entering the hydration reaction amounts to about 18–24% if added to the system as a quartz replacement and 12–18% if added as a cement replacement (Klimesch and Ray, 1998a). NMR studies have revealed that the reaction does not go to completion, and unreacted cement and quartz are evident even after extended autoclaving. In CP-MAS NMR spectra an increasing number of protonated Q^1 and Q^2 species become evident with increasing metakaolin addition, indicating the presence of a high number of silicate anion varieties (Klimesch *et al.*, 1998).

9.4 MICROSILICA-MODIFIED PORTLAND CEMENT

Microsilica, also called **silica fume** or **condensed silica fume**, is a by-product in the production of silicon metal, ferrosilicone, and some other silicone alloys in a submerged-arc electric furnace. Here it condenses from the gaseous phase in the form of very small spherical particles with a mean diameter in the range 0.1–0.2 μm and with a specific surface area between about 10 and 20 m^2/g (BET).

Chemically, microsilica is an amorphous form of silicon dioxide with an SiO_2 content of about 90–98 wt% and small amounts of Al_2O_3, Fe_2O_3, alkalis, and elementary carbon. Besides amorphous SiO_2 it may also contain limited amounts of cristobalite.

The specific gravity of microsilica is around 2.20. Because of its very small particle size and high specific surface area, the bulk unit weight of the loose material is very low, and ranges typically between 240 and 300 kg/m^3. To lower the transportation costs, it is common to ship the material in a predensified form, by which the unit weight may be increased to 540–600 kg/m^3. It is even more convenient to transport and handle the material in the form of a slurry with a high solid content. Blended cements, in which microsilica is combined with Portland clinker and calcium sulfate, are also commercially available. A comparison of different forms of microsilica indicates that densification of this material decreases its chemical reactivity (Sanches de Rojas *et al.*, 1999).

Because of the high reactivity of the amorphous form of SiO_2 and its extreme fineness, microsilica reacts readily with calcium hydroxide in the presence of water, yielding an amorphous C-S-H phase similar to that formed in the hydration of Portland cement (Justines *et al.*, 1990; Justines, 1992; Guindy, 1993; Papadakis, 1999). This pozzolanic reaction progresses much faster than similar reactions with other commonly available pozzolanic materials. Compared with the hydration of pure tricalcium silicate, the conversion of microsilica to C-S-H is slower at shorter reaction times, but faster after about 28 days (Table 9.6). The average length of the polysilicate anion constituting the C-S-H phase is somewhat greater.

Blends of microsilica and slaked lime or hydrated lime exhibit setting

Table 9.6 Rate of microsilica to C–S–H conversion
compared with the hydration rate of C$_3$S.

Hydration time (days)	C$_3$S (%)	Microsilica (%)
1	22	6
3	37	17
7	48	38
28	67	78
84	87	98

C/S = 1.11; *w/s* = 0.70

Source: Justines (1992)

and hardening, if mixed with appropriate amounts of water (Guindy, 1993); however, binders of this kind are rarely employed in practice.

If added to concrete mixes to replace a fraction of Portland cement (typically in amounts of 5–15 wt% per weight of cement), microsilica tends to increase their water requirement. This is due mainly to the capability of this additive to bond significant amounts of water to its large surface by adsorption. The effect of microsilica on rheology depends on a variety of factors (Murata *et al.*, 1994): generally, the rheology of a fresh concrete mix deteriorates with increasing specific surface area of the microsilica, while the strength of the hardened material tends to increase. The rheology of the fresh concrete mix will also depend on the dispersibility of the microsilica used, and improves with increasing degree of dispersion of this additive. Thus undensified silica fume generally performs better than one that has undergone mechanical densification. Silica fume that has been stored for an extended length of time also possesses a lower dispersibility. A significant improvement of the degree of dispersion may be attained by ultrasonic treatment of the microsilica suspended in water.

It is common to add to the fresh mix appropriate amounts of a superplasticizer, by which measure the negative effect of microsilica on rheology may be largely or completely eliminated. As well as reducing the amount of water bound to the surface of the microsilica particles, the superplasticizer also causes an effective dispersion of these particles within the concrete mix. Under such conditions the individual particles of microsilica can occupy spaces between the coarser cement grains that would otherwise be filled with water. This may result in a reduction of the amount of water needed to fill the empty spaces between cement particles, and in the formation of a denser matrix with a better gradation of fine particles. The overall effect of the superplasticizer on rheology will depend on a variety of factors, such as the amount of microsilica in the mix, the quality and amount of superplasticizer added, and the time at which the additive was introduced (Szwabowski and Galaszewski, 1996; Duval and Kadri, 1998). If a naphthalene sulfonate superplasticizer is employed in appropriate amounts, up to about 10% of microsilica does not reduce concrete workability (Duval and Kadri, 1998). The required dosage of the superplasticizer will also depend on the composition of the cement, and especially on its C_3A content. It may be lowered with decreasing C_3A content in the binder (Duval and Kadri, 1998). It has also been reported that the workability of concrete mixes containing microsilica can be improved by applying a two-stage mixing technique (Tamimi, 1996).

In concrete mixes made with Portland cement, the addition of microsilica increases the initial hydration rate of the cement and especially that of the alite and belite phases (Wu and Young, 1984; Huang and Feldman, 1985a, 1985b; Durekovic, 1986; Lilkov *et al.*, 1997). This is explained by a nucleation effect of the microsilica particles present. After several days, however, the hydration settles down to normal rates (Fidjestøl and Lewis, 1998).

The rate at which microsilica converts to C-S-H may vary in different mixes, and will also depend on the water/cement ratio employed. In general, the pozzolanic reaction gets under way within the first few hours after mixing with water, and this process is associated with an enhanced liberation of hydration heat (Sanchez de Rojas and Frias, 1996; Meng and Schiessl, 1997). There is a rapid consumption of microsilica within the first days of hydration, and under favorable conditions up to 50% of the original microsilica may be consumed within the first 28 days (Lilkov *et al.*, 1997). The conversion of microsilica slows down significantly in the later stages of hydration, and even after long curing times a significant proportion of it remains non-reacted (Pietersen and Bijen, 1992). It appears that the rate-controlling factor in this stage is the build-up of an inhibiting layer of reaction products around the individual microsilica particles, which slows down the formation of additional C-S-H. It has also been suggested that – in addition to a C-S-H phase – water-rich gels (C-S-H gel or silica gel) are formed as intermediate products (Meng, 1997).

The amount of free calcium hydroxide in Portland cement–microsilica mixes increases initially, as its formation in the hydration of tricalcium silicate is faster than its consumption in the pozzolanic reaction with microsilica. Later on, however, the amount of free calcium hydroxide may start to decline, when the amount of it consumed in the pozzolanic reaction exceeds the rate by which it is formed in the hydration of tricalcium and dicalcium silicate (Papadakis, 1999). This crossover point – that is, the time at which the rate of $Ca(OH)_2$ consumption exceeds the rate of its formation – will depend on the amount and reactivity of the microsilica present, as well as on the reactivity of the clinker, and can occur after several hours or a few days of hydration, or not at all (especially at low microsilica additions). The amount of residual free calcium hydroxide in mature paste will generally decline with increasing amounts of microsilica in the original mix. It will also decline with decreasing water/solid ratio, as under these conditions the C/S ratio of the formed C-S-H phase tends to increase.

The microsilica present in a mix with Portland cement may convert to C-S-H only if sufficient amounts of calcium hydroxide are formed in the concurrent hydration of tricalcium and – to a lesser extent – dicalcium silicate. In most instances the amount of microsilica should not exceed about 20–25% per weight of the Portland cement to secure complete or nearly complete conversion of it to C-S-H and, at the same time, to preserve some residual free calcium hydroxide in the mature paste.

The texture of cement pastes containing microsilica is characterized by the presence of agglomerates of individual microsilica particles. Such agglomerates are present even when a superplasticizer has been added to the mix (Mitchell *et al.*, 1998), but in smaller numbers. Within these agglomerates a reaction with calcium hydroxide and formation of the C-S-H phase also occurs, but the rate of this process is significantly reduced, especially in the central part, and the reaction is not completed even after 180 days.

Mature Portland cement pastes modified with silica fume contain increased amounts of C-S-H at the expense of portlandite. The C-S-H phase does not differ much from that formed in the hydration of pure Portland cement, but may have a lower average C/S ratio and a higher degree of polymerization, especially at higher microsilica additions (Wu and Young, 1984; Durekovic and Popovic, 1987; Pietersen and Bijen, 1992). It also exhibits a very low A/S ratio (0.04–0.07) (Pietersen and Bijen, 1992). Portlandite is present in the paste in a more dispersed form, and its average crystal size is much smaller than in pastes without added microsilica. Microsilica also reduces the volume of larger capillary pores and causes the formation of a finer and less interconnected pore structure, thus lowering the permeability of the resulting hardened cement paste.

It has also been observed that Portland cement pastes with added silica fume contain an increased number of hollow shell pores, also called Hadley grains, in the size range 1–15 μm (Kjellsen and Atlassi, 1999). They are formed primarily at early ages by dissolution of the cement grains, and in pastes without silica fume usually become filled with fresh hydrates as the hydration progresses. It appears that in pastes with added silica fume the number of empty shells is significantly increased. These hollow shell pores appear to be connected to the "continuous" capillary pore system by much smaller gel pores, and their presence contributes to a reduced permeability of the paste.

In hardened mortar or concrete mixes the added microsilica reduces the porosity of the transitional zone and lowers the concentration of large oriented $Ca(OH)_2$ crystals precipitated at the aggregate surface (Bentur *et al.*, 1988; Tasdemir *et al.*, 1996). The overall thickness of the transitional zone is also reduced. All this may be attributed to the filler effect of the microsilica particles, and to their ability to lower the local water/solid ratio in the immediate vicinity of the aggregate surface.

If used as a partial replacement for Portland cement, in combination with a superplasticizer, microsilica distinctly increases the strength of concrete or mortar mixes, if compared at equal water/cement ratios (Huang and Feldman, 1985b; Bentur *et al.*, 1988; Panchenko and Opoczky, 1992; Babu and Prakash, 1995; Mak and Torii, 1995; Toutanji and El-Korchi, 1995; Wild, 1995; Duval and Kadri, 1998; Papadakis, 1999). Figure 9.3 shows data on the effect of microsilica on strength published recently by Toutanji and El-Korchi (1995). An even more favorable increase of long-term strength is obtained if a belite-rich cement, rather than ordinary Portland cement, is employed as binder (Kato *et al.*, 1997).

The improvement of strength due to microsilica replacement has been attributed either to an improvement of the intrinsic strength properties of the paste formed in the presence of microsilica (Darwin *et al.*, 1988; Cong *et al.*, 1992), or to a strengthening of the bond between the cement paste and the aggregate surface as a direct result of a densification of the interfacial zone by the microsilica present (Bentur *et al.*, 1988; Goldman and Bentur, 1989; Toutanji and El-Korchi, 1995). A direct comparison of strengths obtained on plain pastes and mortars or concretes revealed that whereas the strength of mortar or concrete is distinctly increased by the presence of microsilica, that of plain paste is barely affected (Huang and Feldman, 1985b; Bentur *et al.*, 1988; Rosenberg and Gaidis, 1989; Panchenko and Opoczky, 1992; Toutanji and El-Korchi, 1995) (see also Table 9.7). Thus it appears that the strengthening of the cement paste/aggregate bond is the main factor responsible for the strength increase seen in the presence of added microsilica.

It has been reported that the strength properties of concrete made with

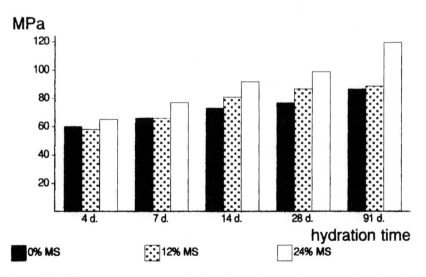

Figure 9.3 Effect on concrete strength of replacing Portland cement with microsilica.
Source: Toutanji and El-Korchi (1995)

Table 9.7 Compressive strength of Portland cement pastes and mortars produced with different degrees of replacement with microsilica (in MPa).

w/s		0% microsilica	16% microsilica	25% microsilica
0.22	Paste	95.2	88.1 (92.5%)	85.3 (89.6%)
0.28	Paste	93.5	93.6 (100.1%)	89.4 (95.6%)
0.34	Paste	81.5	80.3 (98.8%)	82.6 (101.7%)
0.22	Mortar	93.2	103.1 (110.6%)	109.4 (117.4%)
0.28	Mortar	83.1	94.3 (113.5%)	97.8 (117.6%)
0.34	Mortar	73.5	82.1 (111.7%)	90.1 (122.6%)

Curing: 56 days at 29 °C and r.h. < 95%

Source: Toutanji and El-Korchi (1995)

microsilica may be improved even further by a surface treatment of microsilica with sulfuric acid prior to mixing with other concrete mix components (Li and Chung, 1998).

Silica fume may be effectively employed to obtain concrete compositions with extremely high strengths. To obtain such strengths, mixes with a high binder contents and water/cement ratios as low as 0.20 must be produced. The mixes must contain adequate mounts of microsilica in combination with a superplasticizer. To prevent cracking caused by high thermal gradients the use of a Portland cement with a high C_2S content may be required (Kinoshita *et al.*, 1995).

Under excessive exposure to external forces, in concretes produced without microsilica any cracks would develop preferentially around coarse aggregates, owing to the higher porosity and relatively weak bonding existing at the cement paste/aggregate interface. As, however, in the presence of microsilica the interfacial bond becomes stronger, failure of the material gets under way only under higher external stresses, and the cracks that are formed usually traverse the aggregates. As a result the strength of the concrete increases. At the same time the fracture energy of the material decreases dramatically, especially in large maximum size aggregate mixes, and as a result the brittleness index also increases significantly (Tasdemir *et al.*, 1996).

In addition to strength, the addition of microsilica also favorably affects other properties of concrete:

- The modulus of elasticity increases with the amount of microsilica added to the concrete mix (Galeota and Giamatteo, 1989).
- The creep deformation and drying shrinkage are reduced in mixes made with microsilica (Bentur and Goldman, 1989; Tachibana, 1990).
- The permeability of concrete and along with it the migration of aggressive chemicals in the hardened concrete is reduced by the addi-

tion of microsilica to the mix (Perraton *et al.*, 1988; Panchenko and Opoczky, 1992; Alexander and Magee, 1999).

- The resistance of concrete to sulfate attack is significantly improved in mixes containing microsilica (Carette and Malhotra, 1983; Cohen and Bentur, 1988; Berke, 1989; Aköz *et al.*, 1995). This is due to the reduced permeability of the material, to the lower free calcium hydroxide content in the hardened paste, and to the reduced amount of alumina available for ettringite formation, as more of it is incorporated in the C-S-H phase.
- Microsilica is also effective in controlling the alkali-silica reaction in concrete (Aitcin and Regourd, 1985). When microsilica is added to the mix, more alkalis become entrapped within the C-S-H phase, and the concentration of OH^- ions in the pore solution declines accordingly (Diamond, 1983).
- The migration of chloride ions is reduced significantly in concrete mixes made with microsilica (Gantefall, 1986; Marusin, 1988; Perraton *et al.*, 1988; Moukwa, 1993). This in turn increases the resistance of the resulting concrete to seawater and deicing salts.
- The use of microsilica can also substantially increase the electric resistivity of concrete, hence slowing the rate of corrosion of steel reinforcement (Vennesland and Gjorv, 1983).

The addition of microsilica generally improves the frost resistance of concrete. However, at higher microsilica additions the reduced permeability of the material may hamper moisture migration, and may cause a build-up of internal hydraulic pressure and cracking. To prevent this the amount of added microsilica must not be excessively high. The use of an air-entraining admixture may also be necessary, to provide an adequate air-void system.

Moderate amounts of microsilica have no negative effect on the pH of the pore solution and thus on the protection of steel reinforcement against corrosion. However, at high microsilica additions the pH value of the pore solution may drop significantly (below pH = 12), hence increasing the risk of a depassivation of the steel surface (Wiens, 1995).

An exposure to elevated temperature (up to 600 °C) tends to increase the strength of microsilica-modified Portland cement based concrete. This may be attributed to the formation of additional C-S-H in a reaction between residual calcium hydroxide and microsilica (Saad *et al.*, 1996).

If concrete mixes made with high amounts of microsilica are exposed to fire, an excessive internal water vapor pressure may develop, causing cracking, strength reduction, and occasionally explosive spalling. Such a danger exists especially in moist concretes. To prevent this, it may be necessary to add appropriate fibers to the fresh mix in high-risk structures.

The use of microsilica-modified Portland cement concrete may be indicated especially in applications where reduced permeability and high chemical resistance of the material are required. The use of this material is also essential in making concrete of extremely high strength. Compressive strengths exceeding 100 MPa may be obtained under these conditions.

A material closely related to silica fume is **colloidal silica**, which differs from the former by having a significantly higher specific surface area (up to 80 m^2/g). If added to Portland cement mixes this material reacts with the liberated calcium hydroxide at a faster rate than microsilica, and leads to the development of higher early strengths (Chandra and Bergqvist, 1997).

9.5   RICE HUSK ASH BASED CEMENT

Rice husk ash is a product that may be obtained by a controlled ashing of rice husk, a material widely available in rice-producing countries.

The main constituent of the ash is SiO_2, which constitutes about 80–98% of the material. It is present in the original rice husk in the form of hydrated silica, and converts into amorphous silica upon burning at 400–700 °C. At temperatures above 750 °C it may even convert to cristo-balite (a high-temperature modification of SiO_2). The ash may also con-tain a few per cent of K_2O and Na_2O, and up to 20% of residual carbon. Rice husk ash is highly porous. If ashed at 400 °C its surface area (BET) may exceed 100 m^2/g, but this decreases with increasing burning tempera-ture and burning time.

The ultimate quality of the ash depends greatly on the burning con-ditions, and especially on the burning temperature (Sugita *et al.*, 1997). At temperatures above about 600 °C the pozzolanicity of the material will decline owing to the conversion of the primary formed amorphous SiO_2 to crystoballite. At burning temperatures that are too low, the quality of the material will also be poor, owing to the presence of a considerable amount of unburnt carbon. Thus to obtain optimum quality an ashing temperature of 500–600 °C for a short period of time appears most appropriate. The product formed under these condition has a specific surface area of about 10–20 m^2/g. A particularly high-quality ash may be produced by the use of fluidized bed combustors (Hara *et al.*, 1992).

Because of the presence of a highly reactive form of SiO_2 and its high specific surface area, rice husk ash reacts readily with calcium hydroxide in the presence of water, yielding an amorphous C-S-H phase (Zhang *et al.*, 1996; Lin, 1997). In suspensions of rice husk ash in saturated calcium hydroxide solutions at 40 °C the formation of a semi-crystalline calcium silicate hydrate phase of the composition $Ca_{1.5}.SiO_{3.5}.xH_2O$ has also been reported (Yu *et al.*, 1999).

If blended or ground together with quicklime or hydrated lime (in ratios of 70–80 wt% ash to 20–30 wt% lime), rice husk ash yields a binder suit-able for the production of rapid-hardening, water-resistant mortar mixes.

Rice husk ash may also be combined with Portland cement in a way similar to microsilica. Again, the added ash tends to increase the water demand, and it may be necessary to add a superplasticizer to obtain good rheology of the fresh mix. Because of the pozzolanic reaction, pastes made with rice husk have a lower free calcium hydroxide content than comparable neat Portland cement pastes. In concrete mixes the amount of calcium hydroxide in the interfacial zone is reduced, and the density of the zone is increased (Zhang *et al.*, 1996). The strength of cement pastes incorporating rice husk ash is not increased by as much as that of mortar or concrete mixes. This indicates that the main factor contributing to the strength increase of the latter is the densification of the existing interfacial zone (Zhang *et al.*, 1996).

In addition to an increase of strength, the presence of rice husk ash also improves the resistance of concrete to acid attack. The expansion due to alkali-silica reaction and sulfate attack is reduced considerably. Also, the frost resistance has been found to be significantly improved by adding rice husk ash to the concrete mix (Mehta and Folliard, 1995).

REFERENCES

Aitcin, P.C., and Regourd, M. (1985) Use of condensed silica fume to control alkali-silica reaction: a field case study. *Cement and Concrete Research* **15**, 711–719.

Aitcin, P.C. *et al.* (1986) Comparative study on the cementitious properties of different fly ashes, in *Proceedings 2nd ICFSS*, ACI SP-91, pp. 91–114.

Aköz, F. *et al.* (1995) Effect of sodium sulfate concentration on the sulfate resistance of mortars with and without silica fume. *Cement and Concrete Research* **25**, 1360–1368.

Alexander, M.G., and Magee, G.J. (1999) Durability performance of concrete containing condensed silica fume. *Cement and Concrete Research* **29**, 917–922.

Ambroise, J., and Martin-Calle, S. (1993) Pozzolanic behaviour of thermally activated kaolin, in *Proceedings 4th ICFSS*, ACI SP-132, pp. 731–748.

Ambroise, J., Maximilien, S., and Pera, J. (1994) Properties of metakaolin blended cements. *Advances in Cement Based Materials* **1**, 161–168.

Babu, K.G., and Rao, G.S.N. (1994) Early strength behaviour of fly ash concretes. *Cement and Concrete Research* **24**, 277–284.

Babu, K.G., and Prakash, P.V.S. (1995) Efficiency of silica fume in concrete. *Cement and Concrete Research* **25**, 1273–1283.

Bentur, A., and Goldman, A. (1989) Curing effects, strength and physical properties of high strength silica fume concretes. *Journal of Materials in Civil Engineering ASCE* **1**, 46–58.

Bentur, A., Goldman, A., and Cohen, M.D. (1983) Contribution of the transition zone to the strength of high quality silica fume concrete. *Materials Research Society Symposium Proceedings* **114**, 97–103.

Bentur, A., Goldman, A., and Cohen, M.D. (1988) Contribution of the transition zone to the strength of high quality silica fume concretes. *Materials Research Society Symposium Proceedings* **114**, 97–103.

Berke, N.S. (1989) *Resistance of microsilica concrete to steel corrosion and chemical attack*. ACR SP-114, pp. 861–855.

Berry, E.E and Anthony, E.J. (1987) Evaluation of potential uses of AFBC solid waste. *Materials Research Society Symposium Proceedings* **86**, 353–364.

Berry, E.E. *et al.* (1989) Beneficiated fly ash: hydration microstructure and strength development in Portland cement systems. ACI SP-114, pp. 241–273.

Berry, E.E *et al.* (1994) Hydration in high volume fly ash binders: chemical factors. *ACI Materials Journal* **91**, 382–389.

Bildeau, A., and Malhotra, V.M. (1995) Properties of high-volume fly ash concrete made with high early strength ASTM type III cement. ACI SP-153, pp. 1–23.

Bland, A.E. *et al.* (1989) Performance characteristics of concrete produced with fluidized bed combustion ash. *Materials Research Society Symposium Proceedings* **136**, 9–22.

Bouzoubaa, N. *et al.* (1998) Laboratory-produced high-volume fly ash blended cements: physical properties and compressive strength of mortar. *Cement and Concrete Research* **28**, 1555–1571.

Brandstetr, J., Havlica, J., and Odler, I. (1997) Properties and use of solid residue from fluidized bed coal combustion, in *Waste Materials Used in Concrete Manufacturing* (ed. S. Chandra), Noyes Publications, Westwood, NJ, USA, pp. 1–53.

Brough, A.R. *et al.* (1995) Microstructural aspects of zeolite formation in alkali activated cements containing high levels of fly ash. *Materials Research Society Symposium Proceedings* **370**, 199–208.

Buttler, F.G., Decter, M.H., and Squith, G.R. (1983) Studies on the desiccation and carbonation of systems containing Portland cement and fly ash. ACI SP-79, pp. 367–383.

Calderone, M.A., and Gruber, K.A. (1995) High reactivity metakaolin (HRM) for high performance concrete. ACI SP-153, pp. 815–827.

Carette, G.G., and Malhotra, V.M. (1983) Mechanical properties, durability and drying shrinkage of Portland cement concrete incorporating silica fume. *Cement, Concrete and Aggregates* **5**, 3–8.

Carette, G.G. *et al.* (1986) Development of heat curing cycles for Portland cement fly ash concrete for the precast industry. ACI SP-91, pp. 249–272.

Carette, G.G. *et al.* (1993) Mechanical properties of concrete incorporating high volume fly ash from sources in the US. *ACI Materials Journal* **90**, 535–544.

Chandra, S., and Bergqvist, H. (1997) Interaction of silica colloid with Portland cement, in *Proceedings 10th ICCC, Göteborg*, paper 3ii106.

Cohen, M.D., and Bentur, A. (1988) Durability of Portland cement-silica fume pastes in magnesium sulfate attack and sodium sulfate solution. *ACI Materials Journal* **85**, 149–152.

Coleman, N.J., and Page, C.L. (1997) Aspects of pore solution chemistry of hydrated cement pastes containing metakaolin. *Cement and Concrete Research* **27**, 147–154.

Cong, X. *et al.* (1992) Role of silica fume in compressive strength of cement paste, mortar and concrete. *ACI Materials Journal* **89**, 375–379.

Curcio, F. *et al.* (1998) Metakaolin as a pozzolanic microfiller for high-performance mortar. *Cement and Concrete Research* **28**, 803–809.

Darwin, D., Shen, Z., and Harsh, S. (1988) Silica fume, bond strength and the compressive strength of mortars. *Materials Research Society Symposium Proceedings* **114**, 105–110.

Day, R.L., and Ward, M.A. (1988) Sulfate durability of plain and fly ash mortars. *Materials Research Society Symposium Proceedings* **113**, 153–161.

de Silva, P.S., and Glasser, F.P. (1990) Hydration of cements based on metakaolin: thermochemistry. *Advances in Cement Research* **3**, 167–177.

de Silva, P.S., and Glasser, F.P. (1992) The hydration behaviour of metakaolin-Ca(OH)$_2$-sulphate binders, in *Proceedings 9th ICCC, New Delhi*, Vol. 4, pp. 671–677.

de Silva, P.S., and Glasser, F.P. (1992) Pozzolanic activation of metakaolin. *Advances in Cement Research* **4**, 167–178.

Dhir, R.K. *et al.* (1988) Contribution of PFA to concrete workability and strength development. *Cement and Concrete Research* **18**, 227–239.

Dhir, R.K., Zhu, W.Z., and McCarthy, M.J. (1998) Use of Portland PFA cement in combination with superplasticizing admixtures. *Cement and Concrete Research* **28**, 1209–1216.

Diamond, S. (1983) Effect of microsilica (silica fume) on pore solution chemistry of cement pastes. *Journal of the American Ceramic Society* **66**, 682–684.

Drottner, J., and Havlica, J. (1997) Low lime binders based on fluidized bed ash. *Studies in Environmental Science* **71**, 401–410.

Durekovic, A. (1986) Hydration of alite and C$_3$A and chances of some structural characteristics of cement pastes by addition of silica fume, in *Proceedings 8th ICCC, Rio de Janeiro*, Vol. 4, pp. 279–284.

Durekovic, A., and Popovic, K. (1987) The influence of silica fume on the mono/disilicate anion ratio during the hydration of CSF containing cement paste. *Cement and Concrete Research* **17**, 108–114.

Duval, R., and Kadri, E.H. (1998) Influence of silica fume on the workability and compressive strength of high-performance concretes. *Cement and Concrete Research* **28**, 533–547.

Fan, Y. *et al.* (1999) Activation of fly ash and its effects on cement properties. *Cement and Concrete Research* **29**, 467–472.

Feng, N. (1993) Properties of zeolite mineral admixtures concretes, in *Mineral Admixtures in Cement and Concrete* (ed. S.N. Ghosh), ABI Books, New Delhi, pp. 396–446.

Fidjestøl, P., and Lewis, R. (1998) Microsilica as an addition, in *Lea's Chemistry of Cement and Concrete* (ed. P.C. Hewlett), Arnold, London, pp. 675–708.

Fraay, A.L.A., Bijen, J.M., and Haam, Y.M. (1989) The reaction of fly ash in concrete: a critical examination. *Cement and Concrete Research* **19**, 235–246.

Galal, A.F. *et al.* (1990) Hydraulic reactivity and microstructure of artificial pozzolana-lime pastes, in *Proceedings 12th International Conference on Cement Microscopy*, Vancouver, pp. 135–153.

Galeota, D., and Giamatteo, M.M. (1989) Stress-strain relations of normal and lightweight concrete with silica fume under uniaxial compression. ACI SP-114, pp. 991–1011.

Gantefall, O. (1986) Effect of condensed silica fume on the diffusion of chlorides through hardened cement paste. ACI SP-91, pp. 991–998.

Gebler, S.H., and Klieger, P. (1986) Effect of fly ash on physical properties of concrete. ACI SP-91, pp. 1–50.

Goldman, A., and Bentur, A. (1989) Bond effects in high silica fume concretes. *ACI Materials Journal* **86**, 440–444.

Gomes, S. *et al.* (1998) Characterization and comparative study of coal combustion residues from a primary and additional flue gas secondary desulfurization process. *Cement and Concrete Research* **28**, 1605–1619.

Goni, S. *et al.* (1997) Microstructural characterization of the carbonation of mortars made with fly ashes, in *Proceedings 10th ICCC, Göteborg*, paper 4iv004.

Guindy, N.M. (1993) Hydration characteristics of lime-silica pastes. *Journal of Thermal Analysis* **40**, 151–157.

Haque, M.N., and Kayali, O. (1998) Properties of high-strength concrete using a fine fly ash. *Cement and Concrete Research* **28**, 1445–1452.

Hara, N. *et al.* (1992) Suitability of rice husk ash obtained by fluidized-bed combustion for blended cements, in *Proceedings 9th ICCC, New Delhi*, Vol. 3, pp. 72–78.

Havlica, J., Brandstetr, J., and Odler, I. (1998) Possibilities of utilizing solid residues from pressurized fluidized bed coal combustion (PFBC) for the production of blended cements. *Cement and Concrete Research* **28**, 299–307.

He, C., Osbaeck, B., and Makovicky, E. (1995) Pozzolanic reactions of six principal clay minerals: activation, reactivity assessment and technological effects. *Cement and Concrete Research* **25**, 1691–1702.

He, C., Makovicky, E., and Osbaeck, G. (1996) Thermal treatment and pozzolanic activity of Na- and Ca-montmorrilonite. *Applied Clay Science* **10**, 351–368.

Helmuth, R. (1987) *Fly Ash in Cement and Concrete*, PCA, Skokie, IL, USA.

Hobbs, D.W. (1986) Deleterious expansion of concrete due to alkali-silica reaction: influence of PFA and slag. *Magazine of Concrete Research* **38** (137), 195–205.

Hobbs, D.W. (1988) Carbonation of concrete containing PFA. *Magazine of Concrete Research* **40** (143), 69–78.

Hobbs, D.W. (1989) Effect of mineral and chemical admixtures on alkali aggregate reaction, in *Proceedings 8th International Conference on AAR*, Kyoto, pp. 173–186.

Huang, C.Y., and Feldman, R.F. (1985) Influence of silica fume on the microstructural development in cement mortars. *Cement and Concrete Research* **15**, 285–294.

Huang, C., and Feldman, R.F. (1985) Hydraulic reactions in Portland cement – silica fume blends. *Cement and Concrete Research* **15**, 585–592.

Hubbart, F.H., Dhir, R.K., and Ellis, M.S. (1985) Pulverized fuel ash for concrete: compositional characterization of United Kingdom PFA. *Cement and Concrete Research* **15**, 185–198.

Ikeda, K. (1997) Preparation of fly ash monoliths consolidated with a sodium silicate binder at ambient temperature. *Cement and Concrete Research* **27**, 657–663.

Iyer, R.S., and Stanmore, B.R. (1995) Surface area of fly ashes. *Cement and Concrete Research* **25**, 1403–1405.

Jalali, S. (1997) Modeling the overall rate of reaction in lime–fly ash systems, in *Proceedings 10th ICCC, Göteborg*, paper 3ii084.

Jiang, L., and Guan, Y. (1999) Pore structure and its effect on strength of high-volume fly ash pastes. *Cement and Concrete Research* **29**, 631–633.

Jiang, W., and Roy, D.M. (1995) Expansive cement produced from AFBC ash by alkali sulfate activation approach. ACI SP-153, pp. 193–212.

Joshi, R.C., and Lohtia, R.P. (1993) Types and properties of fly ash, in *Mineral*

Admixtures in Cement and Concrete (ed. S.N. Ghosh), ABI Books, New Delhi, pp. 118–157.

Justines, H. (1992) Hydraulic binders based on condensed silica fume and slaked lime, in *Proceedings 9th ICCC, New Delhi*, Vol. 3, pp. 284–290.

Justines, H. *et al.* (1990) A 29Si MAS NMR study of the pozzolanic activity of condensed silica fume and the hydration of di- and tricalcium silicate. *Advances in Cement Research* **3**, 111–116.

Kato, H., Katumoto, R., and Ushiyama, H. (1997) Properties of high strength concrete using belite-rich cement and silica fume. *Semento Konkurito Ronbunshu* **51**, 364–369 [ref. CA 128/220688].

Katz, A. (1997) Fly-ash blended cement activated by a strong base, in *Proceedings 10th ICCC, Göteborg*, paper 3ii083.

Katz, A. (1998) Microscopic study of alkali-activated fly ash. *Cement and Concrete Research* **28**, 197–208.

Khalil, J. M., and Wild, S. (1996) Pore size distribution of metakaolin pastes. *Cement and Concrete Research* **26**, 1545–1553.

Kayat, K.H., and Aitcin, P.C. (1993) Silica fume: a unique supplementary cementitious material, in *Mineral Admixtures in Cement and Concrete* (ed. S.N. Ghosh), ABI Books, New Delhi, pp. 227–265.

Khatib, J.M., and Wild, S. (1996) Pore size distribution of metakaolin paste. *Cement and Concrete Research* **16**, 1545–1553.

Kinoshita, M. *et al.* (1995) Study on application of super high range reducer for ultra high strength silica fume concrete (in Japanese). *Semento Konkurito Ronbunshu* **49**, 192–197.

Kjellsen, K.O., and Atlassi, E.H. (1999) Pore structure of cement silica fume systems: presence of hollow shell pores. *Cement and Concrete Research* **29**, 133–142.

Klimesch, D.S., and Ray, A. (1998a) Hydrogarnet formation during autoclaving at 180 °C in unstirred metakaolin-lime-quartz slurries. *Cement and Concrete Research* **28**, 1109–1117.

Klimesch, D.S., and Ray, A. (1998b) Effect of quartz particle size and kaolin on hydrogarnet formation during autoclaving. *Cement and Concrete Research* **28**, 1317–1323.

Klimesch, D.S. *et al.* (1998) Metakaolin additions in autoclaved cement-quartz pastes: a ^{29}Si and ^{27}Al MAS NMR investigation. *Advances in Cement Research* **10**, 93–99.

Kohno, K. *et al.* (1996) Characteristics of high-volume fly ash concrete: mixture proportions versus strength, Young's Modulus, drying shrinkage and adiabatic temperature rise (in Japanese). *Semento Konkurito Ronbunshu* **50**, 10–17.

Kovacs, R. (1982) Utilization of fly ashes in the cement industry (in German). *TIZ-Fachberichte* 106, 658–665.

Kropp, J., Seeberger, J., and Hilsdorf, H.K. (1986) Chemical and physical properties of cement paste and concrete containing fly ash after hydrothermal exposure. ACI SP-91, pp. 201–218.

Kumar, A. (1993) Rice husk ash based cements, in *Mineral Admixtures in Cement and Concrete* (ed. S.N. Ghosh), ABI Books, New Delhi, pp. 342–367.

Lam, L., Wong, Y.L., and Poon, C.S. (1998) Effect of fly ash and silica fume on compressive and flexural behaviour of concrete. *Cement and Concrete Research* **28**, 271–283.

La Rosa, J.L., Kwan, S., and Grutzek, M.W. (1991) Self generated zeolite cement composites. *Materials Research Society Symposium Proceedings* **245**, 211–216.

La Rosa, J.L., Kwan, S., and Grutzek, M.W. (1992) Zeolite formation in class F fly ash blended cement pastes. *Journal of the American Ceramic Society* **75**, 1574–1580.

Li, W., and Roy, D.M. (1986) Investigations of relations between porosity, pore structure and Cl-diffusion of fly ash and blended cement pastes. *Cement and Concrete Research* **16**, 749–759.

Li, X., and Chung, D.D.L. (1998) Improving silica fume for concrete by surface treatment. *Cement and Concrete Research* **28**, 493–498.

Liebig, E., and Althaus, E. (1998) Pozzolanic activity of volcanic tuff and suevite: effect of calcination. *Cement and Concrete Research* **28**, 567–575.

Lilkov, V., Dimitrova, E., and Petrov, O.E. (1997) Hydration process of cement containing fly ash and silica fume: the first 24 hours. *Cement and Concrete Research* **27**, 577–588.

Lin, L.K. (1997) Characteristics and hydration mechanism of RHA cement paste, in *Proceedings 10th ICCC, Göteborg*, paper 3ii108.

Liu, J.N., Silsbee, M.R., and Roy, D.M. (1997) Strength and hydration of an activated alumino-silicate material, in *Proceedings 10th ICCC, Göteborg*, paper 3ii114.

Lotze, J., and Wargalla, G. (1985) Characteristic data and utilization possibilities of ash from circulating fluidized bed furnace (in German). *Zement-Kalk-Gips* **38**, 239–243, 374–378.

Lu, C. (1992) The research of the reactive product and mineral phase for FKJ cementitious material, in *Proceedings 9th ICCC, New Delhi*, Vol. 3, pp. 319–324.

Mak, S.L., and Torii, K. (1995) Strength development of high strength concretes with and without silica fume under the influence of high hydration temperature. *Cement and Concrete Research* **25**, 1791–1802.

Marusin, S.L. (1988) Influence of superplasticizers, polymer admixtures and silica fume in concrete on chloride permeability. ACI SP-108(2), pp. 19–33.

Masazza, F. (1998) Pozzolana and pozzolanic cements, in *Lea's Chemistry of Cement and Concrete* (ed. P.C. Hewlett), Arnold, London, pp. 471–631.

McCarthy, G.J. *et al.* (1989) Factors affecting the ability of a fly ash to contribute to sulfate resistance of fly ash concrete. *Materials Research Society Symposium Proceedings* **136**, 273–282.

Mehta, P.K. (1985) Influence of fly ash characteristics on the strength of Portland-fly ash mixtures. *Cement and Concrete Research* **15**, 669–644.

Mehta, P.K., and Folliard, K.J. (1995) Rice husk ash – a unique supplementary cementing material: durability aspects. ACI SP-154, pp. 531–541.

Meland, I. (1986) Use of fly ash in cement to reduce alkali-silica reactions. ACI SP-91(1), pp. 591–608.

Meng, B., and Schiessl, P. (1997) The reaction of silica fume at early ages, in *Proceedings 10th ICCC, Göteborg*, paper 3ii105.

Mitchell, Hinczak, I., and Day, R.A. (1998) Interaction of silica fume with calcium hydroxide solutions and hydrated cement pastes. *Cement and Concrete Research* **28**,1571–1584.

Moukwa, M. (1993) Durability of silica fume concrete, in *Mineral Admixtures in Cement and Concrete* (ed. S.N. Ghosh), ABI Books, New Delhi, pp. 467–491.

Monzo, J. *et al.* (1995) Mechanical treatment of flyashes: strength development and workability of mortars containing ground fly ashes. ACI SP-153 (1), pp. 339–353.

Murata, K. *et al.* (1994) The effect of characteristics of silica fume on physical properties of mortar and concrete (in Japanese). *Simento Konkurito Ronbunshu* **48**, 364–369 [ref. CA 123/151172].

Norsy, M.S., Galal, A.F., and Abo-El Enein, S.A. (1998) Effect of temperature on phase composition and microstructure of artificial pozzolana-cement pastes containing burnt kaolinite clay. *Cement and Concrete Research* **28**, 1157–1164.

Odler, I., and Skalny, J. (1992) Potential for the use of fossil fuel combustion waste by the construction industry, in *Materials Science of Concrete III* (ed. J. Skalny), American Ceramics Society, Westerville, OH, USA, pp. 319–336.

Odler, I., and Zysk, K.H. (1990) Hydraulic properties of fluidized bed combustion ashes. *Materials Research Society Symposium Proceedings* **178**, 189–196.

Panchenko, A.I., and Opoczky, L. (1992) Hardening of ordinary and expansive cement with silica fume admixture, in *Proceedings 9th ICCC, New Delhi*, Vol. 4, pp. 646–650.

Papadakis, V.G. (1999) Experimental investigation and theoretical modelling of silica fume activity in concrete. *Cement and Concrete Research* **29**, 79–86.

Paya, J. *et al.* (1995, 1996) Mechanical treatment of fly ashes. *Cement and Concrete Research* **25**, 1469–1479; **26**, 225–235.

Pera, J. *et al.* (1986) Cements on the basis of thermally activated clays: role of iron oxide in the development of strength (in French), in *Proceedings 8th ICCC, Rio de Janeiro*, Vol. 4, pp. 344–350.

Perraton, D., Aitcin, P.C., and Vezina, D. (1988) Permeabilities of silica fume concretes. ACI SP-108(2), pp. 63–84.

Piazza, J.L. (1994) Investigations on the reactivities of bituminous and sub-bituminous coal fly ashes in mortars and cement pastes (in German). PhD Thesis, Technical University Clausthal, Germany.

Pietersen, H., and Bijen, J.M. (1992) The hydration chemistry of some blended cements, in *Proceedings 9th ICCC, New Delhi*, Vol. 6, pp. 281–290.

Popovics, S. (1987) Attempts to improve the bond between cement paste and aggregate. *Materials and Structures* **20**, 32.

Ranogajec, J., and Fisang, L. (1992) The hydration of pozzolanic cement on the basis of fired clay bricks, in *Proceedings 9th ICCC, New Delhi*, Vol. 6, pp. 233–239.

Rattanussorn, M., Roy, D.M., and Malek, R.I.A. (1987) Effect of fly ash incorporation on rheology of cement paste. *Materials Research Society Symposium Proceedings* **86**, 229–237.

Rehsi, S.S. (1993) Portland fly ash cement, in *Mineral Admixtures in Cement and Concrete* (ed. S.N. Ghosh), ABI Books, New Delhi, pp. 158–173.

Roberts, L.R. (1989) Microsilica in concrete, in *Materials Science of Concrete I* (ed. J. Skalny), American Ceramics Society, Westerville, OH, USA, pp. 197–222.

Rosenberg, A.M., and Gaidis, J.M. (1989) New mineral admixture for high strength concrete. *Concrete International: Design and Construction* **11**, 31–35.

Roy, A., Schilling, P.J., and Eaton, H.C. (1995) US patent no. 5435843.

Saad, M. *et al.* (1996) Effect of temperature on physical and mechanical properties of concrete containing silica fume. *Cement and Concrete Research* **26**, 669–675.

Salvador, S. (1995) Pozzolanic properties of fly ash – calcined kaolinite: a comparative study with soak-calcined products. *Cement and Concrete Research* **25**, 102–112.

Samarin, A., Munn, R.L., and Ashby, J.B. (1983) Ash in concrete: Australian experience. ACI SP-79(1), pp. 143–172.

Sanches de Rojas, M.I., and Frias, M. (1996) The pozzolanic activity of different materials: its influence on the hydration heat in mortars. *Cement and Concrete Research* **26**, 203–213.

Sanches de Rojas, M.I., Rivera, J., and Frias, M. (1999) Influence of the microsilica state on pozzolanic reaction rate. *Cement and Concrete Research* **29**, 945–949.

Sarkar, S.I., and Xu, A. (1993) Hydration and properties of fly ash concrete, in *Mineral Admixtures in Cement and Concrete* (ed. S.N. Ghosh), ABI Books, New Delhi, pp. 174–225.

Shayan, A., Diggins, R., and Ivanusec, I. (1996) Effectiveness of fly ash in preventing deleterious expansion due to alkali-aggregate reaction in normal and steam-cured concrete. *Cement and Concrete Research* **26**, 153–164.

Shi, C. (1996) Early microstructure development of activated lime – fly ash pastes. *Cement and Concrete Research* **26**, 1351–1359.

Shi, C., and Day, R.L. (1993) Chemical activation of blended cements made with lime and natural pozzolana. *Cement and Concrete Research* **23**, 1389–1396.

Singh, M., and Garg, M. (1992) Development of cementitious properties of phosphogypsum – fly ash systems, in *Proceedings 9th ICCC, New Delhi*, Vol. 4, pp. 489–494.

Solem, J.K., and McCarthy, G.J. (1992) Hydration reactions and ettringite formation in selected cementitious coal conversion by-products. *Materials Research Society Symposium Proceedings* **245**, 76–80.

Sugita, S. *et al.* (1997) On the semi-industrial production of highly reactive rice husk ash and its effect on cement and concrete properties, in *Proceedings 10th ICCC, Göteborg*, paper 3ii109.

Suzuki, T. (1995) Application of high volume fly ash concrete to marine structures. *Chem. Ecol.* **10**, 249–258.

Swabowski, J., and Galaszewski, J. (1996) Effect of superplasticizers and silica fume on the workability of high performance concrete (in Polish). *Cement-Wapno-Gips (Warsaw)*, 212–216.

Tachibana, D. (1990) High strength concrete incorporating several admixtures, in *Proceedings 2nd International Symposium on High Strength Concrete, Berkeley*, pp. 309–330.

Tamini, A.K. (1996) The effect of mixing technique on microsilica concrete, in *Production Methods and Workability of Concrete* (RILEM Proceedings 32), pp. 27–33.

Tasdemir, C. *et al.* (1996) Effects of silica fume and aggregate size on the brittleness of concrete. *Cement and Concrete Research* **26**, 63–68.

Tenoutasse, N., and Marion, A.M. (1986) Characterization of Belgian fly ashes and their behavior in cement paste. ACI SP-91, pp. 51–76.

Toutanji, H.A., and El-Korchi, T. (1995) The influence of silica fume on the compressive strength of cement paste and mortar. *Cement and Concrete Research* **25**, 1591–1602.

Uchikawa, H. (1986) Effect of blending components on hydration and structure formation, in *Proceedings 8th ICCC, Rio de Janeiro*, Vol. 1, pp. 249–280.

Uchikawa, H., and Okamura, T. (1993) Binary and ternary components blended cements, in *Mineral Admixtures in Cement and Concrete* (ed. S.N. Ghosh), ABI Books, New Delhi, pp. 1–83.

Vennesland, O., and Gjorv, O.E. (1983) Silica concrete protection of corrosion of embedded steel. ACI SP-79, pp. 719–729.

von Berg, W., and Puch, K.H. (1993) Utilization of residues from fluidized bed combustion plants (in German). *WGB Kraftwerkstechnik* **73**, 330–334.

Wang, X., Yang, N., and Zhong, B. (1996) Study of reaction mechanism of fly ash – lime – water system (in Chinese). *Guisuanyan Xuebao* 24, 137–141, 125 [ref. CA 125/17320].

Wei, F., Grutzek, W., and Roy, D.M. (1985) The retarding effect of fly ash upon the hydration of cement pastes: the first 24 hours. *Cement and Concrete Research* **15**, 174–184.

Wild, S., and Khatib, J.M. (1997) Portlandite consumption in metakaolin cement pastes and mortars. *Cement and Concrete Research* **27**, 137–146.

Wild, S., Sabir, B.B., and Khatib, J.M. (1995) Factors influencing strength development of concrete containing silica fume. *Cement and Concrete Research* **25**, 1567–1580.

Wild, S., Khatib, J.M., and Jones, A. (1996a) Relative strength, pozzolanic activity and cement hydration in superplasticized metakaolin concrete. *Cement and Concrete Research* **26**, 1537–1544.

Wild, S. *et al.* (1996b) The potential of fire brick clay as a partial cement replacement material, in *International Conference on Concrete in the Service of Mankind* (eds R.K. Dhir and T.D. Dyler), E & FN Spon, London, UK, pp. 685–696.

Wild, S., Khatib, J.M., and O'Farell, M. (1997) Sulphate resistance of mortar containing ground brick clay calcined at different temperatures. *Cement and Concrete Research* **27**, 697–709.

Wild, S., Khatib, J.M., and Roose, L.J. (1998) Chemical shrinkage and autogenous shrinkage of Portland cement-metakaolin pastes. *Advances in Cement Research* **10**, 109–119.

Wu, Z.Q., and Young, J.F. (1984) The hydration of tricalcium silicate in the presence of colloidal silica. *Journal of Materials Science*, **19**, 3477–3486.

Xu, A., and Sarkar, S.L. (1993) Hydration and properties fly ash concrete, in *Mineral Admixtures in Cement and Concrete* (ed. S.N. Ghosh), ABI Books, New Delhi, pp. 174–225.

Xu, A., and Sarkar, S.L. Active β-C_2S cement from fly ash and kiln dust. ACI SP-153(1), pp. 213–227.

Xu, A. (1997) Fly ash in concrete, in *Waste Materials used in Concrete Manufacturing* (ed. S. Chandra), Noyes Publications, Westwood, NJ, USA, pp. 142–183.

Yu, Q., and Zhang, D. (1991) Effect of $C_{11}A_7$.CaF_2 or C_4A_3S addition on the strength increase of Portland fly ash cement. *Guisnanyan Xuebao* **19**, 565–571 [ref. CA 120/278361].

Yu, Q. *et al.* (1999) The reaction between rice husk ash and $Ca(OH)_2$ solution and the nature of its product. *Cement and Concrete Research* **29**, 37–43.

Zampieri, V.A. (1990) Mineralogy and microstructure of reaction products in metakaolinite-lime mixtures, in *Proceedings 12th International Conference on Cement Microscopy*, pp. 76–95.

Zhang, M.H. (1995) Microstructure, crack propagation and mechanical properties of cement pastes containing high volumes of fly ashes. *Cement and Concrete Research* **25**, 1165–1178.

Zhang, M.H., and Malhotra, V.M. (1995) Characteristics of a thermally activated alumino-silicate pozzolanic material and its use in concrete. *Cement and Concrete Research* **25**, 1713–1725.

Zhang, M.H., Lastra, R., and Malhotra, V.M. (1996) Rice-husk ash paste and concrete: some aspects of hydration and the microstructure of the interfacial zone between the aggregate and paste. *Cement and Concrete Research* **26**, 963–977.

10 Calcium aluminate cement

The term **calcium aluminate cement** (abbreviation CAC) covers a range of inorganic binders characterized by the presence of monocalcium aluminate ($CaO.Al_2O_3$, abbreviation CA) as their main constituent. The chemical composition of calcium aluminate cement may vary over a wide range, with Al_2O_3 contents ranging between about 40% and 80%. Unlike Portland cement it does not contain tricalcium silicate, but may contain limited amounts of dicalcium silicate. Calcium aluminate cement is also called **aluminous cement** (abbreviation AC) or **high-alumina cement** (abbreviation HAC). The trade names Ciment Fondu and SECAR (for cements with lower and higher Al_2O_3 contents respectively) are also being widely used. The following is a selection of overview articles discussing calcium aluminate cement: Bensted (1993); George (1983); Kurdowski *et al.* (1986); Mangabhai (1990); Mohan (1991); Su *et al.* (1992); Holterhoff (1993); Scrivener and Capmas (1998).

The development of calcium aluminate cement was spurred by efforts to overcome the problems associated with sulfate attack on Portland cement based concrete used in the construction of railway tunnels in gypsiferous grounds. The first patent relating to this type of binder was filed in 1908 by Bied in France. The cement was introduced into production in 1913, and became known as Ciment Fondu. After it was recognized that calcium aluminate cement gains strength much faster than Portland cement, the binder was used in World War I by the French military in the construction of gun emplacements and shelters, where this property was of paramount importance. After the war, the cement became widely used in other structural applications; however, its use in this area became limited, after failures of structures built with this cement were reported from different countries. Nowadays calcium aluminate cement is being used in a variety of special applications.

10.1 CALCIUM ALUMINATE PHASES

There are a number of calcium aluminate phases. Most of them exhibit hydraulic properties, and are important for cement chemistry.

Calcium hexaaluminate ($CaO.6Al_2O_3$; abbreviation CA_6) is of little interest for cement chemistry as it does not exhibit hydraulic properties. Its structure is closely related to that of corundum.

Calcium dialuminate ($CaO.2Al_2O_3$; abbreviation CA_2) is a constituent of some high-alumina cements, and exhibits a slow though distinct reaction with water. It is monoclinic (a = 1.2840 nm, b = 0.8862 nm, c = 0.5431 nm, β = 1.06.8°, D = 2920 kg/m³), and has a structure based on a framework of AlO_4 tetrahedra; some oxygen atoms are shared between two and others between three tetrahedra.

Monocalcium aluminate ($CaO.Al_2O_3$; abbreviation CA) is the main constituent of all types of calcium aluminate cement, and is mainly responsible for the properties of this binder. In nearly pure form it may be synthesized by burning an equimolar blend of CaO and Al_2O_3 above about 800 °C. In the course of the process other calcium aluminate phases are also formed as intermediates, most notably $C_{12}A_7$ (Fig. 10.1). CA is monoclinic and pseudohexagonal (a = 0.8700 nm, b = 0.8092 nm, c = 1.5191 nm, β = 90.3°, d = 2945 kg/m³). It exhibits distinct hydraulic properties and reacts rapidly with water. Its crystal structure consists of a three-dimensional framework of AlO_4 tetrahedra sharing corners with Ca^{2+} ions in between. An analogous barium compound, monobarium aluminate ($BaO.Al_2O_3$) is also known, and exhibits similar hydraulic properties as the calcium analog (Borovkova *et al.*, 1992; Raina *et al.*, 1992).

Pentacalcium trialuminate ($5CaO.3Al_2O_3$; abbreviation C_5A_3) is probably an equilibrium phase in the strictly binary system $CaO–Al_2O_3$. It may be produced in a solid-state reaction from CA and C_3A under dry conditions.

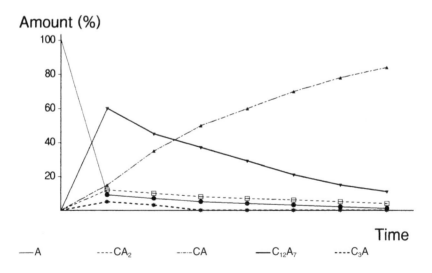

Figure 10.1 Development of Al_2O_3-bearing phases (in mol.%) in the burning of an equimolar blend of CaO and Al_2O_3 at 1200 °C.

In normal moisture air it decomposes rapidly to $C_{12}A_7$ and CA. As C_5A_3 is stable only in the absence of any humidity and oxygen, it has no practical importance, even though it may react readily with water.

Dodeca-calcium heptaaluminate ($12CaO.7Al_2O_3$; abbreviation $C_{12}A_7$) is a common, but in most instances only a minor, constituent of calcium aluminate cements. It is cubic ($a = 1.1983$ nm, $D = 2680$ kg/m^3). The ideal structure of $C_{12}A_7$ consists of an incomplete framework of corner-sharing AlO_4 tetrahedra with a composition of $Al_7O_{16}^{11-}$ and Ca^{2-} ions; one additional O^{2-} ion is statistically distributed between 12 sites. $C_{12}A_7$ produced in air of normal humidity always contains small amounts of water incorporated into its crystal lattice in the form of two OH^- ions replacing the O^{2-} ion. This small amount of water is essential for the structure to be stable. Upon heating, the amount of water bound in this way gradually increases up to about 950 °C but declines upon further heating, until the material melts almost anhydrous at about 1400 °C. The maximum amount of water that may be bound within the structure of $C_{12}A_7$ is about 1.3%, which corresponds to the formula $C_{12}A_7H$. Alternatively, the O^{2-} ion may also be replaced by a pair of halide ions. The fluoride analog $C_{11}A_7.CaF_2$ is a constituent of some special cements, and may form a continuous range of solid solutions with $C_{12}A_7$. In dry oxygen a $C_{12}A_7$ analog that contains excessive oxygen may also be formed. Here the excessive amount of oxygen is present in the form of peroxide (O_2^{2-}) ions, substituting for O^{2-}. $C_{12}A_7$ reacts rapidly with water, and exhibits distinct hydraulic properties.

Tetracalcium trialuminate ($4CaO.3Al_2O_3$; abbreviation C_4A_3) may be synthesized by thermally dehydrating the compound C_4A_3H produced by a hydrothermal process. It is not a constituent of calcium aluminate cements.

Dicalcium aluminate ($2CaO.Al_2O_3$; abbreviation C_2A) can be synthesized only under high-pressure conditions, and is of no practical importance for cement chemistry.

Tricalcium aluminate ($3CaO.Al_2O_3$; abbreviation C_3A) is a regular constituent of Portland cements. It is not present in calcium aluminate cement. Its structure and hydration are discussed in section 2.1.3. It is the most reactive of all calcium aluminates.

10.2 CALCIUM ALUMINATE HYDRATE PHASES

The following calcium aluminate hydrate phases may be formed in the hydration of calcium aluminates.

Monocalcium aluminate hydrate ($CaO.Al_2O_3.10H_2O$; abbreviation CAH_{10}) is a hexagonal phase with a probable ionic structure $Ca_3[Al(OH)_4]_6.18H_2O$ ($a = 1.644$ nm, $c = 0.831$ nm, $D = 1730$ kg/m^3). Most of the water in the structure is only loosely bound, and a loss of water may begin at 80% RH. The compound also loses water upon heating, and this loss is accompanied by a deterioration of crystallinity. By 200–350 °C the material becomes amorphous, and by 1000 °C it converts to monocalcium aluminate (CA). DTA curves are characterized by an exotherm at 130–150 °C, associated with the loss of molecular water, and another one at 290–310 °C due to dehydroxylation. If the water cannot escape easily, the thermal decomposition of CAH_{10} progresses differently, and C_3AH_6 – together with gibbsite (AH_3) – is formed instead. The latter is the typical case, if hardened calcium aluminate cement is heated. CAH_{10} is thermodynamically unstable with respect to C_3AH_6.

Dicalcium aluminate hydrate ($2CaO.Al_2O_3.8H_2O$; abbreviation C_2AH_8) is a hexagonal phase belonging among the AFm phases. It has a layered structure, which may be described by the formula $[Ca_2Al(OH)_6].[Al(OH)_3(H_2O)_3]OH$. The Al is octahedrally coordinated, and the layer thickness is $c = 1.07$ nm ($a = 0.574$ nm, $D = 1950$ kg/m^3). Upon drying or heating the material converts to hydrates with lower water content, including C_2AH_5. C_2AH_8 is also thermodynamically unstable with respect to C_3AH_6.

Tricalcium aluminate hexahydrate ($3CaO.Al_2O_3.6H_2O$; abbreviation C_3AH_6) is the only phase in the system $CaO–Al_2O_3–H_2O$ that is thermodynamically stable at ordinary temperature. All other calcium aluminate hydrates formed in the hydration of calcium aluminate cement convert eventually to C_3AH_6 in the conversion reaction taking place in the hydrated material. C_3AH_6 belongs to the group of hydrogarnet phases whose composition may vary within the compositional region bounded by C_3AH_6, C_3FH_6, C_3AS_3 and C_3FS_3. C_3AH_3 is cubic, and its structure may be expressed by the formula $Ca_3[Al(OH)_6]_2$ ($a = 1.258$ nm, $D = 2530$ kg/m^3). Upon heating the compound converts first (at about 200–250 °C) to $C_{12}A_7H + CH$, and eventually (at about 800 °C) to C_3A.

Tetracalcium aluminate hydrate ($4CaO.Al_2O_3.xH_2O$; abbreviation C_4AH_x) may contain variable amounts of crystalline water, depending on relative humidity and temperature. The form highest in water, C_4AH_{19}, is stable only below 25 °C and at RH > 88%. Upon drying and/or moderate heating it converts first to C_4AH_{13} and subsequently to C_4AH_{11} and even C_4AH_7. The structure of C_4AH_{13} may be expressed by the formula $[Ca_2Al(OH)_6.2H_2O].OH$. In C_4AH_{19} an additional layer of H_2O molecules is present between the principal layers. Tetracalcium aluminate hydrate is not among the hydrate phases formed in the hydration of calcium aluminates constituting calcium aluminate cement; however, it may be formed in the hydration of tricalcium aluminate, a constituent of Portland cement.

In addition to calcium aluminate hydrates, crystalline **aluminum hydroxide** [$Al(OH)_3$; abbreviation AH_3, gibbsite] or amorphous aluminate hydrate (AH_x) is also formed in the hydration of some calcium aluminates.

10.3 THE HYDRATION OF PURE CALCIUM ALUMINATES

In the hydration of monocalcium aluminate (the main constituent of calcium aluminate cement) different hydrate phases may be formed, depending on the temperature of hydration. Up to about 10 °C crystalline CAH_{10} is the sole or major product of hydration, and this phase may be formed, together with C_2AH_8, up to about 27 °C. A significant formation of this phase gets under way several hours after mixing, and this process is accompanied by a distinct hydration heat liberation. There are indications, however, that small amounts of an amorphous calcium aluminate hydrate phase are formed simultaneously in the first hours of hydration at low temperatures (Müller *et al.*, 1992). The formation of CAH_{10} from CA is a through-solution process in which the anhydrous compound is dissolved in the liquid phase, and the hydration product randomly precipitates from the oversaturated solution created in this way (Gulgun *et al.*, 1992).

The primary formed CAH_{10} tends to convert subsequently to C_3AH_6, the only phase in the system C-A-H that is thermodynamically stable. This reaction is also a through-solution process, and gibbsite (γ-AH_3) is formed as a by-product. The conversion of CAH_{10} to C_3AH_6 is not a direct process, and C_2AH_8 is always formed as an intermediate product, even at elevated temperatures (Rashid *et al.*, 1994). The rate of the reaction is temperature dependent. At temperatures close to the freezing point it may take many years until it is completed, whereas at 25 °C a complete conversion occurs within months. The sequence of phase formation taking place in the hydration of CA at ambient temperature is portrayed in Fig. 10.2.

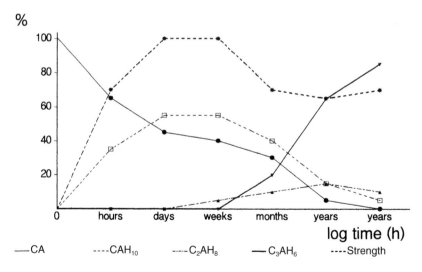

Figure 10.2 Hydration and strength development of monocalcium aluminate (CA): w/s=0.40; 20 °C.

A precondition for the conversion reaction to take place is the presence of free water within the pore system of the hardened material. In contrast, at low humidities, in the absence of free water, a partial decomposition of the hydrated material, associated with a loss of crystalline water, takes place instead; this reaction is also accelerated with increasing temperature.

At temperatures above about $10\,°C$ increasing amounts of C_2AH_8, together with AH_3, are formed at the expense of CAH_{10}, and above about $27\,°C$ CAH_{10} is not formed at all. The conversion of the primary formed hydrates to C_3AH_6 and AH_3 gets under way sooner with increasing temperature, and at the same time the rate of conversion is accelerated. Additional AH_3 is also formed.

Above about $50\,°C$ C_3AH_6 is the sole hydration product formed in the hydration of CA. It has been suggested that the formation of this phase is preceded by a transitory formation of C_2AH_8 (Rashid *et al.*, 1994), but it seems likely that a direct formation of C_3AH_6 may also take place if some C_3AH_6 has already been nucleated (Scrivener and Capmas, 1995).

The hydration of CA is accompanied by a liberation of heat of hydration. In paste hydration at ambient temperature the rate at which heat is liberated is slow initially, but a heat liberation maximum develops at the end of an induction period, after which time the rate of the hydration reaction is significantly accelerated. The time at which the maximum heat output (ranging up to about 800 W/kg; Edmonds and Majumdar, 1988) is reached becomes extended as the temperature increases from $0\,°C$ up to about $30\,°C$, indicating a gradual slow-down of the hydration reaction. At even higher temperatures, however, the hydration is accelerated again as the rate of temperature release increases. The origin of the slow-down of the hydration rate at around $30\,°C$ is not obvious. It is assumed that this phenomenon is due to a slowed-down nucleation of the hydrates formed under these conditions. The temperature of maximum retardation may be lowered and the extent of the retardation may be reduced by adding alumina (Al_2O_3) to the system (Rettel *et al.*, 1997). Following the induction period the hydration of CA is rather fast, and – at sufficiently high water/cement ratios – most of it is consumed within a few days.

The reactions that can take place in the hydration of CA, depending on the temperature, may be summarized as follows:

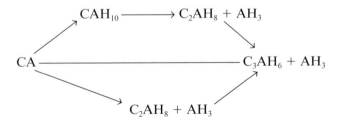

Solid-state ^{27}Al NMR studies have revealed an overall conversion of Al(4) present in the unhydrated CA to Al(6) in the hydrated phases (Müller *et al.*, 1992; Cong and Kirkpatrick, 1993; Kudryavtsev and Kouznetsova, 1997). During the induction period both hydrated Al(4) and Al(6) environments have been detected.

The following enthalpies have been reported for the individual reaction steps (Mohan, 1991):

$$CA \rightarrow CAH_{10} \qquad\qquad 245 \pm 5 \text{ J/g} \qquad\qquad (10.1)$$

$$CA \rightarrow C_2AH_8 + AH_3 \qquad\qquad 280 \pm 10 \text{ J/g} \qquad\qquad (10.2)$$

$$C_2AH_8 + AH_3 \rightarrow C_3AH_6 + AH6 \qquad 120 \pm 50 \text{ J/g} \qquad\qquad (10.3)$$

The amount of water needed for complete hydration of CA to CAH_{10} is rather high, and corresponds to a water/solid ratio of $w/s = 1.13$. For the formation of $C_2AH_8 + AH_3$ the required w/s is 0.63, and for the formation of C_3AH_6 $w/s = 0.46$ is sufficient. These values are distinctly higher than those needed for the hydration of the calcium silicates that constitute Portland cement (C_2S and C_3S).

Conversely, the conversion of CAH_{10} to C_3AH_6 and AH_3

$$3CAH_{10} \rightarrow C_3AH_6 + 2AH_3 + 18H \qquad\qquad (10.4)$$

is associated with a release of free water (60% of the total amount bound within CAH_{10}) and along with it with a distinct decline of the volume of solids (to 47% of its original value).

In the conversion of C_2AH_8 to C_3AH_6

$$3C_2AH_8 \rightarrow 2C_3AH_6 + 9AH_3 \qquad\qquad (10.5)$$

37.5% of the chemically bound water is released, and the volume of solids is reduced to 66% of the original value.

The volume of calcium aluminate hydrates formed in the hydration of CA is smaller than the sum of the volumes of the original CA and the water participating in the reaction. This **chemical shrinkage** for the hydration of monocalcium aluminate amounts to 16 vol.% in the formation of CAH_{10} and to 25 vol.% if C_3AH_6 and AH_3 are formed as products of hydration.

The hydration reaction of other calcium aluminates is similar to that of CA; however, the mutual ratio of the individual hydrate phases formed may be altered. In the hydration of $C_{12}A_7$ the fraction of C_2AH_8 formed is increased at the expense of CAH_{10}, whereas in the hydration of CA_2 increased amounts of AH_3 are formed. There are, however, distinct differences in the rate of hydration between different calcium aluminates; in general, the reactivity increases with increasing C/A ratio. Among the constituents of high-alumina cement, CA_2 is the phase with the slowest hydration; CA hydrates faster, and $C_{12}A_7$ is even more reactive (Das and Daspodda, 1993).

The following reaction enthalpies have been reported for the hydration of $C_{12}A_7$:

$$C_{12}A_7 \rightarrow C_2AH_8 + AH_3 \qquad\qquad 350 \pm 10 \text{ J/g} \qquad\qquad (10.6)$$

$$C_2AH_8 + AH_3 \rightarrow C_3AH_6 + AH_3 \qquad\qquad 110 \pm 25 \text{ J/g} \qquad\qquad (10.7)$$

10.4   TECHNOLOGY OF CALCIUM ALUMINATE CEMENT PRODUCTION

Calcium aluminate cement is produced by burning a raw material blend high in Al_2O_3 and CaO at high temperatures. The oxide composition of the raw blend may vary over a wide range, with Al_2O_3 contents between about 40% and 80%, depending on the type of calcium aluminate cement to be produced. The SiO_2 content of the raw mix must be kept low, to prevent or minimize the formation of gehlenite (C_2AS), a phase that does not exhibit hydraulic properties. If a white color of the cement is desired and/or the cement is to be used for high-temperature applications, the iron oxide content of the raw mix must also be kept low.

Most commonly limestone and bauxite are employed as starting materials, which must be blended in an appropriate ratio. Bauxite is a natural product of the weathering of igneous rock under tropical conditions, and is rich in alumina; it may also contain variable amounts of iron. For the production of high-grade calcium aluminate cements with CaO + Al_2O_3 contents of up to 95%, calcined alumina in combination with high-purity lime or limestone must be employed.

Calcium aluminate cement with a lower or intermediate Al_2O_3 content cannot be produced in rotary or shaft kilns of the type common in the manufacture of Portland clinker. The temperature range between incipient melting and complete fusion of the raw mixes is too narrow to permit successful clinkerization, in which a melt and solid phases must coexist. Moreover, the viscosity of calcium aluminate (-ferrite) melts is significantly lower than that of the calcium silicate-aluminate-ferrite melts formed when Portland clinker is produced.

Calcium aluminate cements with lower Al_2O_3 contents and distinct amounts of Fe_2O_3 are commonly produced by complete fusion of a blend of limestone and bauxite at temperatures of 1450–1600 °C in a modified shaft kiln. At these temperatures the whole material converts to a melt, and is collected at the bottom of the kiln. The melted material is tapped off, and crystallization of the melt takes place in the course of cooling outside the kiln. The cooled product appears as a gray, finely grained compact rock. The burning is usually done under oxidizing conditions, but burning under reducing conditions is also possible.

Subsequently, the high-alumina clinker is crushed and ground to a fine powder in a ball mill. The clinker is very hard to grind, and the process is associated with a high power consumption. Unlike Portland cement, no set controlling constituents, such as gypsum, need to be interground with the burnt material.

High-Al_2O_3 calcium aluminate cements are produced by solid-state sintering of calcined alumina and lime or limestone in rotary kilns. In a solid-state reaction between calcium oxide and aluminum oxide Ca^{2+}, together with O^{2-}, migrates into the Al_2O_3 to form calcium aluminates. Of these $C_{12}A_7$ has the highest growth velocity, irrespective of the C/A ratio of the starting material, and other phases are formed as secondary products from this phase and the excess oxide (Scian *et al.*, 1987). In addition to calcium aluminate phases, the resultant product may also contain distinct amounts of residual non-reacted alumina.

10.5   COMPOSITION OF CALCIUM ALUMINATE CEMENT

Table 10.1 shows the range of chemical composition of calcium aluminate cement. The Al_2O_3 content of this binder may vary over a wide range, depending on the existing phase composition. The CaO content may also vary widely, and it declines with increasing amount of Al_2O_3. Distinct amounts of silica and iron oxide are usually present in cements with lower alumina contents. Bivalent iron is typical for cements produced under reducing conditions. TiO_2, MgO, SO_3, and alkali oxides are usually present only in small amounts, and are of little significance.

The phases present in cement produced by a fusion process are formed by crystallization from the melt. However, equilibrium conditions are not necessarily achieved, and the actual phase composition of the resulting cement will also depend on the conditions of cooling. Some of the melt may even yield a glass phase, especially if the cooling takes place rapidly.

Table 10.1 Composition of calcium aluminate cements.

Al_2O_3	38–82%
CaO	18–40%
Fe_2O_3 + FeO	< 0.3–18%
FeO	< 0.2–6%
SiO_2	< 0.5–6%
TiO_2	< 0.1–4%
MgO	< 0.2–1.5%
Na_2O + K_2O	< 0.4–7%
SO_3	< 0.2–0.4%

The phase composition and texture of high-alumina cement produced by solid-state sintering will also be affected by the production process used, but to a lesser extent.

Monocalcium aluminate (CA) is the main phase of calcium aluminate cements, and the one mainly responsible for their properties. This phase is present in calcium aluminate cements in amounts from 40% upwards. It may incorporated variable, but usually small, amounts of Fe_2O_3 and SiO_2, especially in cements with lower Al_2O_3 contents (Rayment and Majumdar, 1994). $C_{12}A_7$ is also present in most commercial cements. Calcium dialuminate (CA_2) is usually present in cements with high Al_2O_3 contents, which may also contain calcium hexaaluminate (CA_6) and even free aluminum oxide (a-Al_2O_3, corundum). Tricalcium aluminate (C_3A) is not a normal constituent of calcium aluminate cements.

The ferrite phase [$C_2(A,F)$] may constitute up to 20–40% of the cement, and its amount will depend on the Fe_2O_3 content of the raw mix. The A/F ratio in this phase may vary over a wide range, and it may also contain some TiO_2, SiO_2, and MgO. The phase is absent in white, iron-free high-alumina cements.

Most of the SiO_2 is present in high-alumina cement in the form of dicalcium silicate, β-C_2S. At higher SiO_2 contents gehlenite (C_2AS) may also be formed. Both phases are present only in cements with lower Al_2O_3 and distinct SiO_2 contents. The presence of gehlenite is unwelcome, as it does not exhibit hydraulic properties, but limited amounts of this phase may be tolerated.

In cements produced under reducing conditions bivalent iron is present in the form of pleochlorite (a complex solid solution of ferric and ferrous iron, silica, and magnesia, of the possible composition $CA_{20}Al_{22.6}Fe^{3+}_{2.4}Mg_{3.2}Fe^{2+}_{0.3}Si_{3.5}O_{68}$), and in the form of wüstite (FeO). Here it is often in a solid solution with MgO.

Perovskite (TiO_2) may be present in cements with high TiO_2 contents in the raw mix.

Finally, a glass phase may also be present if the melt is cooled rapidly and – as a consequence – the crystallization of the melt is incomplete.

Table 10.2 summarizes the composition and main characteristics of a series of commercial calcium aluminate cements that roughly represent the spectrum of products currently available. Both their oxide composition and their phase composition vary over a wide range, affecting the properties of the product. Calcium aluminate cements with lower Al_2O_3 contents are those most widely used, whereas high-Al_2O_3 products are reserved for special applications.

Calcium aluminate cements may differ distinctly in their color, which is determined mainly by their iron content. The gray color of calcium aluminate cements is caused by the ferrous form of iron, whereas ferric compounds lead to a brown color. The cement gets darker with increasing iron content and – at a constant content of this element – with an increasing

Table 10.2 Composition and properties of commercial calcium aluminate cements.

Trade name		Ciment Fondu	Secar 50	Secar 70	Secar 80
Oxide composition	Al_2O_3	39	50.5	70.5	80.5
	CaO	38.5	36.5	29	18
	SiO_2	5	7	–	–
	Fe_2O_3	12	2	–	–
	FeO	4	1	–	–
Phase composition		CA	CA	CA	CA
		$C_{12}A_7$	αA	CA_2	αA
		C_2S	C_2AS		CA_2
		C_4AF	C_2S		
Fineness (m^2/kg)		300	400	370	900
Specific gravity		3.2	3.1	3.0	3.2
Bulk density (kg/m^3)		1150	980	900	700
Compressive strength (MPa)	6 h	41	40	21	16
	1 d	56	58	30	40
	7 d	67	71	54	65
	28 d	74			

proportion of ferrous species. High-alumina cements made from iron-free starting materials are white.

10.6   HYDRATION OF CALCIUM ALUMINATE CEMENT

The hydration of calcium aluminate cement is dominated by the hydration of the CA and other calcium aluminates that are present. Immediately after mixing with water, CaO and Al_2O_3 start to enter the liquid phase until a maximum CaO concentration of about 20–30 mmol/L is reached (Scian *et al.*, 1994). The CaO/Al_2O_3 molar ratio depends on the composition of the cement, and is usually somewhat greater than 1.0 (Scian *et al.*, 1994). At ambient temperature the initial formation of hydrates is rather slow, but it accelerates significantly after an induction period of several hours. This delay is due to the existence of a nucleation barrier in the hydration of CA (Matusinovic and Curlin, 1993). As the hydration is accelerated, the rate of hydration heat liberation is increased as well, and the concentrations of CaO and Al_2O_3 in the liquid phase start to decrease. In paste hydration setting gets under way at this time. C_2AH_8 and AH_3 are formed first, but only in small amounts. At lower hydration temperatures, up to about 15–20 °C, CAH_{10} eventually becomes the main product of hydration, whereas increasing amounts of C_2AH_8 and AH_3 are formed as the temperature increases. At even higher temperatures increasing amounts of C_3AH_6 are also formed.

In cements that contain significant amounts of $C_{12}A_7$, C_2AH_8 is the predominant phase formed, even at low temperatures. The hydration of this phase is fast and very exothermic. Considerable amounts of an amorphous hydrated material may also be formed, especially at near-zero temperatures.

Eventually all calcium aluminate hydrates convert to C_3AH_6, the only phase that is thermodynamically stable in the system $CaO-Al_2O_3-H_2O$, together with additional AH_3. The rate of this conversion increases distinctly with increasing temperature. At 5 °C it may take years until the conversion is completed, whereas above about 50 °C the process is virtually immediate. At low humidities partial dehydration of the formed hydrates, rather than their conversion to C_3AH_6, predominates. Table 10.3 summarizes the effect of temperature and humidity on the stability of hydrated aluminous cement.

The hydration of the ferrite phase is much slower than that of CA and $C_{12}A_7$, and this reaction scarcely affects the setting and initial strength development of calcium aluminate cements. $C_3(A,F)H_6$ is formed as the final product, with $C_2(A,F)H_x$ and $C_4(A,F)H_x$ as intermediates.

Dicalcium silicate in calcium aluminate cements does not yield a C-S-H phase and free calcium hydroxide, as occurs in Portland cement. Instead, strätlingite (also called gehlenite hydrate, C_2AH_8) is formed as the product of hydration.

The total heat of hydration of calcium aluminate cement is in the range 450–500 J/g, and is similar to that of Portland cement. However, 70–90% of it is liberated within the first 24 hours (at 20 °C), making dissipation of the heat into the environment more difficult than in situations where Portland cement is employed, where the liberation of hydration heat takes place much more slowly. This may be critical, especially in the erection of massive structures, where a significant rise of temperature may take place shortly after mixing. Maximum temperatures of up to 80 °C may be reached in the bulk of the concrete structure. This in turn may accelerate the "conversion" of the calcium aluminate hydrates formed. Wet curing has to be employed to prevent superficial dehydration and dusting of the hardened concrete.

The chemical shrinkage associated with the complete hydration of

Table 10.3 The effect of temperature and humidity on the stability of hydrated aluminous cement.

	Low temperature (< 20 °C)	Elevated temperature (> 20 °C)
Dry conditions	Relative stability	Dehydration $CAH_{10} \rightarrow CAH_4$ $C_2AH_8 \rightarrow C_2AH_5$
Humid conditions	Slow conversion to $C_3AH_6 + AH_3$	Fast conversion to $C_3AH_6 + AH_3$

calcium aluminate cement amounts to about 10–12 ml per 100 g of cement. This value is greater than that common in Portland cement (about 5 ml/100 g).

Calcium aluminate cement reacts in the course of hydration with more water than Portland cement. In a Ciment Fondu cement, the type most widely employed, a water/cement ratio of about 0.7 is needed to attain complete hydration. This amount of water is higher than that needed to obtain satisfactory rheology of the fresh concrete mix, in most instances. If no conversion were to take place in the hardened paste such a high water/cement ratio would be acceptable, as the high water addition would still result in a relatively low porosity of the hardened material. However, a distinct liberation of combined water, associated with an increase of porosity, takes place upon conversion of the primary formed CAH_{10} to C_3AH_6. In mixes with lower water/cement ratios the amount of water present may not be sufficient for complete hydration. Nevertheless, acceptable strengths may still be attained, owing to the low porosity of the incompletely hydrated paste. As will be shown below, such pastes exhibit a lower strength loss in the course of subsequent conversion than those made with a higher *w/c*.

The microstructure of the hydrated calcium aluminate cement paste depends on the initial water/cement ratio, hydration time, and hydration temperature (Halse and Pratt, 1986; Scrivener and Capmas, 1995). In secondary electron imaging on fractured surfaces the CAH_{10} phase appears as hexagonal prisms or fine needles, typically up to 10 μm long and 2 μm wide. C_2AH_8 is visualized in the form of small hexagonal platelets. C_3AH_6 appears in the form of spheroidal particles. Well-recognizable crystal forms can be detected, especially in pastes made with higher water/cement ratios and relatively high porosities, whereas at low *w/c* the paste appears in SEM studies as a rather compact material without readily distinguishable morphological features.

10.7 CONVERSION OF CALCIUM ALUMINATE CEMENT

As the calcium aluminate hydrate phases that are initially formed in the hydration of calcium aluminate cement – CAH_{10} and C_2AH_8 – are thermodynamically unstable, they convert over time to C_3AH_6, the only phase that is stable in the system C-A-H. At the same time AH_3 is formed as a by-product.

Table 10.4 shows the effect of temperature on the rate of conversion of CAH_{10} and C_2AH_8 in a humid environment. The rate of conversion is temperature dependent, and increases steeply as the temperature increases. It also increases with increasing water/cement ratio. An increased temperature during the initial hydration may also accelerate the subsequent rate of

Table 10.4 Effect of temperature on the half-time of conversion of CAH_{10} and C_2AH_8 to C_3AH_6.

Temperature (°C)	CAH_{10}	C_2AH_8
5°C	> 20 y	> 20 y
10°C	19 y	17 y
20°C	2 y	21 mo
30°C	75 d	55 d
50°C	32 h	21 h
90°C	2 min	35 s

Table 10.5 Effect of temperature and humidity on the stability of hydrated aluminous cement.

	Low temperature (< 20 °C)	Elevated temperature (> 20 °C)
Dry conditions	Relative stability	Dehydration $CAH_{10} \rightarrow CAH_4$ $C_2AH_8 \rightarrow C_2AH_5$
Humid conditions	Slow conversion to $C_3AH_6 + AH_3$	Fast conversion to $C_3AH_6 + AH_3$

conversion. The process takes place only under humid conditions, as a certain amount of free water is needed for the recrystallization to occur. Thus the rate of conversion slows down significantly if the relative humidity falls below saturation. Under dry conditions, in the absence of free water, dehydration of the primary formed hydrates, rather than conversion to C_3AH_6, may take place. The effect of temperature and humidity on the stability of aluminous cement is summarized in Table 10.5.

The conversion of aluminous cement is associated with the liberation of some of the water that was originally bound within the crystalline lattice of the calcium aluminate hydrates formed (CAH_{10} and C_2AH_8). Thus the amount of free water, and along with it the volume of the pore system, increases, whereas the amount and volume of the solid constituents becomes smaller. This in turn results in a decline of strength. An additional but less important factor affecting the strength adversely is the fact that the intrinsic bond properties of C_3AH_6 and AH_3 are less favorable than those of CAH_{10} or C_2AH_8. This implies that even at equal porosity a hardened paste containing C_3AH_6 and AH_3 has a lower strength than one consisting of hexagonal calcium aluminate hydrates (Scharf and Odler, 1992).

10.8   PROPERTIES OF CALCIUM ALUMINATE CEMENT

The **rheology** of calcium aluminate cement pastes is broadly similar to that of Portland cement pastes (Banfill and Gill, 1986). They show a Binghamtype behavior with slightly lower yield values and a plastic viscosity. If agitated, they exhibit a structural breakdown that depends on the intensity and duration of mixing. Their flow curves are characterized by a distinct hysteresis loop. Usually, concrete mixes made with calcium aluminate cement are somewhat more free flowing, which makes it possible to use a lower water/cement ratio.

At ambient temperatures the **setting** of calcium aluminate cements starts at about 2–6 hours, and the final set typically occurs within 12 hours. In general, the setting is shortened with increasing $C_{12}A_7$ content in the cement (Menetrier-Sorrentino *et al.*, 1986). Just as with pure CA, the setting time of calcium aluminate cement is temperature dependent, and passes a maximum at about 30 °C. At even higher temperatures it becomes drastically shortened. Unlike the initial induction period, the subsequent hydration is affected positively, rather than negatively, by increasing temperature in the whole temperature range.

While calcium aluminate cement is rather slow-setting, it hardens rapidly, and the attained **strength** is typically higher than that of Portland cement at equal water/cement ratios. At 20 °C most of the final strength is attained within 1–2 days. The hydration may even be accelerated and the strength development increased by mechanical activation (friction milling) of the cement (Scian *et al.*, 1994). Typical strength values of calcium aluminate cements of different composition are shown in Table 10.2. The early strengths of cements high in Al_2O_3 and containing CA_2, CA_6, and a-A are generally lower than those of cements containing $C_{12}A_7$ as well as CA. Cements that contain distinct amounts of $C_{12}A_7$ exhibit a relatively high short-term strength, whereas the long-term strength increases with increasing CA content (Menetrier-Sorrentino *et al.*, 1986).

After reaching a maximum value the strength of hardened calcium aluminate cement starts to decline, owing to the conversion of the primary formed CAH_{10} and C_2AH_8 to $C_3AH_6 + AH_3$. The decline of strength ends with the completion of the conversion reaction. Like the strength of the non-converted material, that of the converted paste is determined primarily by porosity, and thus declines with increasing water/cement ratio. However, the strength after conversion may also be affected by factors that alter the morphology and thus the intrinsic bond properties of the C_3AH_6 and AH_3 formed. In particular, the size of the C_3AH_6 crystals increases with the temperature at which they are formed, which in turn adversely affects the bonding capacity of this phase.

Whereas at high water/cement ratios complete hydration of the cement has occurred by the time the maximum strength is attained, at low water/cement ratios (below approximately 0.70) the amount of water is not sufficient for complete hydration. In the subsequent conversion of the hydrated cement, free water – originally bound within the crystalline lattice of the hydrate phases – is liberated, and the hydration can continue. Under these condition a renewed increase of strength may be observed, or at least a further decline of strength may be avoided, as the newly formed calcium aluminate hydrates fill the pore space created by conversion. In addition to calcium aluminate phases, a hydration of the existing non-hydrated β-C_2S (Midgley, 1980) and carbonation (Revai, 1986) may also contribute to the renewed strength increase.

The **modulus of elasticity** of HAC concrete is somewhat higher than that of hydrated Portland cement concrete of the same strength, whereas the **creep** is of the same magnitude.

Owing to a higher water requirement for hydration, the **porosity** of hardened aluminous cement pastes is lower than that of Portland cement at the same water/cement ratio. This has a favorable effect on **permeability** and along with it on the corrosion resistance and freeze–thaw resistance of the hardened material. However, the conversion to aluminous cement is associated with a distinct increase of both porosity and permeability, which must be considered when this cement is used in practice.

To alter the rate of hydration and the properties of the concrete mix made with calcium aluminate cement, **chemical admixtures** may be employed. However, many of the substances effective as chemical admixtures with Portland cement may be ineffective with calcium aluminate cement, or may affect the hydration of this cement differently.

Lithium salts, such as Li_2CO_3 or lithium citrate, exhibit a strong accelerating effect on the hydration of calcium aluminate cement: this effect is dose dependent, and increases with increasing Li addition (Currell et al., 1987; Baker and Banfill, 1992; Matusinovic and Curlin, 1993). In the presence of Li^+ the induction period is significantly shortened, whereas the process of hardening is scarcely affected. It is assumed that lithium salts act by converting to lithium aluminate hydrate ($Li_2O.2Al_2O_3.11H_2O$) in the cement mix, whose crystals act as nuclei for a heterogeneous nucleation of the hexagonal calcium aluminate hydrates.

A significant accelerating effect on aluminous cement hydration is also shown by OH^- ions, which may be present in the form of alkali hydroxides. Thus the rate of hydration is accelerated at high pH. The effect of OH^- is due to the replacement of H_2O by OH^- in the Al environment, leading to a further center for oxobridge formation (Currell et al., 1987).

Hydroxylic organic compounds, such as mono- or oligosaccharides, citric acid, or gluconic acid, act as retarders, and are effective in very small additions (Bayoux et al., 1992b). They apparently act by poisoning the nuclei of the hydration products. This retarding effect is also associated with a liquefying action. A retardation of hydration may also be obtained by lowering the pH of the cement paste.

Alkaline-earth metal chlorides have been found to either accelerate or retard the hydration of high-alumina cement, depending on the amount added and the temperature (Nilforoushan and Sharp, 1995).

Superplasticizers used in combination with Portland cement are distinctly less effective with high-alumina cement. A product based on sulfonated phenol formaldehyde was found to be more effective than naph-thalene-type and melamine-type resins (Banfill and Gill, 1993). The retarding effect associated with the use of superplasticizers may be overcome by simultaneously adding a lithium salt to the mix (Gill *et al.*, 1986).

Common air-entraining agents may also be employed in combination with high-alumina cement to improve frost resistance of the concrete mix.

The **resistance of calcium aluminate cement to chemical attack** differs from that of cements based on calcium silicates. It is determined by the chemical nature of the binder and the porosity and pore structure of the paste, which in turn determine its permeability.

At low or medium water/cement ratios the porosity and permeability of hydrated non-converted aluminous cement pastes are sufficiently low to confine the corrosive action of any external chemical agents to the surface region of the concrete structure. However, as the porosity increases in the course of conversion, the susceptibility to chemical attack of concrete based on aluminous cement increases. An effective way to prevent this from happening is to use initial water/cement ratios that are too low for complete hydration. Under these conditions the water liberated in the conversion of the hexagonal calcium aluminate hydrate phases, formed initially, reacts with the non-hydrated fraction of the cement, thus preserving a low porosity of the hardened paste. Note that the permeability is the main factor determining the resistance of aluminous cement concrete to chemical agents, and this has to be kept in mind when calcium aluminate cement is used in practice.

Unlike Portland cement, hydrated aluminous cement pastes are resistant to diluted acid solutions with pH > 4. This also includes natural waters containing dissolved CO_2. One factor responsible for the good acid resistance of this cement is the absence of free calcium hydroxide among the formed hydrates, a phase that is easily soluble even in dilute acids. At the same time hydrous alumina is not dissolved at pH values down to 4. At very high pH values, such as those existing in alkali hydroxide solutions, corrosion of hardened aluminous cement may occur. Under these conditions AH_3 is dissolved in the liquid phase in the form of alkali aluminates.

Long-term experience indicates that concrete based on aluminous cement exhibits very good resistance to groundwater containing sulfates. Usually only moderate corrosion, limited to a surface region a few millimeters thick, is observed. This is surprising, as C_3AH_6 may react with calcium sulfate to yield ettringite, the phase whose formation is responsible for damage in concrete based on Portland cement. One possible explanation is the fact that in aluminous cement pastes ettringite is formed in a through-solution reaction rather than a topochemical reaction, and precipitates in the existing pores without causing internal stresses. At the same time the filling of the pore space with ettringite lowers the permeability, and prevents the sulfate ions from migrating into deeper regions.

In seawater, calcium aluminate cement is more durable than Portland cement. Conversion also takes place here, but it is usually very slow, except in the tidal zone or in warm waters (Baker and Banfill, 1992). If seawater is employed as mixing water the initial hydration may be retarded, but the final microstructure of the hardened material is very similar to that made with fresh water. Chloroaluminates are formed in the hydration reaction (Halse and Pratt, 1986).

Atmospheric CO_2 reacts with hardened calcium aluminate cement, yielding the calcium carboaluminate hydrates $C_4A\bar{C}_{0.5}H_x$ and $C_4A\bar{C}H_x$, and ultimately hydrous alumina and calcium carbonate. The rate of this process, and ultimately the thickness of the reacted layer, will increase with increasing porosity of the hardened cement paste and increasing filling of the pores with water, which in turn depends on the existing humidity. The rate of reaction is also accelerated in the presence of free alkali hydroxides. This may be the situation if aluminous cement concrete is in contact with Portland cement concrete, which always contains free alkali hydroxides that may migrate into the aluminous material.

The calcium carbonate and hydrous alumina formed in the reaction with CO_2 fill the existing pores and cause a reduction of permeability in the carbonated region. If concrete with a sufficiently low water/cement ratio is produced, the carbonation remains limited to a narrow region close to the surface, even after conversion. Extensive carbonation may occur, however, in concretes made with excessive water/cement ratios, especially after conversion.

Owing to the pore-filling effect associated with carbonation, this reaction increases the strength of the material. In practice, however, this effect cannot be exploited in concrete mixes with appropriate water/cement ratios, because of the limited thickness of the carbonated region that is formed. Significantly greater carbonation may be achieved, however, if the concrete is cured in a carbon dioxide atmosphere rather than in air. A negative effect of superficial carbonation may be a "dusting" of the concrete surface.

In steel-reinforced concrete structures made with calcium aluminate cement and with a sufficiently low water/cement ratio, the reinforcement is sufficiently protected from corrosion. However, in mixes made with too much water, corrosion of the steel may take place, especially after conversion of the hardened paste has occurred, as the cement paste becomes too porous and too permeable for oxygen of the air. Carbonation of the paste, which progresses especially easily in porous mixes, enhances the corrosion process even further, as the pH of the pore solution drops from its original 10–12 to lower values, making the steel susceptible to corrosion.

In conclusion, the resistance of aluminous cement based concrete to most chemical corrosive agents is outstanding, provided that the mix is produced with a sufficient cement content and with a low water/cement ratio, and is properly compacted and cured.

Like other hydrated cementitious systems, hardened calcium aluminate cement pastes decompose and undergo chemical reactions upon heating:

- At moderately elevated temperatures and in the presence of liquid water the hexagonal calcium aluminate hydrate phases CAH_{10} and C_2AH_8, if still present, convert to C_3AH_6.
- As the temperature approaches and even exceeds 100 °C, non-combined water escapes, together with water loosely bound within the calcium aluminate hydrates and alumina gel.
- At around 300 °C AH_3, which has crystallized as the temperature increased, decomposes, yielding a highly reactive form of anhydrous alumina. At roughly the same temperature the thermal decomposition of C_3AH_6 also gets under way. $C_{12}A_7$ is the first anhydrous calcium aluminate phase detectable as a thermal decomposition product; however, CA_2 and CA are also formed at around 600 °C and 1000 °C. The reactions that occur up to about 800 °C are associated with an increase of porosity and a decline of strength.
- At about 800 °C the anhydrous phases formed in the thermal degradation of the hardened cement paste start to react in solid-state reactions with each other and/or the material employed as aggregate. As a result of these processes the "hydraulic bonding" responsible for strength in the native paste, is gradually replaced by a "ceramic bonding," and the strength of the material starts to increase again, after a strength minimum at about 800 °C.
- At temperatures well above 1000 °C a melt starts to be formed, and this process is associated with a distinct shrinkage and loss of strength. The exact temperature of melt formation will depend on the composition of the cement employed. It will increase with increasing Al_2O_3 content, whereas the presence of iron oxide has a negative effect on the temperature of melt formation.

10.9   STRONTIUM- AND BARIUM-BASED ALUMINOUS CEMENTS

10.9.1   Strontium aluminate cement

Strontium aluminate cement may be produced by solid-state sintering of an approximately equimolar blend of SrO or $SrCO_3$ and Al_2O_3 at about 1500 °C (Borovkova, 1992). The main constituent of the cement is monostrontium aluminate ($SrO.Al_2O_3$), but some hexaaluminate ($SrO.6Al_2O_3$) may also be present. The cement hydrates relatively slowly, yielding a strontium aluminate hydrate of the approximate composition $SrO.Al_2O_3.7–10(H_2O)$ as the product of hydration. In the later course of hydration this phase gradually converts to $6SrO.Al_2O_3.6H_2O$ + hydrous alumina, and this reaction is associated with a loss of strength (Borovkova, 1992). The cement has the potential to be used in the production of refractory concrete for applications at temperatures up to about 2000 °C.

10.9.2   Barium aluminate cement

Barium aluminate cement may be produced by solid-state sintering of an approximately equimolar blend of $BaCO_3$ and Al_2O_3 at 1400–1500 °C (Borovkova *et al.*, 1992; Raina *et al.*, 1992). The sintering temperature may be reduced by adding a suitable mineralizer such as MgO to the starting blend (Raina *et al.*, 1992). The main constituent of the cement is mono-barium aluminate ($BaO.Al_2O_3$), but small amounts of tribarium aluminate ($3BaO.Al_2O_3$) and barium hexaaluminate ($BaO.6Al_2O_3$) may also be present (Borovkova *et al.*, 1992).

At low water/cement ratios, below $w/c = 0.35$, the main crystalline phase formed in the hydration of barium aluminate cement is monobarium aluminate heptahydrate ($BaO.Al_2O_3.7H_2O$), whereas at higher water contents tribarium aluminate hexahydrate ($3BaO.Al_2O_3.6H_2O$), together with hydrous alumina, is also formed.

Barium aluminate cement hydrates quite rapidly, and at sufficiently high water/cement ratios it may hydrate completely within three days. The hardened cement paste exhibits only a small strength loss if heated to high temperatures.

Barium aluminate cement exhibits significantly improved refractory properties compared with calcium aluminate cement, and may be considered for applications at temperatures of up to about 2000 °C. It has also the capacity to absorb radioactive and X-ray radiation very effectively.

10.10   ALUMINOUS CEMENT BASED COMPOSITE BINDERS

Inorganic cementitious systems with special properties may be obtained by combining calcium aluminate cement with other materials. In such blends the aluminous cement may either constitute a main constituent or may be present just in limited amounts, to modify the properties of the binder. Such systems may be produced either by premixing the dry constituents or by adding them separately to the wet mix.

10.10.1   Calcium aluminate cement + Portland cement

Blends of calcium aluminate and Portland cement exhibit different setting behavior and strength development than each of these two binders separately. Even minor amounts of Portland cement added to calcium aluminate cement, just like minor amounts of calcium aluminate cement added to Portland cement, shorten the setting time of the resultant cement blend very significantly. In the range between about 15% and 85% of Portland cement (85–15% calcium aluminate cement), and at water/cement ratios yielding a paste of plastic consistency, the setting time is shortened to just a few minutes (at ambient temperature). Immediately after setting a rapid strength development gets under way, producing measurable strength values in less than an hour. The final strengths, however, are lower than those attained with Portland or calcium aluminate cements alone. Alternatively prehydrated rather than non-hydrated calcium aluminate cement may also be employed, which results in improved workability of the binder (Gu *et al.*, 1994).

In mixes in which calcium aluminate cement is the main constituent, the reaction mainly responsible for fast setting and strength development is a very fast hydration of the calcium aluminate phases. The accelerated hydration responsible for fast setting and strength development is due to the increase of the pH value of the mix, brought about by the Portland cement addition.

In mixes with Portland cement as the main constituent, the setting is due to a rapid formation of ettringite and the hydration of calcium aluminate cement. The hydration of the calcium silicates has little influence on the setting process, but contributes to the subsequent strength development (Gu *et al.*, 1994). In addition to the phases formed in the hydration of pure Portland and calcium aluminate cements, strätlingite (gehlenite hydrate, C_2SAH_8) may also be formed in the hydration process.

Because of the extremely fast setting, and the associated difficulties with mixing, placing and compacting, blending of calcium aluminate cement with Portland cement must be avoided in normal concreting practice. However, blends of Portland cement with additions of calcium aluminate cement may be employed in applications in which extremely fast setting and hardening are required. Further acceleration of the setting and hardening process may be achieved if small amounts of a lithium salt are added to the system.

10.10.2 Calcium aluminate cement + gypsum

If calcium aluminate cement is combined with gypsum in amounts corresponding to 15–40 wt% SO_3 the resulting binder exhibits expanding properties, and may be used as self-stressing cement. Ettringite and hydrous alumina are formed as products of hydration (Xue *et al.*, 1986; Bayoux *et al.*, 1992a):

$$3CA + 3C\bar{S}H_2 + 32H \rightarrow C_3A.3C\bar{S}.32H + 2AH_3 \qquad (10.8)$$

At lower SO_3 contents in the blend, monosulfate may also be formed, or the primary formed ettringite may convert to this phase:

$$3CA + \bar{S}H_2 + 17H \rightarrow C_3A.C\bar{S}.12H + 2AH_3 \tag{10.9}$$

$$6CA + C_3A.3C\bar{S}.32H + 16H \rightarrow 3C_3A.C\bar{S}.12H + 4AH_3 \tag{10.10}$$

The resulting expansion increases with increasing gypsum content and – up to about 40 °C – also with increasing temperature. The self-stress value of the binder increases with increasing CA/CA_2 ratio.

10.10.3 Calcium aluminate cement + Portland cement + gypsum

If limited amounts of calcium aluminate cement and additional amounts of gypsum are combined with Portland cement, the amount of ettringite in the hydration is increased above the amount formed in the hydration of Portland cement alone. This, in turn, is associated with an expansion of the paste. The same effect is achieved if Portland clinker alone (rather than Portland cement), calcium aluminate clinker, and amounts of gypsum higher than common in Portland clinker are ground together. The setting time decreases and the expansion of the binder increases with increasing amounts of calcium sulfate and CAC cement. A binder of this type is commonly called a type M expansive cement (see section 21.3).

10.10.4 Calcium aluminate cement + granulated blast furnace slag

It has been observed that 1:1 blends of calcium aluminate cement with ground granulated blast furnace slag exhibit a steady strength increase without the decline of strength that is common in pure calcium aluminate cement, if hydrated at ambient or moderately elevated temperatures (Majumdar *et al.*, 1990a, 1990b; Quillin, 1993; Quillin and Majumdar, 1994; Osborn and Singh, 1995). The short-term strength of such mixes is lower than that of mixes made with pure calcium aluminate cement, but the long-term strength is higher than that of pure calcium aluminate cement mixes after conversion. In the hydration of such blends the CAH_{10} and C_2AH_8 phases formed at ambient temperature, or the C_3AH_6 phase formed at elevated temperature, are gradually replaced by strätlingite (gehlenite hydrate, C_2AH_8), which is the stable phase in the C-A-S-H system at this temperature (Fentiman *et al.*, 1990; Majumdar *et al.*, 1990b; Richardson and Groves, 1990; Rayment and Majumdar, 1994; Romero *et al.*, 1997). This phase is formed in a reaction between the calcium aluminate hydrate phases and silicate ions originating from the slag. The resulting strätlingite may accommodate distinct amounts of foreign ions, and in particular Fe^{3+}, Mg^{2+}, SO_4^{-2}, and Ti^{4+} (Rayment and Majumdar, 1994). Simultaneously an amorphous phase is formed at the surface of the slag particles: this is probably is a combination of a C-S-H gel and aluminum-rich gels, and even gels of the hydrate phases that have not crystallized yet (Romero *et al.*, 1997). In blends that contain slag with a high MgO content, hydrotalcite-type phases may also be formed in the hydration (Rayment and Majumdar, 1994). The following hydrate phases were found to be formed in blends with different ratios of calcium aluminate cement and slag at different temperatures (90 days) (Romero *et al.*, 1997):

- CAC 80% + slag 20%
    - 5 °C: CAH_{10}
    - 40 °C: $C_3AH_6 + C_2ASH_8$ (little) + $Al(OH)_3$ (little)
- CAC 50% + slag 50%
    - 5 °C: $C_2ASH_8 + CAH_{10}$ (little)
    - 40 °C: $C_2ASH_8 + C_3AH_6$
- CAC 20% + slag 80%
    - 5 °C: C_2ASH_8
    - 40 °C: C_2ASH_8

In the presence of adequate amounts of granulated blast furnace slag the conversion of primary formed CAH_{10} and C_2AH_8 to C_3AH_6, the phase responsible for the loss of strength in calcium aluminate cement, is prevented or reversed to a high degree. A binder of this type appears to be suitable for applications in which long-lasting service of the erected structure is required (Majumdar *et al.*, 1990a, 1990b; Quillin, 1993; Quillin and Majumdar, 1994). The binder also exhibits excellent resistance to sulfates, seawater, and soft water (Osborn and Singh, 1995).

10.10.5   Calcium aluminate cement + blast furnace slag + sodium carbonate

Blends of calcium aluminate cement and granulated blast furnace slag hydrate in a different way if mixed with a solution of sodium carbonate (2M) rather than with plain water. The formation of strätlingite is reduced or prevented, and calcium carboaluminate ($C_4A\bar{C}H_{11}$), as well as CAH_{10}, is formed instead. At high slag contents calcium carbonate (Ca_2CO_3) may also be formed (Romero *et al.*, 1997). Compared with mixes made with plain water, mixes made with the Na_2CO_3 solution exhibit higher strengths at lower and medium slag contents and lower strengths when the slag content is high (Romero *et al.*, 1997).

10.10.6   Calcium aluminate cement + microsilica

Microsilica, an industrial by-product consisting of a highly dispersed amorphous form of SiO_2, does not undergo a pozzolanic reaction if combined with calcium aluminate cement, unlike Portland cement. This is because free calcium hydroxide is not produced in the hydration of microsilica. Moreover, calcium aluminate cement pastes do not contain significant amounts of alkali hydroxides, and the pH of their pore solution is distinctly lower than that of Portland cement pastes. It has been observed, however, that microsilica in amounts of 10–15%, if combined with calcium aluminate cement, prevents or slows down the conversion reaction, and strätlingite is formed instead. In this way the decline of strength observed when primary formed hexagonal calcium aluminate hydrates convert to C_3AH_6 can be prevented (Bentsen *et al.*, 1990; Macdargent, 1992; Collepardi *et al.*, 1995; Ding *et al.*, 1995; Bensted, 1996). The addition of microsilica retards the hydration of monocalcium aluminate, and as a consequence the short-term strength development of this binder is reduced.

10.10.7   Calcium aluminate cement + C_2S or C-S-H gel

In blends of calcium aluminate cement and non-hydrated or prehydrated dicalcium silicate the primary formed CAH_{10} and C_2AH_8 phases tend to convert to strätlingite (C_2ASH_8), rather than to C_3AH_6. This appears to be another way by which the loss of strength associated with the conversion to the latter hydrate phase may be prevented (Rao, 1980).

10.10.8   Calcium aluminate cement + calcium carbonate

The addition of finely ground calcium carbonate (calcite) to calcium aluminate cement results in the formation of calcium aluminocarbonate hydrate ($C_4A\bar{C}H_{11}$) as the product of hydration, instead of CAH_8 and C_3AH_6. At sufficiently high additions of the carbonate conversion of the cement may be avoided, and hence the strength loss associated with it may be prevented (Trivino, 1986). The long-term effectiveness of this measure is questionable, however.

To prevent the strength loss associated with conversion it has also been suggested that the freshly produced concrete should be treated with gaseous carbon dioxide (for example with flue gas; Trivino, 1986).

10.10.9   Calcium aluminate phosphate cement

Calcium aluminate phosphate cementitious systems are obtained by combining calcium aluminate cement with alkali or ammonium phosphates or polyphosphates (Sugama and Carciello, 1991; Ma and Brown, 1992, 1994; Walter and Odler, 1996). They are discussed in section 12.3.

10.10.10   Calcium aluminate cement + sodium silicate

According to Ding and Beaudoin (1996), the addition of small amounts of sodium silicate promotes the formation of strätlingite (C_2SAH_8) in calcium aluminate cement pastes. At the same time the hydration of the cement is strongly retarded. The additive delays, but does not prevent completely, the formation of the hydrogarnet phase in the cement paste.

10.10.11   Calcium aluminate cement combined with calcium sulfoaluminate and microsilica

Mino *et al.* (1991) developed a binder consisting of calcium aluminate cement (68 wt%), a mixture of "amorphous calcium aluminate" (γ) and anhydrite, which they called "calcium sulfoaluminate" (12 wt%), and microsilica (20 wt%), in combination with a superplasticizer based on naphthalene sulfonate. If mixed with water, steel powder and steel fibers in the ratio binder : steel powder : steel fibers : water = 100:150:20:21 and cured for 7 days in water at 50 °C, the hardened material attained a compressive strength of 180 MPa and flexural strength of 33 MPa. The hydration products formed were ettringite, C_3AH_6, and AH_3. The material also contained significant amounts of non-reacted CA and anhydrite. If heated to temperatures between 110 °C and 600 °C the hydrates decomposed, and $C_{12}A_7$ was formed instead. At the same time compressive strengths of up to 280 MPa and flexural strengths of up to 55 MPa were achieved.

10.11   UTILIZATION OF CALCIUM ALUMINATE CEMENT

10.11.1   Civil engineering applications

The fundamental problem associated with the use of calcium aluminate cement instead of Portland cement in civil engineering applications is its long-term durability. The loss of strength associated with the conversion of the primary formed hydrates may be critical, and in fact collapses of structures erected with calcium aluminate cement have been reported from various countries. The problem is compounded by the fact that the porosity of the hardened cement paste is also increased with the conversion, which may cause simultaneous corrosion of the steel reinforcement. The phenomenon of strength loss and steel corrosion may be especially critical in places with high temperature and humidity, such as in indoor swimming pools, as the rate of conversion of the binder is enhanced under these conditions. At present the use of calcium aluminate cement in structural applications is banned in most countries. The appropriateness of this measure may be questioned, however, as good long-term concrete performance may be attained with this binder provided appropriate measures are taken.

The strength of concrete made with high-alumina cement increases rapidly immediately after setting, and about 90% of the maximum strength is achieved within about two days. The time at which maximum strength is reached depends on the particular cement used and the temperature and humidity conditions during service. At ambient temperature and normal conditions this point is reached within a few years. Subsequently the strength starts to decline, owing to the conversion of the primary formed hydrates, until the conversion process is completed. A renewed but usually only moderate strength increase may follow if some non-reacted cement is still present at this point, until even this fraction of cement becomes completely hydrated.

If calcium aluminate cement with a distinct iron content is employed, the conversion is usually accompanied by a change of color of the hardened concrete from blackish-gray to yellow-brown. This change is probably due to an oxidation of the existing ferrous compounds by the oxygen of air, which is facilitated by the high porosity of the converted cement paste. Such a change of color, however, may occur independently of conversion in concrete mixes in which the porosity is excessive owing to bad mixing proportions.

The strength loss due to the conversion must be taken into consideration if calcium aluminate cement is used as a binder in civil engineering applications. The magnitude of the strength loss, in both absolute and relative terms, may vary with the particular cement used, and is greatly influenced by the water/cement ratio employed (George and Montgomery, 1992). The following may be considered to be typical values:

Water/cement ratio	Converted strength (in percentage of initial unconverted strength)
0.30	55–70
0.35	45–60
0.40	35–50
0.50	25–40
0.60	15–30
0.70	5–20

The following measures must be taken if calcium aluminate cement is used in concrete practice:

1. The aggregate used to produce the concrete mix must not contain freeable alkalis, to avoid accelerated carbonation of calcium aluminate hydrates formed in the hydration process by CO_2 in the air. The carbonation is accelerated if alkalis are present in the mix. The following chemical reactions will take place in the presence of alkalis:

$$2KOH + CO_2 \rightarrow K_2CO_3 + H_2O \tag{10.11}$$

$$CaO.Al_2O_3.aq + K_2CO_3 \rightarrow CaCO_3 + K_2O.Al_2O_3.2H_2O \tag{10.12}$$

$$K_2O.Al_2O_3.2H_2O + CO_2 \rightarrow K_2CO_3 + Al_2O_3.aq \qquad (10.13)$$

$$CaO.Al_2O_3.aq + CO_2 \rightarrow CaCO_3 + Al_2O_3.aq \qquad (10.14)$$

As a result of this process the calcium aluminate hydrates will convert to calcium carbonate and hydrous alumina. The process is associated with a lowering of the pH value of the pore solution, which in turn may result in corrosion of the steel reinforcement. Rocks that may include freeable alkalis include granite or mica. As well as alkalis derived from the aggregate, alkalis may also migrate into the paste from an outside source, for example if placed in contact with Portland cement concrete.

An aggregate particularly suitable for use in combination with calcium aluminate cement is limestone. The calcium carboaluminate hydrate $(C_4A\bar{C}H_{11})$ formed in the reaction of the $CaCO_3$ of the limestone with the calcium aluminates of the cement contributes to an improved bond between the hardened cement paste and the aggregate surface.

A concrete surface with superior abrasion resistance may be produced if corundum is used as aggregate.

2. In proportioning the concrete mix the water/cement ratio must be kept as low as possible, the maximum acceptable value being $w/c = 0.40$. The cement content should not be less than 400 kg/m^2. In this way the strength loss due to conversion may be minimized and a sufficiently low porosity of the cement paste may be secured, even after conversion.

3. It has to be kept in mind that intensive hydration heat liberation will start within hours after mixing. This may be especially critical in situations where dissipation of the heat is hampered by the large mass of the concrete. In thin, non-loadbearing members, cooling by spraying with cold water for at least 2–3 days may be sufficient to control this problem, whereas in the case of mass concrete a cooling system must be used. On the other hand, the danger of damaging the fresh concrete mix by frost, even at very low ambient temperatures, is very low if calcium aluminate cement is used as binder.

Legal restrictions, high price, and concerns about long-term performance limit the widespread use of calcium aluminate cement for construction purposes. Its use may be attractive in applications where rapid strength development is required, as in the production of prefabricated concrete elements. In applications in which only a limited service time is planned, calcium aluminate cement may be used without concern.

10.11.2   High-temperature applications

The fact that hardened calcium aluminate cement pastes preserve much of their strength up to high temperature makes this cement suitable for use in high-temperature applications. In principle, calcium aluminate cement systems may be used up to the temperature at which a liquid phase starts to be formed in the material, and that depends on the composition of the cement. In general, the temperature of melt formation increases with increasing alumina content of the cement, whereas its iron content affects this temperature unfavorably. Thus a white, high-Al_2O_3, low-(Fe_2O_3 + FeO) cement should be used for applications in which the material will be exposed to particularly high temperatures. Even higher service temperatures may be achieved by the use of barium rather than calcium aluminate cement, but cements of this type are not commercially available at present.

In addition to the binder, the maximum allowable service temperature of refractory concrete is determined by the aggregate employed. Thus an aggregate capable of sustaining the maximum anticipated temperature must be selected.

10.11.3   Calcium aluminate cement based rapid-setting/hardening binder

By combining calcium aluminate cement with 5–20% of Portland cement or Portland cement with 5–20% of calcium aluminate cement a binder may be obtained that exhibits a very short setting time and rapid strength development. Of these two alternatives the latter is more widely used, not least for economic reasons.

The rate of setting and hardening may be controlled by the mutual ratio of both cements; it increases with increasing proportion of the minor constituent, until a plateau is reached. Further acceleration may be obtained if a small amount of a lithium salt, such as Li_2CO_3, is added to the system.

Cement combinations of this type are employed in applications where a short setting time and rapid hardening of the concrete mix are required. In applying a binder of this type the very short time available for mixing, placing, and compacting must be taken into consideration. Examples of such applications include highway or runway repair works (e.g. filling of potholes), plugging of water leaks, and the production of prefabricated elements without the necessity for steam curing.

10.11.4   Glass fiber reinforced systems made with calcium aluminate cements

Calcium aluminate cement has been suggested as a binder for glass fiber reinforced composites. (Majumdar *et al.*, 1981). This cement appears to be more suitable for this purpose than Portland cement, as the pH of its pore solution is distinctly lower. Also, it does not liberate calcium hydroxide in its hydration, which is responsible for the embrittlement seen in such materials after prolonged times of curing.

10.11.5 Calcium aluminate cement as a constituent of expansive cements

Calcium aluminate cement, in combination with Portland clinker and gypsum, is a constituent of type M expansive cement.

10.11.6 Calcium aluminate cement as a constituent of MDF materials

Calcium aluminate cement may be used in combination with a suitable polymer as a constituent of so-called "MDF cements." These are discussed in section 13.3.

REFERENCES

Baker, N.C., and Banfill, P.F.G. (1992) The use of admixtures in high alumina cement mortar for the marine environment, in *Proceedings 9th ICCC, New Delhi*, Vol. 4, pp. 719–725.

Banfill, P.F.G., and Gill, S.M. (1986) The rheology of aluminous cement pastes, in *Proceedings 8th ICCC, Rio de Janeiro*, Vol. 6, pp. 223–227.

Banfill, P.F.G., and Gill, S.M. (1993) Superplasticizers for ciment fondu: effect on rheological properties of fresh paste and mortar. *Advances in Cement Research* **5**, 131–138.

Bayoux, J.P., Testud, M., and Esponosa, B. (1992a) Thermodynamic approach to understand the $CaO-Al_2O_3-SO_3$ system, in *Proceedings 9th ICCC, New Delhi*, Vol. 4. pp. 164–1693.

Bayoux, J.P. *et al.* (1992b) A study on calcium aluminate cement admixtures, in *Proceedings 9th ICCC, New Delhi*, Vol. 4, pp. 705–711.

Bensted, J. (1993) High alumina cement – present state of knowledge. *Zement-Kalk-Gips* **46**, 560–566.

Bensted, J. (1996) A discussion of the paper "The influence of pozzolanic materials on the mechanical stability of aluminous cement" by M. Collepardi, S. Morosi and P. Picciolli. Cement and Concrete Research **26**, 649–650.

Bentsen, S., Selvod, S., and Sandberg, B. (1996) Effect of microsilica on conversion of high alumina cement, in *Calcium Aluminate Cements* (ed. R.J. Manghabai), E. & F.N. Spon, London, pp. 294–298.

Borovkova, L.B. (1992) An investigation into the hardening of monostrontium aluminate, in *Proceedings 9th ICCC, New Delhi*, Vol. 4, pp. 51–56.

Borovkova, L.B. *et al.* (1992) Hardening of alkaline earth metal aluminates I. Barium monoaluminate (in Russian). *Tsement (Moscow)*, pp. 18–24.

Collepardi, M., Morosi, S., and Piccioli, P. (1995) The influence of pozzolanic materials on the mechanic stability of aluminous cement. *Cement and Concrete Research* **25**, 961–968.

Cong, S., and Kirkpatrick, R.J. (1993) Hydration of calcium aluminate cements: a solid state Al^{27} NMR study. *Journal of the American Ceramic Society* **76**, 409–416.

Currell, B.R. *et al.* (1987) The acceleration and retardation of set of high alumina cement by additives. *Cement and Concrete Research* **17**, 420–432.

Damidot, D. *et al.* (1996, 1997) Action of admixtures on fondue cement. *Advances in Cement Research* **8**, 111–119; **9**, 127–133.

Das, S., and Daspodda, P.K. (1993) Some studies on the hydraulic activity of calcium aluminates. *NML Technical Journal* **35**, 29–38 [ref. CA 124/25756].

Ding, J., and Beaudoin, J.J. (1996) Study of hydration mechanism in the high alumina cement – sodium silicate system. *Cement and Concrete Research* **26**, 799–804.

Ding, J., Fu, Y., and Beaudoin, J.J. (1995) Strätlingite formation in high alumina cement-silica fume systems: significance of sodium ions. *Cement and Concrete Research* **25**, 1311–1319.

Dunster, A.M., and Crammond, N.J. (1997) Alkaline hydrolysis with carbonation in high alumina cement concrete beams, in *Proceedings 10th ICCC, Göteborg*, paper 4iv026.

Edmonds, R.N., and Majumdar, A.J. (1988) The hydration of monocalcium aluminate at different temperatures. *Cement and Concrete Research* **18**, 311–320.

Edmonds, R.N., and Majumdar, A.J. (1989) The hydration of mixtures of monocalcium aluminate and blast furnace slag. *Cement and Concrete Research* **19**, 779–782.

Fentiman, C.H. *et al.* (1990) The effect of curing conditions on the hydration and strength development in Fondu:slag, in *Proceedings International Symposium on Calcium Aluminate Cements, London*, p. 272.

Gaztanaga T. *et al.* (1997) Carbonation of hydrated calcium aluminate cement and/or ordinary Portland cement of varying alkali content, in *Proceedings 10th ICCC Göteborg*, paper 4iv003.

George, C.M. (1983) Industrial aluminous cements, in *Structure and Performance of Cements* (ed. P. Barnes), Applied Science Publishers, London, pp. 415–470.

George, C.M., and Montgomery, R.G.J. (1992) Calcium aluminate cement concrete: durability and conversion. A fresh look at an old subject. *Materiales de Construccion (Madrid)* **42** (228), 51–64.

Gill, S.M., Banfill, P.F.G., and El-Jazairi, B. (1986) The effect of superplasticizers on the hydration of aluminous cement, in *Proceedings 8th ICCC, Rio de Janeiro*, Vol. 4, pp. 322–327.

Goni, S. *et al.* (1996) A new insight on alkaline hydrolysis of calcium aluminate cement concrete I. Fundamentals. *J. Materials Research* **11**, 1748–1754.

Gu, P. *et al.* (1994) A study of the hydration of setting behaviour of OPC-HAC pastes. *Cement and Concrete Research* **24**, 682–694.

Guirado, F., Gali, S., and Chinchon, J.S. (1998) Thermal decomposition of hydrated alumina cement (CAH_{10}). *Cement and Concrete Research* **28**, 381–390.

Gulgun, M.A. *et al.* (1992) Preparation and hydration kinetics of pure $CaAl_2O_4$. Materials Research Society Symposium Proceedings **245**, 199–204.

Halse, Y., and Pratt, P.L. (1986) The development of microstructure in high alumina cement, in *Proceedings 8th ICCC, Rio de Janeiro*, Vol. 4, pp. 317–321.

Holterhoff, A.G. (1993) Calcium aluminate cements. *American Ceramic Society Bulletin* **72** (6), 89–90.

Kudriavtsev, A.B., and Kouznetsova, T.V. (1997) On the possibilities of in situ studies of the hydration of aluminate cements using ^{27}Al NMR and proton relaxation rate measurements, in *Proceedings 10th ICCC Göteborg*, paper 2ii028.

Kurdowski, W., George, C.M., and Sorrentino, F.P. (1986) Special cements, in *Proceedings 8th ICCC, Rio de Janeiro*, Vol. 1, pp. 292–318.

Ma, W., and Brown, P. (1992) Mechanical behaviour and microstructure development in phosphate modified high alumina cement. *Cement and Concrete Research* **22**, 1092–1200.

Ma, W., and Brown, P.W. (1994) Hydration of sodium phosphate modified high alumina cement. *Journal of Materials Research* **9**, 1291–1297.

Majumdar, A.J., Singh, B., and Ali, M.A. (1981) Properties of high alumina cement reinforced with alkali resistant glass fibers. *Journal of Materials Science* **16**, 2597–1607.

Majumdar, A.J., Edmonds, R.N., and Singh, B. (1990) Hydration of SECAR 71 aluminous cement in presence of granulated blast furnace slag. *Cement and Concrete Research* **20**, 7–14.

Majumdar, A.J., Singh, B., and Edmonds, R.N. (1990) Hydration of mixtures of ciment fondue aluminous cement and granulated blast furnace slag. *Cement and Concrete Research* **20**, 197–208.

Manghabai, R.J. (ed.) (1990) *Calcium Aluminate Cements*, E. & F.N. Spon, London.

Macdargent, S. (1992) Hydration and strengths of blends CAC – silica fume and stability of hydrates, in *Proceedings 9th ICCC, New Delhi*, Vol. 4, pp. 651–657.

Matusinovic, T., and Curlin, D. (1993) Lithium salts as set accelerators for high alumina cement. *Cement and Concrete Research* **23**, 885–895.

Menetrier-Sorrentino, D., George, C.M., and Sorrentino, F.P. (1986) The setting and hardening characteristics of calcium aluminate cements: studies of the system $C_{12}A_7$-C_3A and $C_{12}A_7$-CA, in *Proceedings 8th ICCC, Rio de Janeiro*, Vol. 4, pp. 339–343.

Midgley, H.G. (1980) The relationship between cement clinker composition and strength recovery of hydrating high alumina cement during conversion, in *Proceedings 7th ICCC, Paris*, Vol. 3, pp. V68–70.

Mino, I., Sakai, E., and Nishioka, A. (1991) The hydration and calcination mechanism of calcium aluminate-based ultra-high strength cement with calcium sulfoaluminate compound. *Materials Research Society Symposium Proceedings* 179, 193–202.

Mohan, L. (1991) Advances in some special and newer cements, in *Cement and Concrete Science and Technology* (ed. S.N. Ghosh), Vol. I, Part I, ABI Books, New Delhi, pp. 253–312.

Müller D. *et al.* (1992) Progress in the ^{27}Al MAS NMR spectroscopy monitoring the hydration of calcium aluminate cements, in *Proceedings 9th ICCC, New Delhi*, Vol. 6, pp. 148–154.

Nilforoushan, M.R., and Sharp, J.H. (1995) The effect of additions of alkaline earth metal chlorides on the setting behaviour of refractory calcium aluminate cement. *Cement and Concrete Research* **25**, 1523–1534.

Osborn, G.J., and Singh, B. (1995) The durability of concretes made with blends of high-alumina cement and ground granulated blast furnace slag. ACI SP-153, pp. 885–909.

Quillin, K.C. (1993) Blended high alumina cements. *Materials World* **1/2**, 103–105 [ref. CA 119/209127].

Quillin, K.C., and Majumdar, A.J. (1994) Phase equilibria in the CaO-Al_2O_3-SiO_2-H_2O system at 5 °C, 20 °C and 38 °C. *Advances in Cement Research* **6**, 47–56.

Raina, S.J. *et al.* (1992) Barium aluminate – its synthesis and formation kinetics, in *Proceedings 9th ICCC, New Delhi*, Vol. 3, pp. 245–249.

Rao, P.B. (1980) Chemistry of arresting strength retrogression in structural high alumina cements, in *Proceedings 7th ICCC, Paris*, Vol. 3, pp. V51–56.

Rashid, S. *et al.* (1994) Conversion of calcium aluminate cement hydrates conversion re-examined with synchrotron energy-dispersive diffraction. *Journal of Materials Science Letters* **13**, 1232–1234.

Rayment, D.L., and Majumdar, A.J. (1994) Microanalysis of high alumina cement clinker and hydrated HAC/slag mixtures. *Cement and Concrete Research* **241**, 335–342.

Rettel, A., Damidot, D., and Scrivener, K. (1997) Study of the hydration of mixtures of CA and α-Al_2O_3 at different temperatures, in *Proceedings 10th ICCC, Göteborg*, paper 2ii026.

Revay, M. (1992) A summary of experiences on estimating the expected changes of strength of concrete made with high alumina cement, in *Proceedings 8th ICCC, Rio de Janeiro*, Vol. 4, pp. 351–356.

Richardson, I.G., and Groves, G.W. (1990) The microstructure of blastfurnace slag/high alumina cement pastes, in *Proceedings International Symposium on Calcium Aluminate Cements, London*, pp. 282–290.

Romero, L. *et al.* (1997) Activation of blastfurnace slag/high alumina cement pastes: mechanical and microstructural evolution, in *Proceedings 10th ICCC, Göteborg*, paper 3ii096.

Scharf, H., and Odler, I. (1992) Intrinsic bond properties of hydrates formed in the hydration of pure clinker minerals, in *Proceedings 9th ICCC, New Delhi*, Vol. 4, pp. 265–270.

Scian, A.N., Porto-Lopez, J.M., and Pereira, E. (1987) High alumina cements: study of $CaO.Al_2O_3$ formation. *Cement and Concrete Research* **17**, 198–204, 525–531.

Scian, A.N., Porto-Lopez, J.M., and Pereira, E. (1991, 1994) Mechanical activation of high alumina cement. *Cement and Concrete Research* **21**, 51–60; **24**, 937–947.

Scrivener, K., and Capmas, A. (1998) Calcium aluminate cement, in *Lea's Chemistry of Cement and Concrete* (ed. P.C. Hewlett), Arnold, London, pp. 709–778.

Sorrentino, D., Sorrentino, F., and George, M. (1995) Mechanism of hydration of calcium aluminate cement, in *Materials Science of Concrete IV* (ed. J. Skalny), American Ceramic Society, Westerville, OH, pp. 41–90.

Su, M., Kurdowski, W., and Sorrentino, F. (1992) Development in non-Portland cements, in *Proceedings 9th ICCC, New Delhi*, Vol. 1, pp. 317–354.

Sugama, T., and Carciello, N.R. (1991) Strength development in phosphate-bonded calcium aluminate cements. *Journal of the American Ceramic Society* **74**, 1023–1030.

Trivino, E. (1986) Aluminous cement: how to avoid degrading of mechanical resistance, in *Proceedings 8th ICCC, Rio de Janeiro*, Vol. 4, pp. 417–422.

Walter, D., and Odler, I. (1996) Investigation on MgO and CaO/Al_2O_3 polyphosphate cements. *Advances in Cement Research* **8**, 41–46.

Wu, Y., Long, S., and Liao, G. (1992) The mechanism of $BaSO_4$ prevention of calcium aluminate hydrates from crystalline transformation, in *Proceedings 9th ICCC, New Delhi*, Vol. 4, pp. 111–117.

Xue, J. *et al.* (1986) Study on high self stress aluminate cement, in *Proceedings 8th ICCC, Rio de Janeiro*, Vol. 4, pp. 305–311.

11 Calcium sulfate based binders

11.1 PHASES IN THE SYSTEM CAO–SO₃–H₂O

Calcium sulfate exists as anhydrite ($CaSO_4$), hemihydrate ($CaSO_4.\frac{1}{2}H_2O$), and dihydrate ($CaSO_4.2H_2O$). The known phases and forms of calcium sulfate are summarized in Table 11.1.

The A-II phase of anhydrite is the one that is thermodynamically stable up to 1180 °C, and converts to the A-I phase above this temperature. A third phase, anhydrite III, is metastable at any temperature; it exists in two distinct forms, designated α and β.

Calcium sulfate hemihydrate is also a thermodynamically metastable phase known in two forms, designated α and β.

Calcium sulfate dihydrate is stable only up to about 45 °C, and loses water at higher temperatures to be converted to hemihydrate. If the heating is done in air, in the absence of liquid water (that is, under "dry" conditions), three quarters of the water incorporated in its crystalline lattice escapes, the crystalline lattice collapses, and the dihydrate converts to

Table 11.1 Phases in the system CaO–SO₃–H₂O.

Hydration degree	Designation	Symbol	Crystalline form	Thermodynamic stability
Anhydrite ($CaSO_4$)	High-temperature anhydrite	A-I	Cubic	> 1180 °C
	Unsoluble anhydrite	A-II	Rhombic	< 1180 °C
	Soluble α-anhydrite	A-IIIα	Hexagonal	Metastable
	Soluble β-anhydrite	A-IIIβ		
Hemihydrate ($CaSO_4 . \frac{1}{2}H_2O$)	α-hemihydrate	α-HH	Orthorhombic	Metastable
	β-hemihydrate	β-HH		
Dihydrate ($CaSO_4 . 2H_2O$)	Dihydrate (gypsum)	DH	Monoclinic	< ~45 °C

hemihydrate, which possesses a highly distorted crystalline lattice. With more intensive heating the remaining water escapes, and a form of anhydrite with a highly distorted crystalline lattice remains. These hemihydrate and anhydrite forms are called *β*-hemihydrate and anhydrite-III*β*. The dehydration of dihydrate may start as low as about 45 °C, but temperatures above 100 °C are required to attain a reasonable reaction rate.

In the presence of liquid water (that is, under "wet" conditions) the dihydrate is dissolved in the liquid phase, and at temperatures above about 105–110 °C the liquid becomes oversaturated with respect to hemihydrate. A form of hemihydrate called *α*-hemihydrate crystallizes from the liquid phase under these conditions. At even higher temperatures *α*-hemihydrate may lose the rest of its water and converts to anhydrite-III*α*. Table 11.2 summarizes the main differences between the two forms of hemihydrate. The differences between *α*- and *β*-anhydrite III are similar.

At temperatures above about 300 °C both anhydrite-III*α* and anhydrite-III*β* convert to the thermodynamically stable phase anhydrite-II (also called "insoluble anhydrite"), which converts further to anhydrite-I if heated above 1180 °C. In a parallel reaction – starting at about 600 °C – a thermal dissociation of calcium sulfate to calcium oxide and sulfur trioxide may also take place:

$$CaSO_4 \rightarrow CaO + SO_3 \tag{11.1}$$

At ambient or moderately elevated temperature calcium sulfate hemihydrate or anhydrite react with liquid water and convert to calcium sulfate dihydrate. Anhydrite-III may also react readily with water vapor, but only to hemihydrate, which subsequently may convert to dihydrate, if allowed to hydrate with liquid water. The hydration of hemihydrate and anhydrite is a through-solution reaction, associated – at appropriate water/cement ratios – with setting and hardening. The setting and hardening process is accompanied by an overall macroscopic expansion of the paste caused by the manner of the growth of dihydrate crystals. The extent of this expansion may be controlled by suitable additives.

Table 11.2 Differences between *α*- and *β*-calcium sulfate hemihydrate.

	α-hemihydrate	*β-hemihydrate*
Crystal size	Small (10–20 μm)	Very small (1–5 μm)
Porosity of particles	Not porous	Porous
Specific surface area	Small	Large
Crystal lattice defects	Few	Many
Water requirement	Lower	Higher
Rate of strength development	Slower	Faster
Final strength	Higher	Lower

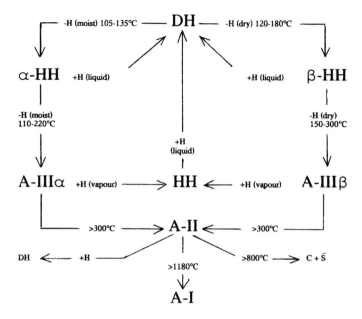

Figure 11.1 Hydration and dehydration reactions in the system CaO–SO$_3$–H$_2$O. DH, dihydrate (CaSO$_4$. 2H$_2$O); HH, hemihydrate (CaSO$_4$.½H$_2$O); A = anhydrite (CaSO$_4$).

The rate of hydration differs in the various calcium sulfates: whereas calcium sulfate hemihydrate and anhydrite-III hydrate rapidly, the hydration of anhydrite-II progresses very slowly.

The possible hydration and dehydration reactions in the system CaO–SO$_3$–H$_2$O are summarized in Fig. 11.1.

11.2 GYPSUM PLASTERS

"Gypsum plasters" is the general designation for inorganic binders that contain anhydrous calcium sulfate or calcium sulfate hemihydrate as their main or sole constituent. The setting and hardening of these binders is associated with their hydration to calcium sulfate dihydrate. The water requirements and reactivity vary greatly in different members of this group of binders, and this in turn affects their rate of strength development and ultimate strength. Gypsum plasters are not suitable for loadbearing applications, owing to the high creep of the hardened material under load. Also, calcium sulfate, and thus also all binders yielding calcium sulfate dihydrate as the product of the hydration process, exhibit a limited yet distinct solubility in water, and thus should not be excessively exposed to liquid water. The pH of the pore solution in hardened pastes made from gypsum plasters is around 7. Thus it must be expected that a steel rein-forcement, if used in combination with this type of binder, will undergo corrosion. On the other hand, gypsum plasters are well compatible with glass or cellulose fibers.

11.2.1   Hemihydrate gypsum plaster (plaster of Paris)

Plaster of Paris is produced by a "dry" dehydration (calcination) of cal-cium sulfate dihydrate, which is available either as natural gypsum or as a by-product of the chemical industry. Increasing amounts of dihydrate are also produced as flue gas gypsum in the desulfurization of flue gas in power plants that use sulfur-containing fossil fuels as source of energy. A number of technologies are available to produce plaster of Paris.

The main or sole constituent of plaster of Paris is calcium sulfate hemi-hydrate. If natural gypsum is used as the starting material the plaster may also contain some anhydrite, calcite, or clay minerals. The binder consists of individual particles whose size and shape corresponds to those of the starting dihydrate. They exhibit a distinct internal porosity and a relatively large specific surface area as a result of the dry dehydration process.

Pastes made from plaster of Paris exhibit fast setting and hardening, and the hydration process is typically completed within hours. Small amounts of chemical additives may be added to the calcined material to modify its properties. These may include set accelerators (such as soluble sulfates or prehydrated plaster), set retarders (such as citric acid or ker-atin), or modifiers of rheology (such as methyl cellulose or carboxy methyl cellulose).

The strength of the hardened paste depends mainly on its overall poros-ity, which is determined by the initial water/binder ratio. To a lesser extent the strength may also be affected by the size and shape of the formed dihy-drate crystals, which may be altered by the hydration temperature and by chemical additives (Rössler and Odler, 1989).

Very high-strength hardened gypsum pastes can by obtained by mechanical compaction of flue gas desulfurization gypsum (Stoop *et al.*, 1996). By applying pressures between 50 and 500 MPa the porosity of the material can be reduced to 21–26 vol.%, resulting in compressive strengths of up to 96 MPa. To achieve acceptable results a few per cent of water must be added to the original desulfurization gypsum powder prior to compacting. The microstructure of such a material is characterized by the presence of granular crystals, aligned boundary to boundary without much intergrowth.

Plaster of Paris is the calcium sulfate based binder that is produced and used in the largest amounts by far. Its main uses include:

- production of gypsum based boards of different kinds;
- gypsum based wall plaster compositions;

- protection of steel and concrete structures from fire damage (fire protective boards or plasters);
- thermal protection (lightweight boards and mortar compositions).

11.2.2 Anhydrous gypsum plaster (soluble anhydride)

Anhydrous gypsum plaster contains β-anhydrite (A-IIIβ) as its main constituent. It is produced in a similar way to hemihydrate gypsum plaster, but at higher calcination temperatures. The binder is very hygroscopic, and tends to absorb water vapor rapidly from the air to convert to hemihydrate. The hemihydrate formed in this way hydrates rapidly to dihydrate, if mixed with water. The properties of this binder are similar to those of hemihydrate gypsum plaster.

11.2.3 Anhydrite binder

Anhydrite binder contains anhydrite (A-II), called also "insoluble anhydrite," as its main constituent. It may be produced by high-temperature calcination (at about 600 °C) of natural or by-product gypsum. If natural or by-product anhydrite are available as starting materials, both of which contain the A-II phase as their main or sole constituent, it is sufficient to grind them to an appropriate fineness. Owing to the very low reactivity of anhydrite-II, about 0.5–1.0% of a suitable additive, such as K_2SO_4, $KAl(SO_4)_2$, $ZnSO_4$, $FeSO_4$, or a combination of them, has to be added to the ground anhydrite to accelerate the hydration process.

If mixed with water the anhydrite converts to dihydrate, and this process is associated with setting and hardening. The hydration reaction progresses rather slowly even in the presence of accelerating additives, but the various additives may affect the rate of hydration – and along with it the rate of strength development – to different degrees (Israel, 1996). The strength of the anhydrite binder is brought about by the intergrowth and bonding of unconverted anhydrite grains with plate-like gypsum crystals. The water requirements of anhydrite binders are rather low, and this in turn enhances the ultimate strength of the hardened material. The use of anhydrite binders is indicated in applications in which a slow rate of strength development is acceptable or desired.

11.2.4 Estrich binder (Estrich gypsum)

Estrich gypsum is produced by heating calcium sulfate based starting materials to temperatures of about 1000–1100 °C. Under these conditions a partial thermal dissociation of the existing or primary formed anhydrite-II takes place, yielding sulfur trioxide and calcium oxide. The formed SO_3 escapes, whereas CaO remains dispersed in the product, thus causing an acceleration of its setting and hardening. Estrich gypsum has properties

similar to those of an anhydrite binder that has not been thermally activated, but it exhibits a distinctly greater, yet still rather slow, rate of hydration. Estrich gypsum has been widely used as flooring plaster, especially in
Germany. By-product anhydrite is usually used as the starting material in
its production.

11.2.5   Alpha-hemihydrate gypsum plaster (autoclave gypsum)

a-Hemihydrate gypsum plaster contains the a form of calcium sulfate
hemihydrate as its sole constituent. It may be produced by a "wet" partial
dehydration of the dihydrate (that is, natural or by-product gypsum), in
which three quarters of the water originally present in the crystalline lattice of dihydrate remains as a constituent of the liquid phase.

To convert calcium sulfate dihydrate to a-hemihydrate liquid water must
be present, and the system must have a temperature that is higher than that
of boiling water at atmospheric pressure. To meet these conditions the
reaction must take place under autoclave conditions at a water vapor partial pressure exceeding 760 torr ($= 101.3$ kPa). A temperature of about
120–140 °C is typically employed under large-scale production conditions.
The product obtained in this way is usually called autoclave gypsum.

It is also possible to produce a-hemihydrate gypsum plaster by heating
to the boiling temperature suspensions of dihydrate in concentrated or saturated salt solutions, such as those of NaCl, $CaCl_2$, $MgCl_2$, NH_4Cl,
$MgSO_4$, $MgNO_3$, or NH_4NO_3 (Wirsching, 1962; Fischer and Uschmann,
1964; Zürz *et al.*, 1991). Such solutions have boiling temperatures higher
than 100 °C even at atmospheric pressure, which makes a recrystallization of $CaSO_4.2H_2O$ to $CaSO_4.1/2H_2O$ possible without the use of an
autoclave.

A third possibility of converting calcium sulfate dihydrate into a-hemi-
hydrate consists in dehydrating the dihydrate by sulfuric acid (Nogishi et
al., 1981; Zürz *et al.*, 1991). At an appropriate combination of acid concentration, temperature, and reaction time the dihydrate suspended in the
acid converts quantitatively to a-hemihydrate. Uner more drastic conditions (that is, at a higher acid concentration, higher temperature, and/or
longer reaction time) the dehydration does not stop at the hemihydrate
level, and anhydrous calcium sulfate is formed instead.

Unlike plaster of Paris, a-hemihydrate gypsum plaster consists of well-
developed individual hemihydrate crystals. The size and shape of these
crystals, and along with them the water requirement and strength of the
resulting product, will depend greatly on the conditions used in the production process, such as the reaction temperature, reaction time, and quality and concentration of the salt solution or concentration of the sulfuric
acid. However, suitable modifying agents added to the starting dihydrate
suspension in small amounts may modify the crystal morphology and gen-

erally improve the properties of the a-hemihydrate gypsum that is produced. A variety of organic acids exhibit such action, including fumaric acid, maleic acid, malic acid, and pyruvic acid.

Alpha-hemihydrate gypsum plaster has a distinctly lower water requirement than plaster of Paris and – at an equal consistency of the starting mix – exhibits a distinctly higher final strength. It is used in applications in which a high strength of the hardened material is essential, as in medical and dental applications, and in the production of molds for the ceramic industry.

11.3 WATER-RESISTANT GYPSUM BINDERS

The performance of gypsum based binders, and in particular their resistance to water, may be improved by combining calcium sulfate hemihydrate with limited amounts of some other constituents.

An effective solution to this problem is to combine calcium sulfate hemihydrate with limited amounts of Portland clinker or Portland cement (Odler and Balzer, 1992; Ostrowski, 1992; Bentur *et al.*, 1994; Gutti *et al.*, 1996; Deng *et al.*, 1998; Kovler, 1998a, 1998b). The improved water resistance achieved by this measure is brought about by the formation of the C-S-H phase in the hydration of dicalcium and tricalcium silicates of the clinker and to a lesser extent also by the formation of tricalcium aluminate trisulfate hydrate ($C_3A.3C\bar{S}.32H$, ettringite) in a reaction of calcium sulfate with the tricalcium aluminate and calcium aluminate ferrite phase present in the clinker. Both reaction products have a very low water solubility, and significantly improve the overall water resistance of the system.

The reaction of calcium sulfate with the tricalcium aluminate phase and the formation of ettringite may cause an expansion of the Portland-clinker-modified calcium sulfate hemihydrate paste. The extent of this expansion may vary over a wide range: it generally increases with increasing amounts of ettringite formed (and thus with an increasing clinker addition up to a pessimum value), and is distinctly greater in humid than in dry conditions (Odler and Balzer, 1992; Gutti *et al.*, 1996).

To limit such undesired expansion and keep it within acceptable limits it has been suggested that a low-C_3A, sulfate-resistant clinker rather than ordinary Portland cement clinker should be used in the preparation of gypsum + clinker-based cementitious mixes (Odler and Balzer, 1992). Other recommendations include the addition of small amounts of calcium chloride (Gutti *et al.*, 1996), sucrose (Ostrowski, 1992), and urea (Ostrowski, 1992) to the mix. The two latter additives prevent the change of the Si coordination from tetrahedral to octahedral, and thus hindering the crystallization of thaumasite ($C_3S\bar{C}\bar{S}H_{15}$), an AFt phase structurally closely related to ettringite, which – at low temperatures – is formed together with ettringite and forms with it a solid solution. Another approach recommended for limiting the expansion consists in the application of a mechanical load while the mix hardens (Danko, 1992).

The strength, water resistance, and expansivity of the hardened material may be improved further by combining hemihydrate gypsum with both Portland clinker/cement and microsilica (Odler and Balzer, 1992; Bentur *et al.*, 1994; Kovler, 1998a, 1998b). Other recommended combinations include: gypsum with Portland cement, ground granulated blast furnace slag, and an organic retarder (Singh *et al.*, 1990; Singh and Garg, 1996); with fly ash (Lin *et al.*, 1995); with Portland clinker and a natural zeolite (Naidenov, 1991); and with fly ash, hydrated lime, and Portland cement (Singh and Garg, 1995). Microsilica, the slag, the zeolite, and the fly ash react with free calcium hydroxide that has been formed in the hydration of Portland clinker, yielding additional amounts of the C-S-H phase and thus reducing the amounts of soluble constituents in the hardened paste. Table 11.3 shows the effect of the addition of sulfate-resistant clinker and microsilica on the strength, expansion, and solubility of hardened hemihydrate gypsum.

Other approaches for improving the water resistance of gypsum binders include combinations of calcium sulfate hemihydrate with microsilica and carbide sludge (Ajrapetov *et al.*, 1996) or with a formaldehyde-based synthetic resin (Rumyantsev and Adrianov, 1995).

The use of gypsum binders with improved water resistance is indicated in applications in which limited exposure of the hardened material to liq-

Table 11.3 Compressive strength, expansion and mass of gypsum-based binder paste after 365 days of curing in air and under water.

Paste composition		Compressive strength (MPa)		Expansion (mm/m)		Mass (%)
		A	W	A	W	paste
HH 100%		11.5	D	0.5	D	0
HH 75%	OPC 25%	18.1	0	19.7	29.1	84.1
HH 50%	OPC 50%	23.1	14.6	15.0	18.5	99.0
HH 75%	SRPC 25%	20.6	0	5.3	7.1	86.8
HH 50%	SRPC 50%	25.2	15.4	3.9	16.6	104.5
HH 65% MS 10%	SRPC 25%	34.6	7.6	1.1	5.8	89.7
HH 40% MS 10%	SRPC 50%	40.2	22.0	2.2	8.3	106.3

Mass: all data in per cent of original mass of paste specimen (*w/s* = 0.50)
A: air curing   W: underwater curing
HH: hemihydrate gypsum
OPC: ordinary Portland cement
SRPC: sulfate-resistant Portland cement
MS: microsilica
D: disintegration of samples
Source: Odler and Balzer (1992)

uid water is to be expected. As the setting and hardening of the binder are not impeded by soluble wood constituents, binders of this kind are also suitable for use in the production of particle boards (Deng *et al.*, 1998).

11.4 OTHER INORGANIC BINDERS CONTAINING CALCIUM SULFATE

A variety of inorganic binders contain variable amounts of calcium sulfate, usually in the form of dihydrate (gypsum). In the hydration of such cements calcium sulfate reacts with another constituent of the cement containing Al_2O_3, calcium hydroxide, and water, yielding ettringite as the product of reaction:

$$3C\bar{S}H_2 + A + 3CH + 24H \rightarrow C_3A.3C\bar{S}.32H \qquad (11.2)$$

The ettringite that is formed exhibits distinct cementing properties, but may cause expansion at a different rate and to a different degree, depending on the Al^{3+} source involved in the reaction. Cements that contain significant amounts of calcium sulfate include expansive cements (see section 21), regulated set cement (see section 5.2), supersulfated cement (see section 8.4), sulfobelite cement (see section 4.2), and sulfoalite cement (see section 4.3). Limited amounts of calcium sulfate are also present in Portland cement (see section 2) and blended cements (see section 7).

Ettringite cements contain an aluminate donor such as monocalcium aluminate, tricalcium aluminate, tetracalcium trialuminate sulfate, or tetracalcium aluminate ferrite, together with calcium sulfate dihydrate and in some instances also calcium hydroxide. In the hydration of such mixes ettringite is formed as the main or sole reaction product. If allowed to react unrestricted, the hydrating paste exhibits a significant expansion; however, hardened pastes with strengths comparable to those of other cements may be produced if the hydration is allowed to take place under mechanically restricted conditions. This may occur in completely closed steel molds, by which measure undesired expansion of the paste may be effectively prevented (Odler and Yan, 1994).

An ettringite cement paste that does not expand noticeably in the course of hydration may be produced by using a high-Al_2O_3 C-A-S glass as the Al_2O_3 source (45% CaO, 35% Al_2O_2, 20% SiO_2) in combination with gypsum (20%)(Yan and Odler, 1995; Yan and Yang, 1997). In this binder no other crystalline phases, besides ettringite, are formed in the hardening process, and it is assumed that the SiO_2 present in the glass converts into amorphous SiO_2 gel or into a C-S-H phase. The strengths obtained are comparable to those of Portland cement, yet the modulus of elasticity is significantly lower.

Song *et al.* (1997) studied the combination of anhydrite II plus blast furnace slag, using K_2SO_4, $Al_2(SO_4)_3$, or $AlK(SO_4)_2$ as accelerators. After one day the anhydrite was almost completely converted to dihydrate but the slag was almost unreacted. At longer hydration time some ettringite was also formed.

REFERENCES

Ajrapetov, G.A., Panchenko, A.I., and Nechushkin, A.Yu. (1996) Multicomponent clinkerless waterproof gypsum (in Russian). *Stroitel'nye Materialy*, **1996** (1), 28–29 [ref. CA 124/350759].

Bentur, A., Kovler, K., and Goldman, A. (1994) Gypsum of improved performance using blends with Portland cement and silica fume. *Advances in Cement Research* **6**, 109–116.

Danko, G.Y. (1992) On hardening peculiarities of gypsum-cement materials, in *Proceedings 9th ICCC, New Delhi*, Vol. 4, pp. 475–421.

Deng, Y., Furuna, T., and Uehara, T. (1998) Improvement of the properties of gypsum particle-board by adding cement. *Journal of Wood Science* **44**, 98–102.

Fischer, K.W., and Uschmann, W. (1964) The production of a-hemihydrate gypsum in technical salt solutions (in German). *Silikattechnik* **15**, 361–366.

Gong, J. *et al.* (1997) Modification of mixed cementing material of gypsum with conventional Portland cement. *Hunan Daxue Xuebao, Ziran Kexheban* **24**, 78–83 [ref. CA 128/260823].

Gutti, C.S. *et al.* (1996) The influence of admixtures on the strength and linear expansion of cement stabilized phosphogypsum. *Cement and Concrete Research* **26**, 1083–1094.

Israel, D. (1996) Investigations into the relationship between the degree of hydration, flexural tensile strength and microstructure of setting anhydrite. *ZKG International* **49**, 228–234.

Kovler, K. (1998a) Strength and water absorption for gypsum-cement-silica fume blends of improved performance. *Advances in Cement Research* **10**, 81–92.

Kovler, K. (1998b) Setting and hardening of gypsum-Portland cement-silica fume blends. *Cement and Concrete Research* **28**, 423–438, 523–531.

Lin, F. *et al.* (1995) Study of hydration-hardening mechanism and water resisting property of desulfogypsum–fly ash binder (in Chinese). *Guisuanyan Xuebao* **23**, 219–226 [ref. CA125/35787].

Naidenov, V. (1991) *Cement and Concrete Research* **21**, 898–904, 1023–1027.

Nogishi, H., Osuga, Y., and Sekiya, M. (1981) Metastability of gypsum dihydrate in sulfuric acid. *Gypsum-Lime* **175**, 23–26.

Odler, I., and Balzer, M. (1992) Investigations in the system Portland clinker–gypsum plaster–condensed silica. *Materials Research Society Symposium Proceedings* **245**, 95–102.

Odler, I., and Yan, P. (1994) Investigations on ettringite cements. *Advances in Cement Research* **6**, 165–171.

Ostrowski, C. (1992) Mechanism of deterioration of gypsum cement binder materials, in *Proceedings 9th ICCC, New Delhi*, Vol. 5, pp. 335–341.

Rössler, M., and Odler, I. (1989) Relationship between pore structure and strength of set gypsum pastes (in German). *Zement-Kalk-Gips* **42**, 96–100, 419–424.

Rumyantsev, B.M., and Adriamov, R.A. (1995) Improvement of water resistance of gypsum products (in Russian). *Izvestiya Vysshikh Uchebnykh Zavedenii, Stroitel'stvo* **1995**(12), 51–60.

Singh, M., and Garg, M. (1995) Phosphogypsum–fly ash cementitious binder: its hydration and strength development. *Cement and Concrete Research* **25**, 752–758.

Singh, M., and Garg, M. (1996) Relationship between mechanical properties and porosity of water-resistant gypsum binder. *Cement and Concrete Research* **26**, 449–456.

Singh, M., Garg, M., and Rehsi, S.S. (1990) Durability of phosphogypsum based water resistant anhydrite binder. *Cement and Concrete Research* **20**, 271–176.

Song, J.T. *et al.* (1997) Hydration in the system anhydrite II–blastfurnace slag (in Korean). *Yoop Hakhoechi* **34**, 861–869 [ref. CA 128/274178].

Stoop, B.T.J., Larbi, J.A., and Heijen, W.M.M. (1996) Compaction of FGD gypsum. *ZKG International* **49**, 158–164.

Sychova, L., and Anufriev, V. (1997) Clinkerization of anhydrite with phosphogypsum and calcium fluoride, in *Proceedings 10th ICCC, Göteborg*, paper 1i032.

Wirsching, F. (1962) The production of α-hemihydrate gypsum plaster from by-product gypsum in salt solutions (in German). *Zement-Kalk-Gips* **15**, 439–441.

Yan, P., and Odler, I. (1995) An ettringite cement based on a C-A-S glass and gypsum. *Advances in Cement Research* **7**, 125–128.

Yan, P., and Yang, W. (1997) Cementitious behaviour of $CaO-Al_2O_3-SiO_2$ glass (in Chinese). *Guisuanyan Xuebao* **25**, 480–484.

Zürz, A. *et al.* (1991) Autoclave-free formation of α-hemihydrate gypsum. *Journal of the American Ceramic Society* **74**, 1117–1124.

12 Phosphate cements

Phosphate cements are the main representatives of **acid-base cements**. In these the setting/hardening process is brought about by a reaction between a more acid and a more basic compound. The product is then a salt or a hydrated salt produced in this reaction. The acid reactant may be an inorganic acid or an acid salt (for example monoammonium phosphate, or acid phosphates and polyphosphates of ammonium and alkali metals), or even an organic chelating agent such as polyacrylic acid [(CH$_2$–CH(COOH)-)$_n$] or eugenol [CH$_2$=CH.CH$_2$.C$_6$H$_3$(OCH$_3$)H]. The basic constituent is usually a weakly basic or amphoteric metal oxide with a moderately small ionic radius (MgO, CaO, ZnO), but may also include other substances that are more basic than their acid counterpart. Examples of such materials are tetracalcium phosphate [Ca$_4$(PO$_4$)$_2$O], wollastonite (CaO.SiO$_2$), or an acid-decomposable aluminosilicate-type glass. Most reactions of this type take place at ambient temperatures, but some require an elevated temperature. Some of the hardened products produced by such reactions have been called chemically bonded ceramics because of their unusual properties, which were unattainable in the past by reactions taking place at these temperatures.

12.1 MAGNESIUM PHOSPHATE CEMENT

The constituents of magnesium phosphate cement are magnesium oxide (calcined magnesia, MgO) and a water-soluble acid phosphate, most often diammonium hydrogen phosphate [(NH$_4$)$_2$HPO$_4$]. The magnesium oxide must be available in a powdered form, whereas the phosphate may be either dissolved in the mixing water or dry-mixed with the MgO powder first. In the first instance, which is more common, the phosphate is employed in the form of a saturated or a highly concentrated solution (solubility 57.5 g (NH$_4$)$_2$HPO$_4$ in 100 ml H$_2$O at 10 °C), which is mixed with the oxide in a ratio needed to yield a mix of appropriate consistency. In the subsequent hydration process **struvite** (NH$_4$MgPO$_4$.6H$_2$O) is the main product of reaction, and the one mainly responsible for setting and hardening:

$$MgO + (NH_4)_2HPO_4 + 5 H_2O \rightarrow NH_4MgPO_4.6H_2O + NH_3 \quad (12.1)$$

Other reaction products that may also be formed in limited amounts in unwanted side-reactions are dittmarite ($NH_4MgPO_4.H_2O$), schertelite [($NH_4)_2Mg(HPO_4)_2.4H_2O$], newberite ($MgHPO_4.3H_2O$), and magnesium phosphate [$Mg_3(PO_4)_2.4H_2O$]. An amorphous or poorly crystalline phase may also be formed in mixes with very low water contents (Hall *et al.*, 1998). Gaseous ammonia is released in the hardening reaction.

Upon mixing with water, solid diammonium phosphate dissolves very quickly in the liquid phase, whereas the dissolution rate of magnesium oxide is much slower. Initially, colloidal gel particles of the reaction product are formed around the MgO particles and gradually fill the interparticle space, causing setting and hardening. Eventually, the primary formed gelous material converts into an interlocking network of struvite crystals. The size of these depends greatly on the starting water/solid ratio, and increases from about 0.5 μm to 10 μm as the water content in the mix increases from 5% to 12% (Hall *et al.*, 1998).

To obtain a paste of plastic consistency typically about 30–50 ml of the saturated phosphate solution is mixed with 100 g of calcined magnesia. Under these conditions significant amounts of non-reacted MgO may be found in the hardened material after the hardening reaction has been completed. It is not obvious, however, whether the presence of non-reacted magnesia is essential for good bonding.

At ambient temperature the initial rate of reaction is rather fast and associated with an intensive heat liberation. Typically, setting occurs in a few minutes, unless retarders are added to the mix. Measurable strength is usually attained within the first hour after mixing, and more than half of the ultimate strength is attained within 4 hours. The ultimate (28 day) strengths of well-produced magnesium phosphate cement mortars may exceed 50 MPa.

The actual reaction rate also depends on the quality of the calcined magnesia employed, and especially on its thermal history and particle size. In general, the reaction rate slows down with increasing temperature of calcination. An increasing fineness of the calcined magnesia results in a shortening of the setting time and an increase of the early strength.

The setting time of magnesium phosphate cement may be conveniently controlled by adding a set retarder to the mix, such as sodium tetraborate (borax, $Na_2B_4O_7.10H_2O$), boric acid (H_3BO_4) or its salts, fluorides, silicofluorides, sodium silicate (Na_2SiO_3), or sodium potassium tartrate ($NaKC_4H_4O_6$). Up to about 10% of retarder (borax) may be added to the mix without adversely affecting the ultimate strength. It is believed that in the presence of borate ions a surface layer of magnesium borate is formed at the surface of magnesia grains, thus delaying the setting reaction. An additional beneficial effect of the borate based retarders is a moderate increase of the ultimate strength of the paste.

In pastes made with magnesia produced at a relatively low calcination temperature the residual MgO may partially convert to $Mg(OH)_2$ after long curing times, and this reaction may be associated with an undesired expansion of the material. Thus the used magnesia should be "dead burnt": that is, calcined at or above 1500 °C.

Phosphates that may be employed in magnesium phosphate cements instead of diammonium hydrogen phosphate include monoammonium dihydrogen phosphate ($NH_4H_2PO_4$), monobasic and dibasic acid phosphates of the alkali metals (NaH_2PO_4, KH_2PO_4, Na_2HPO_4, K_2HPO_4), and sodium pyrophosphate ($Na_4P_2O_7$). A distinct advantage of the use of alkali phosphates instead of ammonium phosphate consists in the fact that noxious ammonia is not released in the hardening reaction. On the other hand, the strengths attained with these starting materials are generally lower than those of a cement based on diammonium phosphate. The reaction products formed under these conditions are non-crystalline or microcrystalline phases that have not been more closely characterized, rather than struvite or other crystalline magnesium ammonium phosphates.

In magnesium polyphosphate cements MgO is allowed to react with alkali or ammonium polyphosphates $[(NH_4)_{n+2}P_nO_{3n+1}]$ with chain lengths of up to 50 phosphate units. In mixes based on ammonium polyphosphates both the setting time and the strength have been reported to increase with increasing chain length. A blend of different crystalline ammonium magnesium phosphates was formed as reaction product (Muntean and Gutul, 1992). In contrast, mixes based on sodium polyphosphates acquired a putty-like consistency first, and exhibited measurable strengths only after longer curing times. No crystalline reaction products could be detected by X-ray diffraction (Dimotakis *et al.*, 1992; Walter and Odler, 1996). A gradual polyphosphate chain degradation was observed in the cement mixes by ^{31}P NMR (Dimotakis *et al.*, 1992).

The main area of application of magnesium phosphate cement is mortars or concrete mixes for rapid repair works, such as repairing of potholes in concrete pavings. Here, one of the advantages is the good bonding of magnesium phosphate cement mixes to mature concrete based on Portland cement. Bond strengths exceeding 4 MPa may be achieved. Magnesium cement mortar/concrete also exhibits an acceptable abrasion resistance, good frost resistance, and a coefficient of thermal expansion similar to that of Portland cement mixes (Seehra *et al.*, 1993). The rheological characteristics of magnesium phosphate cement mixes are similar to those based on Portland cement.

Magnesium phosphate cements have also been recommended for containment of chemical and radioactive waste (Wagh *et al.*, 1993).

Upon heating above about 200 °C, hardened magnesium phosphate cement exhibits a gradual loss of chemically bound water and a moderate shrinkage, but preserves most of its strength at least up to 1000 °C, and thus may also be considered for high-temperature applications. Magnesium phosphate cements also used to be considered as dental cements, but are rarely employed in this area nowadays (Sharp and Winbow, 1989).

Finally, magnesium phosphate cement may be considered as a suitable binder in glass fiber reinforced composites, as the pH value of its pore solution is significantly lower than that of most other cements.

12.2 CALCIUM PHOSPHATE CEMENTS

In calcium phosphate cement blends two calcium phosphates of different basicity are usually allowed to interreact, to yield hydroxyapatite [$Ca_5(PO_4)3OH$, abbreviation OHAp), the phase responsible for setting and hardening.

Table 12.1 shows the phases existing in the system $CaO–P_2O_5–H_2O$. Out of these only monocalcium phosphate (anhydrous or monohydrate) is reasonably soluble in water, whereas the solubility of all other calcium phosphates is very limited. At any pH above 4.8 hydroxyapatite is the phase that is least soluble, and hence the one that is thermodynamically stable in this system.

A unit cell of **hydroxyapatite** may be expressed by the formula $Ca_{10}.(PO_4)_6.(OH)_2$. Four of the calcium sites are coordinated with eight oxygens each, whereas the remaining six are seven-coordinated. Six of these seven bonds are with oxygen, whereas the seventh is with an hydroxyl ion. The theoretical Ca/P ratio of this stoichiometric compound is 1.67, but because of a possible calcium deficiency, the stability region of this phase may extend down to Ca/P = 1.50. Such a deficiency may be associated with the formation of up to one vacancy per unit cell on one of the seven-coordinated calcium sites, and a charge balance in such an instance is maintained by the removal of one of the hydroxyl oxygens. The remaining proton becomes bound to an oxygen that is also coordinated with a phosphate. Thus the composition of hydroxyapatite having a Ca/P ratio of 1.50 may be expressed by the formula $Ca_9.(PO_4)_5.(HPO_4).OH$. The structure of hydroxyapatite is closely related to that of fluorapatite, in which the OH$^-$ ion is replaced by F [$Ca_5.(PO_4)_3.F$].

Table 12.1 Phases in the system $CaO–P_2O_5–H_2O$.

Phase	Abbreviation	Chemical formula	C/P
Monocalcium phosphate	MCP	$Ca(H_2PO_4)_2$	0.5
Dicalcium phosphate	DCP	$CaHPO_4$	1.0
Dicalcium phosphate dihydrate	DCPD	$CaHPO_4 . 2H_2O$	1.0
Octacalcium phosphate	OCP	$Ca_8H_2(PO_4)_6.5\ H_2O$	1.33
Tricalcium phosphate	TCP	$Ca_3(PO_4)_2$	1.5
Hydroxyapatite	HAP	$Ca_5(PO_4)_3OH$	1.67
Tetracalcium phosphate	TeCP	$Ca_4(PO_4)_2O$	2.0

In water suspensions all calcium phosphates have the potential to convert to hydroxyapatite. Such a reaction, however, would be associated with the liberation of free phosphoric acid, and a decline of pH with all calcium phosphates that are more acid than hydroxyapatite. It would stop as soon as the singular point with hydroxyapatite was obtained, where an equilibrium would be established:

$$5CaHPO_4 + H_2O \rightarrow Ca_5.(PO_4)_3.OH + 2H_3PO_4 \tag{12.2}$$

On the other hand, in suspensions of tetracalcium phosphate – the only calcium phosphate that is more basic than hydroxyapatite – the conversion would be associated with the liberation of free lime, and with an increase of pH:

$$3Ca_4.(PO_4)_2.O + 2H_2O \rightarrow Ca_{10}.(PO_4)_6.(OH)_2 + Ca(OH)_2 \tag{12.3}$$

Thus a complete conversion to hydroxyapatite, without any changes in the pH value of the liquid phase, may take place only if tetracalcium phosphate is combined with another calcium phosphate phase that is more acid than hydroxyapatite.

Calcium phosphate cements are combinations of tetracalcium phosphate and a phosphate more acid than hydroxyapatite, preferably anhydrous dicalcium phosphate or dicalcium phosphate dihydrate. Upon contact with water these two phosphates dissolve until the singular point for both compounds is reached, which corresponds to a pH value of 7.4. At this point the liquid phase is oversaturated with respect to hydroxyapatite, which starts to precipitate, causing setting and hardening:

$$CaHPO_4.2H_2O + Ca_4(PO_4)_2O \rightarrow Ca_5(PO_4)_3OH + 2H_2O \tag{12.4}$$

$$CaHPO_4 + Ca_4(PO_4)_2O \rightarrow Ca_5(PO_4)_3OH \tag{12.5}$$

Since, at the singular point composition, the liquid phase is also oversaturated with respect to tricalcium phosphate and octacalcium phosphate, these phases may also precipitate as metastable intermediates. However, as both of them are more soluble than hydroxyapatite, they eventually also convert to this phase.

The presence of tetracalcium phosphate is essential for this reaction to take place, as this is the only phase in the $CaO–P_2O_5–H_2O$ system that is more basic than hydroxyapatite. Moreover, the structural similarity of tetracalcium phosphate to hydroxyapatite may also stimulate the reaction. Of the calcium phosphates that are more acid than hydroxyapatite, the best results are obtained with the dicalcium phosphates, and thus only these are commonly used in calcium phosphate cements. In pure water and at ambient temperature reactions (12.4) and (12.5) progress very slowly;

however, both the reaction rate and the resulting strength may be controlled by the experimental conditions employed.

Reasonably fast setting and strength development may be achieved if diluted phosphoric acid or solutions of some organic acids are used as the mixing liquid, instead of plain water. An addition of fluorine ions to the liquid phase also has an accelerating effect. Ammonium and alkali phosphates exhibit a catalytic effect on the reaction.

The exact setting time of the mix may be adjusted by varying the pH value of the liquid, which may be done by varying the ratio XH_2PO_4/X_2HPO_4 (where $X = NH_4^+$, Na^+, or K^+) (Vanis and Odler, 1997). In general, the setting time is extended as the pH of the liquid phase increases.

The setting time may also be shortened and the final strength increased by adding crystalline hydroxyapatite to the mix as a seeding material. Additions of up to 40 wt% have been recommended for this purpose, but additions of just a few per cent may be sufficient.

Finally, the properties of the mix will also depend on the fineness of the starting materials. It has been reported that strengths exceeding 50 MPa have been obtained in mixes made from coarse tetracalcium phosphate and fine dicalcium phosphate, whereas the mix did not even set if the tetracalcium phosphate was fine and the dicalcium phosphate coarse (Chow *et al.*, 1991).

To increase the strength of the hardened material even further, Muroyame *et al.* (1998) recommended adding acrylamide to the starting mix, and curing the hardened material at 37 °C to polymerize the monomer.

In addition to using blends of tetracalcium and dicalcium phosphate for preparing suspensions that set and harden with time, such blends may also be used for producing compacts that may attain high strengths if placed into appropriate solutions. Flexural strengths of up to 28 MPa have been reported in compacts stored in a saturated potassium phosphate solution (Vanis and Odler, 1997).

The main area of application of calcium phosphate cements is in dentistry and medicine.

12.3   ALUMINUM PHOSPHATE BINDER

The aluminum phosphate binder consists of a combination of orthophosphoric acid (H_3PO_4) or monoaluminum phosphate [$Al(H_2PO_4)_3$] and aluminum oxide (fused alumina, Al_2O_3). Upon heating to elevated temperatures the two constituents interact, and aluminum orthophosphate ($AlPO_4$) is formed as the final product of the reaction at temperatures above about 500 °C. This is a cementing process that results in the development of a hardened material. Aluminum phosphate binder is widely employed in refractory concrete applications, and is discussed in more detail in Chapter 23.

Silsbee *et al.* (1991) studied materials that had been produced by combining calcined gibbsite or an X-ray amorphous form of alumina with phosphoric acid and by subsequent heat treatment at temperatures between 60 and 300 °C. The structure of the resultant material was characterized by the presence of residual unreacted material (gibbsite crystals) embedded in an amorphous aluminum phosphate matrix. The amorphous fraction exhibited some heterogeneity on a very fine scale. Splitting tensile strengths between about 10 and 20 MPa were attained.

Ma and Brown (1992a) developed a cement–inorganic polymer composite material by combining calcium aluminate cement with sodium phosphate based inorganic polymers, such as $(NaPO_3)_n$ and $(NaPO_3)_n.Na_2O$, and shear-mixing the mix with a low water/solid ratio. The resulting mix exhibited an accelerated hydration rate and a greatly improved strength, as compared with plain high-alumina cement (flexural strength up to 34 MPa after one day). The phosphate-containing reaction products formed under these conditions were X-ray amorphous.

In separate work Walter and Odler (1996) studied mixes of calcium aluminate cements with sodium polyphosphates having chain lengths between $n = 4$ and $n = 30$. The obtained strengths were distinctly improved by the addition of the polyphosphate, but the material exhibited a significant shrinkage in the course of curing.

12.4   ZINC PHOSPHATE CEMENT

Zinc phosphate cement consists of finely powdered zinc oxide suspended in phosphoric acid. The setting and hardening of this cement results from a chemical reaction between these two constituents, in which zinc phosphate tetrahydrate is formed as the product of reaction:

$$3ZnO + 2H_3PO_4 + H_2O \rightarrow Zn_3(PO_4)_2.4H_2O \qquad (12.6)$$

The setting of the cement is rapid, and measurable strength may be obtained within minutes. The hardening reaction is completed within 24 hours, yielding compressive strengths of up to 100 MPa. To improve the strength properties even further an addition of small amounts of other oxides (MgO, SiO_2) to the original mix may be indicated. The mixing liquid needs to be buffered. Zinc phosphate cement is mainly used in dentistry.

12.5   CALCIUM PHOSPHOSILICATE CEMENTS

Some highly dispersed powders in the $CaO–P_2O_5–SiO_2$ system have been found to exhibit hydraulic properties. Compacts made from powders of this kind and preignited to 700 °C exhibited flexural strengths of up to 70 MPa after being placed into water for appropriate lengths of time (Hu *et al.*, 1988a, 1988b; Roy, 1988; Steineke *et al.*, 1988; Steineke, 1991). Optimum results were reported with a mix having a $CaO:P_2O_5:SiO_2$ ratio of 55.5:33.3:11.1 wt%. The preignited powders contain hydroxyapatite, dicalcium silicate, and possibly amorphous C-S and C-P phases as their constituents. The hardening process seems to be associated with a fast hydration of the dicalcium silicate phase, which appears to be present in a highly reactive form. According to Feng and Yang (1997) the PO_4^{-3} ions that are present mostly remain preserved in the hardening reaction, and only a small fraction of them convert to $P_2O_7^{-4}$. However, ^{29}Si and ^{31}P NMR studies have revealed the presence of some polymeric tetrahedra in Q^3 and Q^4 form, which were probably formed in a reaction between silicates and phosphates contained in the material (Feng and Yang, 1996).

C-P-S powders preignited to 1300 °C contain neither hydroxyapatite nor dicalcium silicate, but only tricalcium phosphate as the sole crystalline phase. Both pastes and compacts made from such powders exhibit setting and hardening that may be enhanced significantly by the catalytic action of ammonium, sodium or potassium phosphates present in the mixing water or solution in which the compacts are stored (Vanis and Odler, 1996). The product of the hardening reaction is hydroxyapatite in combination with a C-S-H and/or C-S-P-H gel.

Finally, powders made from $CaO–SiO_2–P_2O_5–CaF_2$ glasses of appropriate composition also exhibit hydraulic properties, if ammonium phosphate is added to the mixing water (Kokubo and Yoshihara, 1991). The products of the hardening reaction are calcium ammonium phosphate $(CaNH_4PO_4.H_2O)$ and hydroxyapatite.

REFERENCES

Abdelrazig, B.E.I. (1988) Phase changes on heating ammonium magnesium phosphate hydrates. *Thermochimica Acta* **129**, 197–215.

Abdelrazig, B.E.I. (1988) The microstructure and mechanical properties of mortars made from magnesia-phosphate cement. *Cement and Concrete Research* **18**, 415–425.

Abdelrazig, B.E.I. (1989) The microstructure and mechanical properties of mortars made from magnesia polyphosphate cement. *Cement and Concrete Research* **19**, 147–158.

Abdelrazig, B.E.I. *et al.* (1984) Chemical reactions in magnesium phosphate cements. *Proceedings of the British Ceramic Society* **35**, 141–154.

Brown, P.W. (1992) Phase relationship in the ternary system $CaO-P_2O_5-H_2O$ at 25 °C. *Journal of the American Ceramic Society* **75**, 17–22.

Brown, W.E., and Chow, L.C. (1986) A new calcium phosphate water setting cement, in *Cement Research Progress* (ed. W.E. Brown), American Ceramic Society, Westerville, OH, USA, pp. 352–279.

Brown, P.W., Sample D., and Hecker, N. (1991) The low temperature formation of synthetic bone. Materials Research Society Symposium Proceedings 179, 41–48.

Brown, P.W., Hocker, N., and Hoyele, S. (1991) Variations in solution chemistry during the low-temperature formation of hydroxyapatite. *Journal of the American Ceramic Society* **74**, 1848–1954.

Chow, L.C. *et al.* (1991) Self setting calcium phosphate cement. *Materials Research Society Symposium Proceedings* **179**, 3–24.

Dimotakis, E.D., Klempered, W.G., and Young, J.F. (1992) Polyphosphate chain stability in magnesia-polyphosphate cements. *Materials Research Society Symposium Proceedings* **245**, 205–210.

El-Jazairi, B. (1987) The properties of magnesia phosphate cement based materials for rapid repair of concrete. *Proceedings 3rd International Conference on Structural Faults and Repair, London*, pp. 9–13.

Feng, X., and Yang, N. (1996) Studies on the hydration properties of chemically bonded ceramics in $CaO-SiO_2-P_2O_5-H_2O$ system using TMS-GS and ^{29}Si, ^{31}P NMR. *Guisuanyan Tongbao* **15**, 11–17 [ref. CA 125/93795].

Feng, X., and Yang, N. (1997) Variation of the anion polymerization degree of $CaO-SiO_2-P_2O_5-H_2O$ CBC materials during hydration. *Cement and Concrete Research* **27**, 407–413.

Frantzis, P., and Baggott, R. (1996) Effect of vibration on the rheological characteristics of magnesia phosphate and ordinary Portland cement slurries. *Cement and Concrete Research* **26**, 387–395.

Hall, D.H., Stevens, R., and El-Jazairi, B. (1998) Effect of water content on the structure and mechanical properties of magnesia phosphate cement mortar. *Journal of the American Ceramic Society* **81**, 1550–1556.

Hirano, M. (1993) Development of calcium phosphate based cement (in Japanese). *Nyu Seramikkusu* **6**, 61–65 [ref. CA 120/13714].

Hu, J., Agraval, D.K., and Roy, R. (1988a) Investigation of hydration phases in the system $CaO-SiO_2-P_2O_5-H_2O$. *Journal of Materials Research* **3**, 772–780.

Hu, J., Agraval, D.K., and Roy, R. (1988b) Study of strength mechanism in newly developed chemically bonded ceramics in the system $CaO-SiO_2-P_2O_5-H_2O$. *Cement and Concrete Research* **18**, 103–108.

Jiang, W., and Yang, N. (1992) Phosphate-bonded dental cements, in *Proceedings 9th ICCC, New Delhi*, Vol. 3, pp. 338–344.

Kokubo, T., and Yoshihara, S. (1991) Bioactive bone cement based on $CaO-SiO_2-P_2O_5$ glass. *Journal of the American Ceramic Society* **74**, 1739–1741.

Ma, W., and Brown, P.W. (1992a) Cement–inorganic polymer composites, microstructure and strength development, in *Proceedings 9th ICCC, New Delhi*, Vol. 4, pp. 424–429.

Ma, W., and Brown, P.W. (1992b) Mechanical behaviour and microstructural development in phosphate modified high alumina cement. *Cement and Concrete Research* **22**, 1192–1200.

Monma, H. (1988) On hydraulic calcium phosphates, in *Proceedings MRS International Meeting on Advanced Materials, Tokyo*, pp. 15–26.

Muntean, M., and Gutul, M. (1992) MgO–ammonium polyphosphate system, in *Proceedings 9th ICCC, New Delhi*, Vol. 3, pp. 345–350.

Muroyame, K. *et al.* (1998) Effect of addition of arylamide on hydraulic hardening of calcium phosphate. *Maki Materiaru* **5**, 31–36 [ref CA 128/285603].

Park, C.K., Silsbee, M.R., and Roy, D.M. (1998) Setting reaction and resultant structure of zinc phosphate cement in various orthophosphoric acid cement-forming liquids. *Cement and Concrete Research* **28**, 141–150.

Petri, M., and Odler, I. (1991) Physico-mechanical properties of hardened Mg-phosphate cement. *Materials Research Society Symposium Proceedings* **179**, 83–86.

Popovic, S., Rajendran, N., and Penko, M. (1987) Rapid hardening cements for repair of concrete. *ACI Journal* **84**, 64–73.

Roy, D.M. (1988) Recent advances in phosphate chemically bonded ceramics, in *Proceedings MRS International Meeting on Advanced Materials, Tokyo*, Vol. 13, pp. 213–228.

Sarkar, A.K. (1990) Phosphate cement based fast setting binders. *Ceramics Bulletin* **691**, 234–238.

Seehra, S.S., Gupta, S., and Kumar, S. (1993) Rapid hardening magnesium phosphate cement for quick repair of concrete pavements: characterization and durability aspects. *Cement and Concrete Research* **23**, 254–266.

Sharp, J.H., and Winbow, H.D. (1989) Magnesia-phosphate cements, in *Cement Research Progress* (ed. W.E. Brown), American Ceramic Society, Westerville, OH, USA, pp. 233–264.

Silsbee, M.R. *et al.* (1991) Low temperature (< 300 °C) phosphate ceramics from reactive aluminas. *Materials Research Society Symposium Proceedings* **179**, 49–58.

Steineke, R.A. (1991) Development of chemically bonded ceramics in the CaO-SiO_2-P_2O_5-H_2O system. *Cement and Concrete Research* **21**, 66–72.

Steineke, R.A. *et al.* (1988) Reactions and bonding sol-gel derived chemically bonded ceramics in the system CaO–P_2O_5–SiO_2, in *Proceedings MRS International Meeting on Advanced Materials, Tokyo*, Vol. 13, pp. 229–238.

Sugama, T., and Kukacka, L.E. (1983a) Magnesium monophosphate cements derived from magnesium diammonium phosphate solutions. *Cement and Concrete Research* **13**, 407–416.

Sugama, T., and Carciello, N.R. (1991) Strength development in phosphate bonded calcium aluminate cement. *Journal of the American Ceramic Society* **74**, 1023–1030.

Sugama, T., and Kukacka, L.E. (1983b) Characteristics of magnesium polyphosphate cements derived from ammonium polyphosphate solutions. *Cement and Concrete Research* **13**, 499–506.

Sugama, T. *et al.* (1995) Mullite microsphere-filled lightweight calcium phosphate cement slurries for geothermal wells: setting and Properties. *Cement and Concrete Research* **25**, 1305–1310.

Vanis, P., and Odler, I. (1997) Investigations on calcium phosphate cements, in *Proceedings 10th ICCC, Göteborg*, paper 2ii062.

Wagh, A.S. *et al.* (1993) Mechanical properties of ammonium phosphate cements and their zeolite composites. *Ceramic Transactions* **37**, 139–152.

Walter, D., and Odler, I. (1996) Investigations of MgO and CaO/Al_2O_3 polyphosphate cements. Advances in Cement Research **8**, 41–46.

Xie, L., and Monroe, E.A. (1991) Calcium phosphate dental cements. *Materials Research Society Symposium Proceedings* **179**, 25–39.

Zürz, A., Odler, I., and Detki, B. (1991) Investigation on phosphate cements hardening at room temperature. *Materials Research Society Symposium Proceedings* **179**, 69–82.

13 Cementitious systems modified with organic polymers

Organic polymers are high molecular weight products of polymerization or polycondensation of pertinent low molecular weight starting building units (so-called **monomers**). Substances that consist of only a limited number of monomeric units are called **oligomers**, and may be intermediate products of polymerization or polycondensation.

Thermoplasts are organic polymers that consist of long-chain molecules that are mutually interconnected just by weak Van der Waals forces. The chain length of the individual molecules constituting the polymer is usually not identical, and may vary over a certain range. The polymers may be polymerized to different degrees, and thus their medium molecular weight may vary in different products. This is one of the factors that determine the physico-mechanical properties of the material. Upon heating to elevated temperatures thermoplasts tend to attain plastic properties with a viscosity that declines with increasing temperature.

Thermosets, by contrast, have a three-dimensional structure, and are typically formed in a reaction between two different starting compounds. Upon heating they do not melt and tend to preserve the characteristics of a solid, but may undergo thermal degradation, especially in air.

Elastomers are organic polymers that exhibit distinct elastic properties.

Many organic polymers, commonly called plastics, are widely used in different applications as structural materials. Others may be used, finely dispersed in water, as polymeric **dispersions**. Low molecular weight polymers may exist in the form of liquids whose viscosity increases with the degree of polymerization. Upon further polymerization they also gradually attain the characteristics of a solid.

Polymers may be produced either just from one or from two or more different monomers. The latter are called **co-polymers**.

Some organic polymers may be combined with inorganic cements to yield materials with unique properties. A precondition for such application is a sufficient stability of the polymer in a high-pH environment, typical for most cementitious systems.

There are three fundamentally different ways in which organic polymers may be combined with inorganic cements:

- The polymer may be employed to fill the pore space of the hardened cement paste.
- A polymeric dispersion may be interblended with the fresh cement paste to yield a combined cementitious-polymeric system.
- A new material may be created in a chemical interaction between the polymer and constituents of the inorganic binder.

13.1   POLYMER-IMPREGNATED CEMENTITIOUS MATERIALS

In polymer-impregnated cementitious materials the prehardened cement paste, mortar or concrete is impregnated with a low molecular weight, low-viscosity polymer precursor, which, after having filled the existing pore space, is brought to polymerization. The resultant material is a hardened cementitious body, in which the original pores are filled with the polymer.

The impregnation may be performed on complete hardened structures or on prefabricated concrete elements after the hydration of the binder has been completed. The quality of the binder is of little relevance, but Portland cement is employed most often.

Prior to impregnation, the existing pore space must be emptied, and any water present in the pores must be removed. This may be done most effectively by heating the concrete to temperatures above 100 °C, especially if elements of limited size are to be impregnated; however, more gentle methods, such as drying in air, may also be considered.

Complete impregnation of the material can be achieved only in concrete/mortar elements of limited size. To achieve this, the element must be placed in a chamber that is subsequently evacuated to remove the air present in the pores. The chamber is then filled with the impregnating liquid, and air pressure is applied to force the liquid into the pores. Pressures of around 50–100 kPa are appropriate.

Larger elements or whole concrete structures that cannot be treated in this way may be impregnated by applying the impregnating liquid on their surface, for example by brush. In such cases the impregnation is limited to a surface region, the depth of which will depend on the fineness of the pore system of the cement paste and the viscosity of the liquid. In many instances such an approach is sufficient to achieve the desired improvement of the performance of the structure or the concrete element.

A variety of polymer precursors may be used as impregnating liquids, as long as their viscosity is sufficiently low to allow effective impregnation. Precursors of polymethyl methacrylate or of polystyrene are those used most commonly. A cross-linking agent may also be added to the precursor liquid, to modify the properties of the resultant polymer.

The polymerization of the precursor present in the pore system to produce the final polymer may be done in several ways.

The approach most widely employed is to heat the concrete element, typically to temperatures of about 80–100 °C. The polymerization process may be accelerated significantly by adding to the impregnating liquid substances that decompose at elevated temperatures, to yield free radicals that act as catalysts.

It is also possible to perform the polymerization by exposing the impregnated concrete to ionizing radiation. Under these conditions the polymerization takes place at ambient temperature, and the method may also be applied for very large elements or monolithic structures.

The impregnation of hardened concrete or mortar with polymers increases the strength of the material significantly: fourfold increases are not uncommon. At the same time the material attains a more brittle character. The creep of the material is also reduced significantly, which is related to its reduced capacity to take up water. For the same reason the material exhibits an exceptionally high frost resistance. The resistance against chemical attack is also improved significantly, as the corrosive agent can exhibit its action only at the surface of the material, and is prevented from entering its pore system.

Upon exposure to fire and high temperatures the polymer component undergoes thermal degradation and oxidation. Toxic gaseous substances may also be released. The process is associated with a strength loss of the material.

Polymer-impregnated concrete and mortar elements are being employed in applications where an exceptionally high performance of the material is required, such as in highly corrosive environments.

13.2   POLYMER-MODIFIED CEMENT MIXES

Polymer-modified cement mixes are produced by mixing together an inorganic cement (most commonly Portland cement), water, and a dispersion of an organic polymer. If concrete aggregate or sand is also added, polymer-modified concrete or mortar mixes are obtained. Compared with similar polymer-free mixes, the addition of water may be reduced or omitted entirely, as the polymeric dispersion supplies part or all of the water needed for hydration and for obtaining adequate rheology of the mix.

In a polymeric dispersion (also called latex) the polymer is present in the form of small, separated droplets that are dispersed in the water phase, which may also contain dispersing agents to stabilize the whole system. A 50% concentration of the polymer may be considered fairly typical.

Upon mixing and subsequent hardening a three-dimensional polymeric network develops within the material, which is intimately combined with the three-dimensional structure of the hardened cement paste. A variety of polymer dispersions may be combined with inorganic cements, as long as the polymeric material is sufficiently resistant to sustain the high-pH environment of the cement paste. These may be: thermoplasts, such as polyvinyl acetate, polyvinyl chloride or polyacrylate; thermosets, such as epoxides, polyesters, or polyurethanes; and also elastomers, such as natural rubber latex or a butadiene-styrene copolymer. Polymer additions between 5% and 20% may be considered typical.

If two-component resinous materials are employed, such as epoxy resins, the resin precursor must be mixed with a curing agent to get the polymerization reaction under way. After mixing, the precursor/curing agent blend has only a limited pot life, and its viscosity changes with time as the polymerization reaction progresses. However, one-component pot-life-free epoxies have been developed in which this drawback is eliminated. Such products contain a curing agent that does not react at normal temperatures, and the chain reaction starts only if the mix is heated up to 80 °C. A slow polymerization reaction of the epoxy precursor may also be initiated by the catalytic action of alkalis present in the binder (Ohama *et al.*, 1992).

The addition of the polymeric dispersion usually improves the rheology of the fresh mix. Very often, however, at the same time the amount of air entrapped in the mix in the course of mixing is increased. An addition of a defoamer (of silicone type) may be needed to prevent this from happening.

The initial rate of hydration of the cement in the mix may be adversely affected by the presence of the polymeric dispersion. This effect can be explained by the formation of a diffusional barrier around the non-hydrated cement particles (Grosskurth, 1991; Ollitrault-Fichet *et al.*, 1998). It has also been suggested that the polymer acts as a retarder, first by suppressing the nucleation of the cement hydration products and thus increasing the induction time, and second by restraining the growth of these phases during the acceleration stage (Zeng *et al.*, 1966).

The addition of polymeric dispersions significantly affects the physico-mechanical properties of the hardened material, which also depend on the existing temperature (Grosskurth, 1991). At temperatures below the glass transition temperature the cement/polymer composite behaves like a brittle solid, independent of the amount of added polymer. At temperatures above the glass transition temperature the most marked effect is a significant increase of the fracture energy, accompanied by a decline in the modulus of elasticity. This effect is particularly pronounced if an elastomer is employed as the polymeric component. At a constant water/cement ratio the flexural strength increases with increasing polymer content in the mix (Schulze, 1999), whereas the compressive strength is usually altered only little but may sometimes decline, owing to air entrapment. The creep of the hardened cement pastes tends to increase with increasing polymer additions. Figure 13.1 shows the effect of the addition of different polymers on the flexural strength of the polymer-modified material. Table 13.1 summarizes the properties of a cement paste combined with different amounts of a butadiene-styrene co-polymer.

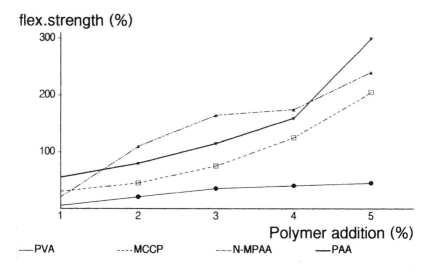

Figure 13.1 Relative increase of flexural strength in polymer-cement composites made with different polymers. PVA, polyvinyl alcohol; MCCP micro-crystalline cellulose polymers; N-MPAA, N-methylol acrylamide; PAA, polyacrylamide
Source: Gopinathan (1992)

Table 13.1 Properties of hardened Portland cement pastes modified with a butadiene-styrene copolymer dispersion.

Property	Units	ml dispersion/100 g cement			
		0	10	30	50
Compressive strength	MPa	112	92	72	58
Flexural strength	MPa	8.4	10.9	13.4	14.0
Modulus of elasticity	GPa	30.5	25.3	19.1	14.5
Hardness	MPa	402	328	248	175
Fracture of energy	Nm	44	82	193	281
Bond strength	MPa	0.68	1.04	1.16	1.28

Curing: 365 d in humid air

Source: Odler and Liang (1992)

An important feature of polymer-modified mixes is their improved bonding to old concrete or steel surfaces, which makes them particularly suitable for repair work (Schulze, 1999). The addition of polyvinyl alcohol, polyvinyl acetate or epoxy resin to the fresh mix appears to be most effective in this respect (Stark *et al.*, 1996). An improved bond may also be achieved between the fibers and cement matrix in fiber-reinforced composites, resulting in improved strength properties (Fu *et al.*, 1996).

Polymer-modified mortar/concrete mixes also tend to exhibit an improved resistance to chemical corrosion and abrasion, and a high water-tightness (Shaker *et al.*, 1997). It has also been reported (Wang *et al.*, 1998) that the addition of acrylic latex to a fresh concrete mix is effective at reducing the corrosion of steel reinforcement. A similar effect may be achieved by applying a polymer coating on the rebars.

Upon heating, thermal degradation or even burning-out of the polymeric component may take place, followed – at even higher temperatures – by the decomposition of the hydrated cement paste. Polymer-modified mortars and concretes are particularly suitable for heavy-duty uses, especially for applications where they are expected to be exposed to frequent mechanical impacts or chemical corrosion, but not to elevated temperatures. They are also widely used for repair works, especially where a good bond to the damaged underlying old concrete surface and/or uncovered steel reinforcement is essential.

Efforts have also been made to combine inorganic cements with inorganic polymers, and in particular calcium aluminate cement with sodium polyphosphates (Walter and Odler, 1996). The strengths obtained with such combined binders are generally higher than those of the calcium aluminate cement alone. In the range between 4 and 30 phosphate units the obtained strength does not seem to be affected by the chain length of the polyphosphate used.

13.3 MDF CEMENT

The term "MDF cement" is misleading, as this is not a cement but a cementitious, polymer-modified and specially processed material. The designation "MDF" stands for "macrodefect-free," and it originated at a time when it was believed that the absence of macrodefects was solely responsible for the high strengths of this material.

The starting constituents in making an MDF material are an inorganic cement, an organic polymer, and mixing water. Small amounts of glycerine have been found to improve the processing, and may also be added. Two cement/polymer combinations have been found to be particularly suitable, and are employed nowadays almost exclusively:

- calcium aluminate cement in combination with a polyvinyl alcohol–polyvinyl acetate co-polymer, and
- Portland cement in combination with polyacrylamide.

The calcium aluminate cement used should preferably have a high Al_2O_3 content. A product with the brand name Secar 71 by Lafarge, with an Al_2O_3 content of around 70%, has been found to be particularly suitable, but other calcium aluminate cements may also be used. The polyvinyl alcohol–polyvinyl acetate co-polymer is the product of an incomplete hydrolysis of polyvinyl acetate. The selection of an appropriate product is critical. The simultaneous presence of hydroxy and acetyl groups within the polymer appears to be essential. It has been suggested, however, that both the strength and the water resistance of the resultant MDF cement will increase with decreasing degree of hydrolysis of the polymer (Branca *et al.*, 1996; Santos *et al.*, 1999) and with decreasing molecular weight (Santos *et al.*, 1999). The polymer droplets in the dispersion should be small enough (that is, below 100 μm) to ensure rapid and effective distribution of the polymer within the MDF mix during processing. A PVA-PVAl product widely and successfully used in producing MDF materials is one with the brand name Gohsenol KH-172 by Nippon Gohsel.

There do not seem to be any particular requirements for the composition of the Portland cement employed. Ordinary Portland cement performs better than slag-modified Portland cement (Santos *et al.*, 1999). The polyacrylamide polymer should also be available in a well-dispersed form, to facilitate processing.

In addition to polyvinyl alcohol/acetate and polyacrylamide, some other polymers have also been employed as the organic constituents of MDF cement, including polypropylene glycol (Hsu and Juaang, 1992) and hydroxypropyl-methyl cellulose (Drabik *et al.*, 1992, 1998). As well as Portland and calcium aluminate cements, sulfoaluminate-ferrite-belite cement (in combination with hydroxypropyl-methyl cellulose) has also been employed as constituent of an MDF material (Drabik *et al.*, 1992, 1997, 1998).

The following ratios of the individual constituents may be considered typical for MDF materials:

- inorganic cement   75–85 wt%
- organic polymer   5–10 wt%
- water   8–15 wt%
- glycerine   0.3–0.6 wt%

It is also possible to add also various inorganic fillers to the mix, such as finely ground silica, fly ash, metallic powders to increase electrical and thermal conductivity, or silicon carbide to increase abrasion resistance (Poon and Groves, 1988; Young, 1991; Kim *et al.*, 1992a, 1992b). They may be introduced in amounts of up to 50 wt% without any dramatic affect on the final strength of the resulting MDF material. MDF materials have also been developed that are reinforced with small amounts of organic fibers or whiskers (Alford and Birchall, 1985b; Kim *et al.*, 1992a, 1992b; Malek and Salama, 1992).

The production of MDF materials consists of a series of steps:

1   premixing;
2   high-shear mixing;

3 forming;
4 curing.

Premixing

The components are premixed in a conventional low-sheer blender.

High-shear mixing

This procedure is usually performed in a two-roll mill, a device commonly used in the plastics and rubber industry. The apparatus consists of two rolls that rotate counter to each other. The gap between them is adjustable, and may vary between about 0.5 and 2 mm. Owing to a difference in their speeds of rotation, shear forces are generated in the material that is allowed to pass through the gap between the rolls. In this way shear rates exceeding 1000/s may be realized.

High-shear mixing is an important step in making MDF cement. Originally it was believed that high-shear mixing just helps to eliminate macrodefects (represented by large pores and air voids) from the paste, and the polymer acts as a processing aid that helps to achieve this goal. This belief was based on the well-known fact that the flexural strength of porous bodies is a function not only of overall porosity but also of the critical flaw size, which in cement pastes is characterized by the size of the largest pores present. In a further development it was recognized, however, that high-shear mixing, in addition to eliminating large pores, also helps to induce chemical interactions between the polymer and the inorganic cement (Tan *et al.*, 1996), and it was recognized that much of the mechanical strength of MDF materials is due to these mechanically induced chemical changes.

The way in which the material is high-shear-mixed largely determines its ultimate physico-mechanical properties (McHugh and Tan, 1992; Lewis and Kriven, 1993; McHugh *et al.*, 1996; Tan *et al.*, 1996). Within a reasonable range the strength will increase with increasing shear rate and increasing mixing time. However, the shear rate cannot be increased with impunity, because with an increasing rate of mixing the amount of heat generated by viscous damping also increases, and increased temperature of the mix may cause the material to stiffen before the processing is completed. An additional increase of temperature also takes place as a consequence of exothermic chemical reactions taking place within the material. A cooling of the rolls of the two-roll mill may extend the time available for high-shear mixing and thus affect positively the quality of the MDF material produced.

During high-shear mixing the consistency of the mix is gradually altered. The material ultimately attains a rubbery consistency, with viscoelastic properties.

Forming

Different methods may be used to obtain the desired shape of the material. Sheets with thicknesses between about 0.5 and 2 mm may be produced by calendering. These may be densified further by pressing between two plates, which may be heated. Other possible processing methods include extrusion and injection molding or pressing.

Curing

After forming, the material may be cured in humid air at ambient or elevated temperature. A final drying at about 80 °C increases the ultimate strength significantly.

The hardening and strength development of MDF materials is the result of chemical reactions taking place in the material during processing and curing. These include both an interaction between the inorganic cement and the polymer, and cross-linking reactions within the polymer itself (Opoczky and Horvath, 1994). Van der Waals forces may also be involved (Rodrigues and Joekes, 1998).

In the system comprising calcium aluminate cement and polyvinyl alcohol/acetate the cement hydration is limited, as the polymer slows down the hydration rate. Non-hydrated particles of the cement become constituents of the developed structure, held together by a polymeric-inorganic matrix. The phase C_2AH_8 is formed as the main hydration product of the cement (Lewis and Kriven, 1993).

In the system comprising Portland cement and polyacrylamide more hydrated material is formed, but again a significant part of the cement remains non-hydrated. The cement hydration products are similar to those formed in the absence of the polymer, but they exhibit a denser packing.

Because of the high pH of the liquid phase brought about by the hydration of the cement, the polymers also undergo chemical reactions. In the polyvinyl alcohol/acetate co-polymer a hydrolysis of the acetate groups takes place. The liberated acetate groups react with cations in the liquid phase, and in particular with Ca^{2+} ions, and calcium acetate is precipitated (Harsh *et al.*, 1992):

$$-CH_2\text{-}\underset{\underset{\text{CO}}{\overset{\displaystyle |}{\underset{|}{\overset{\displaystyle |}{\text{O}}}}}{\overset{|}{\text{C}}}H\text{-}CH_2\text{-}\underset{OH}{\overset{|}{C}}H\text{-}CH_2\text{-}\underset{OH}{\overset{|}{C}}H\text{-}CH_2\text{-}$$

-CH₂-CH-CH₂-CH-CH₂-CH-CH₂-   -CH₂-CH-CH₂-CH-CH₂-CH-CH₂
    OH   O   OH   +OH-   OH   OH   OH
            CO   →   +
            CH₃   CH₃-CO-O-   (13.1)

$$2\ CH_3\text{-}CO\text{-}O\text{-} + Ca^{2+} \rightarrow CH_3\text{-}CO\text{-}O\text{-}Ca\text{-}O\text{-}CO\text{-}CH_3 \qquad (13.2)$$

In the case of polyacrylamide the amide groups undergo hydrolysis, and carboxy groups are formed instead (Silsbee *et al.*, 1991a; Hu, 1992):

$$-CH_2-CH-CH_2- \quad +H_2O \quad -CH_2-CH-CH_2- \; + \; NH_3 \qquad (13.3)$$
$$\downarrow \qquad\qquad \rightarrow \qquad\qquad \downarrow$$
$$CO \qquad\qquad\qquad\qquad CO$$
$$\downarrow \qquad\qquad\qquad\qquad\qquad \downarrow$$
$$NH_2 \qquad\qquad\qquad\qquad OH$$

It is generally believed that the chemically modified polymer interacts with the cations liberated in the cement hydration, but the exact nature of this interaction is not obvious. It has been suggested (Rodgers *et al.*, 1985; Young, 1991; Harsh *et al.*, 1992) that $(Al(OH)_4)^-$ ions released by the calcium aluminate cement cross-link the polyvinyl alcohol chain in the following way:

$$-CH_2-CH-CH_2-CH-CH_2-$$
$$\downarrow \qquad\qquad \downarrow$$
$$O \qquad\qquad O$$
$$\searrow \qquad \swarrow$$
$$Al$$
$$\swarrow \qquad \searrow$$
$$O \qquad\qquad O$$
$$\downarrow \qquad\qquad \downarrow$$
$$-CH_2-CH-CH_2-CH-CH_2-$$

However, this assumption was questioned by Desai *et al.* (1992). A similar cross-linking involving polyvinyl alcohol and Ca^{2+} ions is considered unlikely (Igarashi and Takaheshi, 1992).

As to polyacrylamide, it has been postulated that after removal of the NH_2 group the following bonds may be formed, involved in cross-linking reactions (Hu, 1992; Silsbee *et al.*, 1991a):

$$-CH_2-CH-CH_2- \qquad\qquad -CH_2-CH-CH_2-$$
$$\downarrow \qquad\qquad\qquad\qquad\qquad \downarrow$$
$$CO \qquad\qquad\qquad\qquad\qquad CO$$
$$\downarrow \qquad\qquad\qquad\qquad\qquad \downarrow$$
$$O \qquad\qquad\qquad\qquad\qquad O$$
$$\downarrow \qquad\qquad\qquad\qquad\qquad \downarrow$$
$$Ca \qquad\qquad\qquad\qquad\qquad Al$$
$$\downarrow \qquad\qquad\qquad\qquad\qquad \| \|$$

In addition to the participation of ionic species in cross-linking reactions, it is also believed that additional bonds develop between the -OH or –CO–OH groups of the polymer and ionic species at the cement grain surface (Drabik *et al.*, 1992; Harsh *et al.*, 1992; Hu, 1992). The bonding forces formed in this way are quite strong, and thus the failure of MDF material tends to occur within the polymer phase or clinker grains rather than at their interface (Hu, 1992).

On the micrometer scale the structure of MDF materials consists of densely packed residual unreacted cement grains embedded in a polymeric matrix. An important role in the development of strength is played by an interphase region that surrounds the unreacted grains and is responsible for the high degree of bonding between cement particles and the polymeric matrix. In this region crystallites of C_2AH_8 (5–8 mm in size) reside in an amorphous matrix (Lewis and Kriven, 1993; Lewis *et al.*, 1994). The following composition was estimated for a mature calcium aluminate cement-based MDF material and the original mix (Lewis and Kriven, 1993; Lewis *et al.*, 1994):

Component	Composition (wt%)	(vol.%)
Original mix		
High alumina cement	84.3	65.2
Polyvinylalcohol	5.9	12.3
Glycerol	0.6	1.4
Water	9.3	21.1
Mature MDF material		
Unreacted cement ($CA + CA_2$)	78.7	66.6
Binder phase		
C_2AH_8	10.1	13.2
$Al(OH)_3$	4.4	4.6
Plasticized PVA	6.8	15.6
Non-volatiles		
Volatiles (in binder phase)		
Plasticized PVA	6.8	15.6
H_2O	5.5	14.0

The distribution of constituents of the binder phase between the interphase region and the bulk matrix was estimated as follows (Lewis *et al.*, 1994):

Constituent (vol.%)	Interphase region	Matrix
C_2AH_8	39.5	
$Al(OH)_3$	13.8	
PVA	9.9	36.8
	63.2	36.8

MDF materials exhibit a relatively small but distinct porosity (Alford *et al.*, 1982; Poon and Groves, 1987; Silsbee *et al.*, 1991b; Young, 1991; Igarashi and Takaheshi, 1992; Roy, 1992). In general the porosities found in Portland cement based MDF materials are larger than those of calcium aluminate cement based products. From mercury porosimetry values of up to 20 vol.% have been reported for the former and below 5 vol.% for the latter. Values found by nitrogen adsorption are usually very low. The maximum flaw sizes in MDF materials are reduced typically to 10–100 μm (Russell *et al.*, 1991).

Hardened MDF pastes exhibit strengths that – if compared with conventional cementitious materials – must be considered extremely high. Out of the two main types of MDF materials, those based on calcium aluminate cement are about twice as strong as those based on Portland cement. Flexural strengths exceeding 100 MPa may be achieved in the Portland cement/polyacrylamide system and strengths exceeding 200 MPa in the calcium aluminate cement/polyvinyl alcohol/acetate system. Such high strengths can be attributed to several factors, including a high degree of compaction, a low overall porosity, an absence of large pores acting as macrodefects, and the extremely favorable intrinsic strength properties of the cement/polymer composite material, in which an important role is played by the interphase region (Tan *et al.*, 1996). The typical physico-mechanical properties of MDF cement may be summarized as follows (Chatterjee, 1992):

- flexural strength   150–200 MPa
- compressive strength   300 MPa
- Young's modulus   50 GPa
- critical stress intensity factor   3 MPa.m$^{1/2}$
- Poisson's ratio   0.2
- Density   2500 kg/m^3
- coefficient of thermal expansion 9.7×10^{-6} m/mK

For a further improvement of mechanical properties MDF materials may be reinforced with fibers. Carbon and alumina fibers gave the best results (Park, 1998). The optimum fiber amount was 10–15% by volume.

A serious drawback of MDF materials is their sensitivity to water. Upon immersion in water they exhibit a significant uptake of water and an expansion, associated with a loss of strength, mainly within the first two weeks. The residual strength may be as low as 20% of the original strength of the oven-dried material, and the linear expansion may exceed 20 mm/m (Poon and Groves, 1988; Igarashi and Takaheshi, 1992). This loss of strength must be attributed to a swelling and softening of the organic polymer phase constituting the MDF material. It has also been suggested that the reduction of strength seen under moist conditions is due to a weakening of the bonding between the polymer and the cement matrix by van der Waals forces and a base/acid interaction between the polymer and water (Rodrigues and Joekes, 1998). If kept under water, the material may lose weight, in spite of water uptake and swelling. This loss is due to a partial dissolution of the cement and the polymer in the surrounding water (Young, 1991). Upon redrying the material retains a significant fraction, but usually not all, of its original dry strength.

Just as in samples kept under water, an uptake of water and expansion of the material associated with a loss of strength may also be observed in samples kept in humid air. Here the amount of absorbed water and the extent of expansion increase with increasing relative humidity. Weight increases of up to 10% and linear expansions up to 0.8% may be observed under these conditions. The actual extent of water uptake and expansion will also depend on the processing and curing conditions. Samples hot-pressed at 90 °C have been reported to show a lower expansion than those processed at only 40 °C (Igarashi and Takaheshi, 1992). The loss of strength may exceed 50%, and will also increase with relative humidity and curing time.

Along with an uptake of water, swelling and loss of strength, MDF cements also exhibit a distinct increase of creep deformation upon exposure to water. This effect increases with increasing relative humidity.

Under the same experimental conditions all these moisture effects generally increase with increasing polymer content in the MDF cement.

Efforts have been made to eliminate or at least reduce the adverse effect of moisture on MDF materials. A significant improvement was achieved by the use of organosilane or titanate coupling agents or by incorporating an isocyanate compound to cross-link the polyvinyl alcohol chains through urethane bonding (Young, 1991).

It has also been suggested (Rodrigues and Joekes, 1998) that the water resistance of MDF materials might be improved by an *in situ* reticulation of the polymer. This would entail combining with Portland cement an aqueous solution of partially hydrolyzed polyvinyl alcohol and sodium silicate, which would act as the reticulation agent.

Drabik *et al.* (1998) achieved a reduction of water susceptibility of an MDF system consisting of a combination of a sulfoaluminoferrite belitic clinker and hydroxypropylmethyl cellulose by adding sodium polyphosphate to the original mix. This improvement was attributed to the formation of Al(Fe)–O–C(P) cross-links in the hardened material.

The resistance of MDF cement to elevated temperature is limited by the presence of the organic constituent. At around 500 °C the polymer is burnt out, and along with this the strength of the material declines distinctly.

Efforts have also been made to produce MDF-like materials in which the role of the organic polymer is taken over by an inorganic polymer, specifically sodium polyphosphate (Ma and Brown, 1992). In this way flexural strengths of up to about 30 MPa were achieved in combination with calcium aluminate cement. The resulting phosphate-containing reaction products were found to be amorphous.

MDF materials have been produced in limited amounts so far. A variety of possible applications is being considered for these materials, including:

- armor: personnel armor, vehicle armor, fragmentation protection etc.;
- loadbearing structural elements: floors, ceilings, partitions etc.;
- nuclear waste containment;
- transportation of gases and liquids: pipes, containers;
- plastics fabrication: tools for plastics fabrication, compressive moulds;
- brake-lining matrixes (replacement for phenolic resins used currently);
- electrical engineering applications.

Compared with metals, a distinct advantage of MDF materials is that they can be formed at low temperatures, and they do not require machining.

13.4 OTHER POLYMER–CEMENT COMPOSITE MATERIALS

Hasegawa *et al.* (1995) developed a new cementitious material by combining calcium aluminate cement with a methanolic solution of a phenol resin precursor. The water-free mix must be processed by high-shear mixing in a twin-roller mill in a way similar to that used in the production of MDF materials. The rheology of the resin–cement paste may be improved by adding to the system small amounts of a modifier such as alcohol-soluble polyamide and a plasticizer. A typical mix proportion is as follows:

Material	Parts per weight
Calcium aluminate cement	100.0
Phenol resin precursor (methanol-soluble resol)	12.3
Modifier	1.8
Plasticizer	2.3

The phenol resin precursor acts as a processing aid, as a filler to occupy the space between the calcium aluminate cement particles, and as a source of water for the hydration of the cement. The methanol also initially affects the rheology of the mix positively, but most of it evaporates in the course of processing. In the subsequent curing at 200 °C the resin precursor undergoes polycondensation, yielding a high molecular weight phenol-formaldehyde resin and liberating free H_2O (about 2 wt% of total). The latter in turn reacts with the cement, causing its partial hydration. The resulting material consists essentially of a phenol-formaldehyde resin matrix with enclosed particles of partially hydrated calcium aluminate cement. A formation of some kind of chemical bond between the matrix and cement particles appears likely. The porosity of the material is very low, and typically does not exceed 3 vol.%. The following strength properties have been reported for the hardened material:

- flexural strength   120–220 MPa
- compressive strength   300 MPa
- tensile strength   54–100 MPa
- bending modulus of elasticity   32–45 GPa

Upon heating, the strength of the material is maintained up to about 250 °C, but it declines at higher temperatures owing to thermal degradation of the resinous part. Unlike MDF materials, the material is stable upon exposure to water, and does not exhibit significant strength reduction even after immersion in water for three months. Similar materials may also be obtained if calcium aluminate cement in the starting mix is replaced by Portland or Portland–blast furnace slag cement, but the resulting strengths are distinctly lower.

REFERENCES

Alford, N., and Birchall, J.D. (1985a) Properties and potential applications of macro-defect free cement. *Materials Research Society Symposium Proceedings* **42**, 265–276.

Alford, N., and Birchall, J.D. (1985b) Fibre toughening of MDF cement. *Journal of Materials Science* **20**, 37–45.

Alford, N., Groves, G.W., and Double, D.D. (1982) Physical properties of high strength cement paste. *Cement and Concrete Research* **12**, 349–358.

Birchall, J.D., Howard, A.J., and Kendall, K. (1981) Flexural strength and porosity of cement. *Nature* **289**(29), 388–390.

Branca C. *et al.* (1996) The influence of polyvinylalcohol on macro-defect free cement composites. *Materials Engineering* **7**, 55–60.

Chatterjee, A.K. (1992) Special and new cements, in *Proceedings 9th ICCC, New Delhi*, Vol. 1, pp. 177–212.

Desai, P.G., Young, J.F., and Wool, R.P. (1992) Cross-linking reactions in macro-defect free cement composites. *Materials Research Society Symposium Proceedings* **245**, 179–184.

Drabik, M., and Slade, R.C.T. (1995) Interaction of soluble polymers and hydrated cement in the model subsystem $C_4A_3\bar{S}$-HPMC-(poly-P)-H: investigation by nuclear magnetic spectroscopy. *British Ceramic Transactions* **94**, 242–245.

Drabik, M. *et al.* (1992) Chemistry and porosity in modelled MDF cement materials, in *Proceedings 9th ICCC, New Delhi*, Vol. 3, pp. 386–392.

Drabik, M. *et al.* (1997) MDF-related systems based on sulfobelitic clinker/hydroxypropylmethyl cellulose/polyphosphate glass composition, in *Proceedings 9th ICCC, Göteborg*, paper 3iii011.

Drabik, M., Zimermann, P., and Slade, R.C.T. (1998) Chemistry of MDF materials based on sulfoaluminateferritebelitic clinkers: syntheses and tests of moisture resistance. *Advances in Cement Research* **10**, 129–133.

Fu, X., Lu, W., and Chung, D.D.L. (1996) Improving the bond strength between carbon fiber and cement by fiber surface treatment and polymer addition to cement mix. *Cement and Concrete Research* **26**, 1007–1012.

Gopinathan, K., and Ramano Rao, D.V. (1992) Influence of some water soluble polymers on the mechanical properties of cement-polymer composites, in *Proceedings 9th ICCC, New Delhi*, Vol. 5, pp. 538–543.

Grosskurth, K.P. (1991) Morphology and long term behavior of polymer cement concrete. *Materials Research Society Symposium Proceedings* **179**, 273–281.

Harsh, S., Naidu, Y.C., and Ghosh, S.N. (1992) Chemical interaction between PVA and hydrating HAC: infrared spectroscopic and thermoanalytical investigations, in *Proceedings 9th ICCC, New Delhi*, Vol. 3, pp. 406–412.

Hasegawa, M. *et al.* (1995) A new class of high strength, water and heat resistant polymer-cement composite solidified by an essentially anhydrous phenol resin precursor. *Cement and Concrete Research* **25**, 1191–1198.

Hsu, K.C., and Juaang, Y.T. (1992) Preparations and characteristics of chemically bonded ceramics (in Chinese). *Cailiao Kaksue* 24, 231–239 [ref. CA 120/142248].

Hu, S. (1992) The increasing strength mechanism of the role of interfacial bond in MDF cement, in *Proceedings 9th ICCC, New Delhi*, Vol. 3, pp. 393–399.

Igarashi, H., and Takaheshi, T. (1992) The influence of moisture on the expansion of macro-defect-free cements. *Materials Research Society Symposium Proceedings* **245**, 173–178.

Kim, T.H., Park, Y.W., and Choi, S.H. (1992a) Mechanical and microstructural characterization of MDF cement–SiC whisker composites, in *Proceedings 9th ICCC, New Delhi*, Vol. 3, pp. 373–379.

Kim J.H., Choi, S.H., and Han, K.S. (1992b) Mechanical properties of MDF cement composites with SiC powder, in *Proceedings 9th ICCC, New Delhi*, Vol. 3, pp. 380–385.

Lewis, J.A., and Kriven, W.M. (1993) Microstructure-property relationships in macro-defect-free cement. *MRS Bulletin* **18**, 72–77.

Lewis, J.A., Boyer, M., and Bentz, D.P. (1994) Binder distribution in macro-defect-free cements: relation between percolative properties and moisture absorption kinetics. *Journal of the American Ceramic Society* **77**, 711–716.

Ma, W., and Brown, P.W. (1992) Cement–inorganic polymer composites, microstructure and strength development, in *Proceedings 9th ICCC, New Delhi*, Vol. 4, pp. 424–429.

Malek, R.I.A., and Salama, M.N. (1992) Structure/performance characteristics of Aramide fibers reinforced MDF cement, in *Proceedings 9th ICCC, New Delhi*, Vol. 5, pp. 499–504.

McHugh, A.J., and Tan, L.S. (1992) Processing–properties interactions in macro defect free cements. *Materials Research Society Symposium Proceedings* **245**, 185–190.

McHugh, A.J. *et al.* (1996) Processing–structure–properties interaction in polymer-cement composites, in *Proceedings MAETA Workshop on High Flexural Polymer-cement Composites, Sakata*, pp.59–67.

Montaro, L. *et al.* (1990) Influence of added polymer emulsions on the short-term physical and mechanical characteristics of plastic mortars. *Cement and Concrete Research* **201**, 62–68.

Odler, I., and Liang, L. (1992) Physico-mechanical properties of cementitious systems modified by elastomer dispersions, in *Proceedings 7th International Congress on Polymers in Concrete, Moscow*, pp. 201–212.

Ogawa, H., Okuda, A., and Kawachi, T. (1997) Fundamental properties of polymer-modified cement mortars using latent cure type epoxy resin (in Japanese). *Semento Konkurito Ronbunshu* **511**, 894–899 [ref. CA 129/ 220815].

Ohama, V., Demura, K., and Endo, T. (1992) Strength properties of epoxy-modified mortars without hardener, in *Proceedings 9th ICCC, New Delhi*, Vol. 5, pp. 512–511.

Ollitrault-Fichet, R. *et al.* (1998) Microstructural aspects in a polymer-modified cement. *Cement and Concrete Research* **28**, 1687–1693.

Omaha, Y. (1989) Principle of latex modification and some properties of latex-modified mortars and concretes. *ACI Materials Journal* (Nov.-Dec.), 611–618.

Opoczky, L., and Horvath, I. (1994) Fundamental research for the preparation of macrodefect free cement (Hun.). *Epotöanyag* **46**, 2–8 [ref. CA 121/211457].

Park, C.K. (1998) Characterization of fiber reinforced macro-defect-free cementitious materials. *Journal of the Ceramic Society of Japan* **106**, 268–271.

Poon, C.S., and Groves, G.W. (1987) The effect of latex on macrodefect-free cement. *Journal of Materials Science* **21**, 2148–2152.

Poon, C.S., and Groves G.W. (1988) The microstructure of macrodefect-free cement with different polymer contents and the effect on water solubility. *Journal of Materials Science* **23**, 657–660.

Rao, B.K. *et al.* (1992) Improving ductility of concrete by incorporating natural rubber latex, in *Proceedings 9th ICCC, New Delhi*, Vol. 5, pp. 550–556.

Rodgers, S.A. *et al.* (1986) High strength cement pastes. *Journal of Materials Science* **20**, 2853–2860; *Materials Research Society Symposium Proceedings* **42**, 45–51.

Rodrigues, F.A., and Joekes, I. (1998) Macro-defect free cements: a new approach. *Cement and Concrete Research* **28**, 877–885.

Roy, D.M. (1992) Advanced cement systems, including CBC, DSP, MDF, in *Proceedings 9th ICCC, New Delhi*, Vol. 1, pp. 357–380.

Russell, P. *et al.* (1991) Moisture resistance of macrodefect-free cement. *Ceramic Transactions* **16**, 501–519.

Santos, R.S. *et al.* (1999) Macro-defect free cements: influence of poly(vinyl alcohol), cement type and silica fume. *Cement and Concrete Research* **29**, 747–751.

Schulze, J. (1999) Influence of water-cement ratio and cement content on the properties of polymer-modified mortars. *Cement and Concrete Research* **29**, 909–915.

Shaker, F.A., El-Dieb, A.S., and Reda, M.M. (1997) Durability of styrene-butadiene latex modified concrete. *Cement and Concrete Research* **27**, 711–720.

Silsbee, M.R., Roy, D.M., and Adair, J.H. (1991a) The chemistry of MDF cements produced from polyacrylamide-cement-water pastes. *Materials Research Society Symposium Proceedings* **179**, 129–144.

Silsbee, M.R., Roy, D.M., and Perez-Pena, M. (1991b) A view of macro-defect-free (MDF) pastes. *Materials Research Society Symposium Proceedings* **179**, 145–156.

Skenderovic, B. (1992) Effect of addition of different quantities of polymeric dispersions on hydration of Portland cement and characteristics of mortars, in *Proceedings 9th ICCC, New Delhi*, Vol. 5, pp. 525–530.

Stark, J., Fimming, A., and Dempfwolf, F. (1996) The adhesive bond between epoxide resin modified mortar and the base material (in German). *Bautenschutz-Bausanierung* **19**, 34–41.

Sujjavenich, S., and Lundy, J.R. (1998) Development of strength and fracture properties of styrene-butadiene copolymer latex modified concrete. *ACI Materials Journal* **95**, 131–143.

Tan, L.S. *et al.*(1996) Evolution of mechano-chemistry and microstructure of a calcium aluminate–polymer component. Part II. Mixing rate effect. *Journal of Materials Research* **11**, 1739–1747.

Wakesugi, M. (1996) Organic polymer and cement. *Semento Konkurito Ronbunshu* **594**, 150–152 [ref. CA 125/20301].

Walter, D., and Odler, I. (1996) Investigation of MgO and CaO/Al₂O₃ polyphosphate cements. *Advances in Cement Research* **8**, 41–46.

Wang, X., Chen, G., and Wu, K. (1992) Experimental investigations and fracture properties of ordinary concrete and polymer cement concrete, in *Proceedings 9th ICCC, New Delhi*, Vol. 5, pp. 557–563.

Wang, S.X. *et al.* (1998) Corrosion inhibition of reinforcing steel by using acrylic latex. *Cement and Concrete Research* **28**, 649–653.

Young, J.F. (1991) Macro-defect-free cement: a review. *Materials Research Society Symposium Proceedings* **179**, 101–121.

Young, J.F. (1996) Organo-cement composites (MDF cements): current status, in *Proceedings NAETA Workshop on High Flexural Polymer-cement Composites, Sakata*, pp. 1–12.

Young, J.F., and Berg, M. (1992) Macro-defect-free cement: a novel organo-ceramic composite. *Materials Research Society Symposium Proceedings* **271**, 609–619.

Zeng, S., Short, N.R., and Page, C.L. (1996) Early-age hydration kinetics of polymer-modified cement. *Advances in Cement Research* **8**, 1–9.

14 Densely packed cementitious materials

Densely packed cementitious materials are commonly also called also "DSP materials," which is an acronym for "Densified Systems containing homogeneously arranged ultrafine Particles." The alternative term "reactive powder concrete" is also used. In principle these are cementitious materials formed by the hydration of densely packed particles of an inorganic binder in combination with ultrafine particles of a second, less reactive or non-reactive filler. Some additional particulate materials may also be present.

Inorganic binders, which are materials produced by mechanical comminution of coarser intermediates such as clinker, gypsum, or granulated blast furnace slag, consist of particles with sizes ranging between about 1 and 100 μm. Even if such powders are packed to a maximum degree, empty spaces remain that cannot be filled with solids, owing to the absence of particles that are finer than those produced in the comminution process. Upon mixing with water these empty spaces must be filled with the liquid phase, and their combined volume determines the theoretically lowest amount of water needed for mixing. This amount of water may be reduced if the empty spaces between the densely packed cement particles are filled with a second material, whose particles are small enough to fit. Based on this principle, cementitious systems can be created that exhibit an extremely high degree of packing, a low porosity, and – as a consequence – a very high strength and extremely low permeability. The binder most commonly used in the production of DSP materials is Portland cement. The best results are obtained with a cement with a low C_3A content and high silica modulus (Richard and Cheyrezy, 1995). Also, the cement should not be ground too finely, as this increases its water demand. Alternatively, slag-Portland cement may be used (Young and Jennings, 1991).

Silica fume (microsilica) is the ultrafine material most commonly used in making DSP materials. Its main functions are:

- to fill the voids between the next larger class of particles, i.e. particles of cement;

- to improve the rheology of the fresh mix, by virtue of the perfect sphericity of the microsilica particles;
- to form additional hydrated material in a pozzolanic reaction with calcium hydroxide liberated in the hydration of the cement.

The characteristic of silica fume that is most important in making good-quality DSP materials is the absence of particle aggregates. To avoid or minimize the presence of such aggregates, only silica fume products that have not been mechanically compacted may be used (Aldridge *et al.*, 1992). Microsilica slurries must also be excluded, as the quantity of water contained in them would exceed the total quantity required for making the DSP mix. Best results have been reported with silica fume coming from the zirconia industry, as it has the lowest tendency to aggregation, and is free from impurities (Richard and Cheyrezy, 1995). The particle size and specific surface area of the silica fume are secondary factors, and products with specific surface areas ranging between 10 m^2/g and 20 m^2/g may be considered acceptable. The most injurious impurities that may be present in silica fume products are residual carbon and alkalis (de Larrard, 1989).

Typically, the silica fume/cement ratio in DSP materials is about 0.20–0.25. This ratio corresponds to the best space-filling performance, but it is too high for a full conversion of the silica fume to C-S-H in a pozzolanic reaction.

Besides microsilica, other ultrafine materials may be used as constituents of DSP systems, as long as their particles are small enough to fill the existing spaces left between the particles of cement. Aldridge *et al.* (1992) produced a DSP material using ultrafine rutile (TiO_2) instead of microsilica as filler. The resulting material had properties comparable to those of a DSP product made with microsilica, in spite of the absence of a pozzolanic reaction in this combination of starting materials.

To achieve good space-filling between the cement particles, the particles of microsilica (or rutile) must be effectively dispersed in the liquid phase. This may be achieved by introducing a superplasticizer to the system, in addition to cement and the microfiller. Superplasticizers based on melamine or naphthalene may be used for this purpose. Even better dispersion can be achieved with polyacrylate based dispersing agents (Richard and Cheyrezy, 1995), but these tend to retard the hydration process. To be sufficiently effective, the superplasticizer must be added in relatively large amounts, typically 1.5–2.0% per weight of cement.

The strength of DSP material may be further improved by adding to the starting mix – in addition to microsilica – medium-size particles, with average diameters between about 2.5 and 9 μm (Cheong *et al.*, 1997). Such materials may include finely ground granulated blast furnace slag, fine fractions of fly ash, paper sludge ash, or calcined bauxite. At a dosage of cement : microsilica : medium-size particle = 7:2:1 all these additives are able to lower the water/solid ratio and increase the compressive strength of

MPa

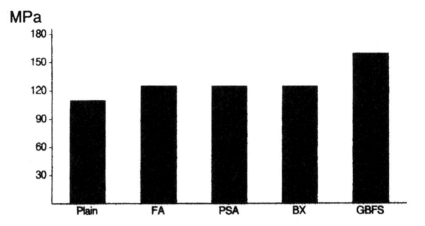

Figure 14.1 Effect of medium-size particles on compressive strength of DSP material. Mixing ratio cement : microsilica : medium-size particles = 7:1:2.
FA, fly ash; PSA, paper sludge ash; BX = calcined bauxite; GBFS =
granulated blast furnace slag.
Source: Cheong *et al.* (1997)

the DSP material; granulated blast furnace slag is the most effective in this
respect (see Fig. 14.1).

In addition to cement and microsilica, it is also possible to add larger,
non-reactive particles to the starting mix. The maximum particle size of
this "aggregate" fraction should not exceed 600 μm. Particles with diameters below 150 μm should be absent, to prevent interference with the
largest cement particles. If particles of this size class are added to the system, the volume of the paste (that is, microsilica + cement + water) must
be at least 20% greater than the void index of the aggregate in the non-
compacted state (Richard and Cheyrezy, 1995). Under these conditions
the aggregate particles will not form a rigid skeleton within the hardened
material, but instead will constitute a set of inclusions trapped in a continuous matrix.

Fine steel particles with diameters not exceeding 800 μm may also be
used as aggregate in DSP products: this results in an even higher strength
of the hardened material (Richard and Cheyrezy, 1995). By limiting the
size of the aggregate particles it is possible to avoid the formation of
microcracks in the material due to external mechanical forces, autogenous
shrinkage, and differential thermal expansion between aggregate and the
paste.

The appropriate water/solid ratio in producing DSP materials ranges
between about 0.12 and 0.22. Under these conditions a dense but fairly
fluid paste is obtained, thanks to the combination of an optimum particle
size distribution of the solids that are present and the chemical action of
the dispersant.

In principle DSP mixes may be processed (including mixing, placing, and compacting) in a similar way as is common in the production of conventional high-performance concrete. Conventional low-shear mixers may be used, but long mixing times – in the range of 10–20 min – are required to achieve an acceptable dispersion of the particles within the mix. A more effective approach is separate dispersion of the microsilica particles in water using a high-shear mixer, before adding cement and other solid constituents to the resulting suspension. Still more effective, however, is combined high-shear mixing of all constituents.

Significant amounts of air may be trapped in the fresh DSP mix during mixing. This air is difficult to remove because of the thixotropic nature of the mix. The presence of air voids in the hardened material is unwelcome, because they adversely affect the strength properties. To remove the entrapped air, vacuum deaeration of the mix after mixing and prior to casting may be necessary. Alternatively, the entrapped air may be removed by deaeration of the mix after it has been cast, or by applying mechanical pressure to the cast mix for a few seconds (Richard and Cheyrezy, 1995).

If mechanical pressure is applied for a longer time to a mix that has not yet set, even some of the original mixing water (up to 20–25%) may be removed. The actual amount removed will depend on the magnitude of the pressure applied (up to 50 MPa) and on the length of pressing, but will also depend on the geometry and size of the DSP specimens studied.

If the mechanical pressure is applied during the setting phase, typically 6–12 hours after mixing, a fraction of the porosity formed within the sample as a result of chemical shrinkage can also be eliminated (Richard and Cheyrezy, 1995).

The progress of cement hydration and the consumption of the microsilica can be monitored by ^{29}Si NMR measurements. By using this method it has been found that – at room temperature – only limited amounts of the cement undergo hydration, and most of it remains in the material in its non-hydrated form. The ultimate degree of hydration varies between about 40% and 60% (Cheyrezy *et al.*, 1995). A C-S-H phase, similar to that formed in conventional Portland cement hydration, is formed in the hydration process. In DSP materials containing silica fume the C/S ratio of the C-S-H phase varies between about 0.9 and 1.3, depending on the degree of hydration and silica fume content, whereas a ratio of 1.6 has been found in samples without silica fume (Lu *et al.*, 1993). The fraction of longest silicate chains (above octamer) in the resulting C-S-H phase increases as the C/S ratio declines. No cross-linked species were found. Some free calcium hydroxide is also liberated in the hydration process, whereas ettringite cannot be detected in the hardened material (Cheyrezy *et al.*, 1995).

As for the cement, most of the silica fume remains non-reacted, and does not undergo a pozzolanic reaction. In one particular case only 20% of the cement and 9% of the silica fume had been consumed within 28 days of hydration at 20 °C (Zanni *et al.*, 1996). These low degrees of reaction must be attributed to the very low water/solid ratios employed in producing DSP materials.

Because of the low initial water/solid ratio the cumulative pore volume of hardened DSP materials is significantly reduced. At the same time there is a shift to smaller radii (Touse, 1989; Cheyrezy *et al.*, 1995). Both factors affect favorably the strength properties and the permeability of the material.

The hydration process, and especially the pozzolanic reaction, may be accelerated and the microstructure of the hardened sample may be modified by curing the DSP mix at elevated temperatures: that is, above 80 °C. Under these conditions the bound water content increases and along with it the residual porosity declines up to a curing temperature of about 200 °C. At the same time the fraction of freed calcium hydroxide, which reacts further with silica fume, increases, and virtually no free portlandite can be detected in the material cured at 200 °C and above (Cheyrezy *et al.*, 1995). About 65% of the microsilica is consumed within 8 hours at 200 °C (Zanni *et al.*, 1996). If the curing temperature is increased even further (up to 400 °C) the amount of free water and the porosity begin to increase again, as xonotlite begins to be formed as the main hydration product at the expense of the semi-crystalline calcium silicate hydrate that is formed at lower temperatures (Sun and Young, 1993; Cheyrezy *et al.*, 1995; Zanni *et al.*, 1996). The presence of tobermorite, gyrolite, kilchoanite, and $8CaO.5SiO_2$ has also been reported in some preparations autoclaved at higher temperatures (Sun and Young, 1993). Figure 14.2 shows the effect of curing on the compressive strength of a DSP composition that contained 20% of finely ground blast furnace slag as well as microsilica (Cheong *et al.*, 1997).

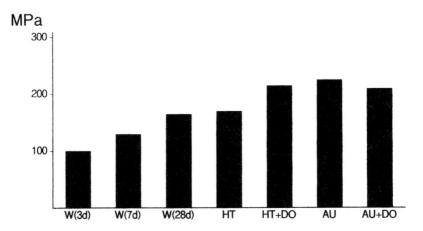

Figure 14.2 Effect of curing conditions on compressive strength of DSP material. W, water curing, 20 °C; HT, hydrothermal curing, 80 °C; AU, autoclave curing, 180 °C; DO, dry oven curing, 200 °C.
Source: Cheong *et al.* (1997)

The physico-mechanical behavior of DSP matrices is purely linear-elastic, with a fracture energy not exceeding about 30 J/m^2. A significant increase of ductility may be achieved by adding steel fibers to the system (1.5–3.0 vol.%). At the same time the presence of fibers improves also tensile strength (Richard and Cheyrezy, 1995).

The following may be considered to be typical physico-mechanical properties of DSP materials (Richard and Cheyrezy, 1995):

DSP materials cured at ambient temperature or heat-treated at temperatures below 90 °C, produced without application of mechanical pressure:

- Compressive strength: 170–230 MPa
- Flexural strength: 30–60 MPa
- Fracture energy: 20–40 J/m^2
- Young's modulus: 50–60 GPa

DSP material exposed to mechanical compaction and heat treated at 250–400 °C

- Compressive strength
    using quartz sand: 490–680 MPa
    using steel aggregate: 650–810 MPa
- Flexural strength: 45–140 MPa
- Fracture energy: 2–20 J/m^2
- Young's modulus: 65–75 GPa

The effect of fiber addition and compaction density on compressive strength becomes apparent from the following data (Richard and Cheyrezy, 1995; relative density is defined as $d = d_0/d_s$, where d_0 designates the density of concrete at demolding, and d_s designates the solid density of the granular mixture, assumed to be compact):

	$d = 0.84$	$d = 0.88$
Non-fibered material cured at 20 °C	145 MPa	180 MPa
Fibered material cured at 20 °C	175 MPa	210 MPa
Non-fibered material cured at 90 °C	215 MPa	255 MPa
Fibered material cured at 90 °C	250 MPa	–

As well as high strength, DSP materials also exhibit low permeability, high corrosion resistance, and volume stability. It has been reported, however, that samples containing undispersed agglomerates of silica fume may expand excessively if exposed to repeated wetting and drying, owing to an alkali-silicate reaction (St John *et al.*, 1996).

If exposed to high temperatures a decline of strength of the material may be observed as the hydrates gradually decompose. In wet samples spalling of the material or a steam explosion may occur owing to the low permeability of the material (Wise and Joner, 1991).

The strength properties, especially at high temperatures, may be improved by reducing the C/S ratio in the starting mix to about 0.8 and drying the material at 400 °C after it has been hydrothermally treated. Such a material exhibited flexural strengths up to 80 MPa, which declined gradually to about 40 MPa after exposure to 800 °C (Wise and Joner, 1991). This decline of strength was attributed to the formation of wollastonite at the expense of amorphous, low-C/S calcium silicate phases.

High-alumina cement may be used instead of Portland cement as a binder in DSP systems, but such a material exhibits chemical shrinkage that is significantly greater than that existing in DSP materials based on calcium silicate. To eliminate this handicap, a DSP material has been developed in which the starting mix contains – besides calcium aluminate cement, silica fume and superplasticizer – a blend of amorphous calcium aluminate and anhydrite. In the hydration process ettringite is formed in the material, in addition to C_3AH_6 and AH_3, and the formation of this phase reduces the overall shrinkage (Mino *et al.*, 1991). A flexural strength of 33 MPa was attained under these conditions. A further increase of strength may be obtained by calcination of the material at temperatures up to 600 °C. In such systems the phase mainly responsible for the attained strength is $C_{12}A_7$, which has been formed in the decomposition of C_3AH_6. Amorphous Al_2O_3 and non-reacted residual CA are also present in significant amounts. Flexural strengths of up to 55 MPa may be achieved by this approach (Mino *et al.*, 1991).

It has been suggested that DSP mixes should be employed instead of high-performance concrete (Richard and Cheyrezy, 1995). Because of the high ductility of DSP material its use (in its fibered modification) is envisioned even for structures not incorporating traditional passive reinforcement. It has been estimated that in the most favorable cases the structures may be up to three times lighter than equivalent structures erected with conventional concrete. The combined effect of reinforcement elimination, reduction of permanent load by lightening the structure, and reduction of the quantity of material used may result in major cost reductions.

Other possible applications of DSP materials include:

- replacement of steel in a variety of mechanical parts (machine components, cutting tools, molds, etc.);
- hardening of military structures and equipment (the material exhibits an excellent projectile impact resistance);
- storage of nuclear and chemical waste.

REFERENCES

Aldridge, L.P., Ross, J.C., and Cassidy, D.J. (1992) DSP with rutile substituting for microsilica. *Materials Research Society Symposium Proceedings* **245**, 223–228.

Cheong, H.M., Lim, J.R., and Cho, D.W. (1997) Properties and microstructure of ultra high strength cement composites, in *Proceedings 10th ICCC, Göteborg*, paper 2ii037.

Cheyrezy, M., Maret, V., and Fronin, L. (1995) Microstructural analysis of RPC (reactive powder concrete). *Cement and Concrete Research* **25**, 1491–1500.

de Larrard, F. (1989) Ultrafine particles for making of very high strength concrete. *Cement and Concrete Research* **19**, 161–172.

Lu, P., Sum, G., and Young, J.F. (1993) Phase composition of hydrated DSP cement pastes. *Journal of the American Ceramic Society* **76**, 1003–1007.

Mino, I. *et al.* (1991) The hydration and calcination mechanism of calcium aluminate-based ultra-high strength cement with calcium sulfoaluminate compound. *Materials Research Society Symposium Proceedings* **179**, 193–202.

Pao, G. *et al.* (1998) Experimental study on the microaggregate effect in high strength and super-high-strength cementitious composites. *Cement and Concrete Research* **28**, 171–176.

Richard, P., and Cheyrezy, M. (1995) Composition of reactive powder concretes. *Cement and Concrete Research* **25**, 1501–1511.

St John, D.A., McLeed, L.C., and Milestone, B. (1996) Durability of DSP mortars exposed to conditions of wetting and drying. American Concrete Institute SP-159, pp. 45–56.

Sun, G.K., and Young, J.F. (1993) Hydration reactions in autoclaved DSP cements. *Advances in Cement Research* **5**, 163–169.

Touse, S.A. (1989) Pore structure of low porosity DSP cement paste. *Materials Research Society Symposium Proceedings*, **137**, 449–456.

Wise, S., and Joner, R.K. (1991) High-temperature processing and application of silica modified Portland cement CBC composites. *Materials Research Society Symposium Proceedings* **179**, 89–100.

Young, J.F., and Jennings, H.M. (1991) Advanced cement based materials, in *Cement and Concrete Science and Technology* (ed. S.N. Ghosh), Vol. I, Part I, ABI Books, New Delhi, pp. 346–372.

Zanni, H., Cheyrezy, M., and Maret, V. (1996) Investigation of hydration and pozzolanic reaction in reactive powder concrete (RPC) using ^{29}Si NMR. *Cement and Concrete Research* **26**, 93–100.

15 Miscellaneous inorganic binders

15.1 MAGNESIUM OXYCHLORIDE AND OXYSULFATE CEMENTS

Magnesium oxychloride cement, also called Sorel cement, is obtained by mixing powdered magnesium oxide (calcined magnesia, MgO) with a concentrated solution of magnesium chloride ($MgCl_2$).

The calcined magnesia is produced by calcination of magnesite ($MgCO_3$) at temperatures of around 750 °C. The conditions of calcination affect the reactivity of the formed MgO, and this in turn influences both the reaction rate and the properties of the reaction product. Underburning produces an excessively reactive product, whereas overburning results in an insufficient reactivity and excessively long setting times of the mix.

The setting and hardening of the binder takes place in a through-solution reaction. Initially, a gelous hydration product is formed, and this process is followed by the precipitation of crystalline reaction products. The reaction proceeds in four distinct steps (Menetrier-Sorrentino *et al.*, 1986):

1 *Rapid dissolution of MgO.* This period lasts a few minutes, and is associated with a sharp but short-lasting exothermic peak. The concentration of Mg^{2+} in the liquid phase reaches 0.1–0.25 mol/L at this stage. The pH value of the solution increases by 1 to 2 units.

2 *Induction (nucleation) period.* In the initial, dissolution stage the liquid phase became oversaturated with respect to the magnesium oxychloride hydrate phases. Nevertheless, precipitation of the latter does not take place yet, owing to delayed nucleation. The concentration of Mg^{2+} and Cl^- ions in the liquid phase does not change significantly during this period.

3 *Period of massive precipitation of reaction products.* In this stage the nucleation barrier is overcome, and a massive precipitation of the reaction products gets under way. The concentration of Cl^- in the liquid phase declines rapidly. At the same time the concentration of Mg^{2+} does not change significantly, as additional amounts of MgO dissolve,

to supply the solution with Mg^{2+}. A setting and hardening of the cement paste takes place. The overall rate of the reaction is controlled by the rate of MgO dissolution. At this state of hydration the process is accompanied by the release of distinct amounts of hydration heat. In mixes in which two different hydrates are formed successively, two distinct thermal peaks may be observed.

4 *Period of slowed-down reaction.* The formation of additional amounts of magnesium oxychloride hydrates is gradually slowed down, as the concentration of Cl^- declines and as the system strives toward a final equilibrium state.

There are known four magnesium oxychloride phases in the system $MgO-MgCl_2-H_2O$:

- $2Mg(OH)_2.MgCl_2.4H_2O$, called the 2.1.4 phase;
- $3Mg(OH)_2.MgCl_2.8H_2O$, called the 3.1.8 phase;
- $5Mg(OH)_2.MgCl_2.8H_2O$, called the 5.1.8 phase;
- $9Mg(OH)_2.MgCl_2.5H_2O$, called the 9.1.5 phase.

Of these, the 3.1.8 and 5.1.8 phases may exist at ambient temperature, whereas the 2.1.4 and 9.1.5 phases are stable only at temperatures above 100 °C (Cole and Demediuk, 1955; Demediuk *et al.*, 1955). The 5.1.8 phase tends to precipitate from an oversaturated solution faster than the 3.1.8 phase. Two additional hydrate phases exist in the system $MgO-MgCl_2-H_2O$ at ambient temperature: $Mg(OH)_2$ and $MgCl_2.6H_2O$.

In magnesium oxychloride cement pastes with $MgO/MgCl_2$ molar ratios over 5, and in the presence of sufficient amounts of water, the 5.1.8 phase and some $Mg(OH)_2$ are usually formed as products of the reaction (Menetrier-Sorrentino *et al.*, 1986). However, if the MgO used is too reactive, the reaction product may consist of larger quantities of $Mg(OH)_2$ and smaller quantities of the 3.1.8 phase, which is formed with a lower rate than $Mg(OH)_2$. The 3.1.8 phase, instead of the 5.1.8 phase, is also formed if small amounts of high-alumina cement are added to the original mix (Deng and Zhang, 1966). This "phase conversion" is due to the precipitation of calcium aluminate hydrates at the MgO grain surface, causing a slow-down of its dissolution. In this way, conditions for the precipitation of the 3.1.8 phase are created.

At an $MgO/MgCl_2$ molar ratio below 5 the final reaction product will be the 3.1.8 phase alone or in combination with the 5.1.8 phase. Owing to its higher formation rate, in the initial stages of the hydration process the 5.1.8 phase will be formed in overproportionate amounts. However, after reaching a maximum about 24 hours after mixing, the amount of this phase will start to decline as it becomes converted to the more stable 3.1.8 phase (Deng and Zhang, 1966).

The hardened paste consists of $Mg(OH)_2$ and relatively large acicular magnesium oxychloride hydrate crystals, which form a three-dimensional crystal network. Unreacted MgO in variable amounts may also be present, depending on the reactivity and fineness of the calcined magnesia employed.

Magnesium oxychloride cement exhibits a fast initial strength development, and final strengths that are comparable to those of Portland cement. For a given porosity its strength properties are superior to those of Portland cement. The intrinsic bonding characteristics of the 5.1.8 and 3.1.8 phases are comparable (Deng and Zhang, 1996).

A serious shortcoming of hardened magnesium oxychloride cement is its limited water resistance. If exposed to water both the 5.1.8 and the 3.1.8 phases are dissolved, and Mg^{2+}, Cl^- and OH^- ions enter the liquid phase. The liquid phase may become oversaturated with respect to $Mg(OH)_2$, and crystalline magnesium hydroxide precipitates, whereas chloride ions remain dissolved (Menetrier-Sorrentino *et al.*, 1986; Zhang and Deng, 1994). The process is associated with a loss of strength of the hardened material. Different additives have been recommended to be added to the original mix to reduce the water sensitivity of magnesium oxychloride cement. They include various phosphates, borax, and organic resins. In the presence of phosphoric acid the insoluble $Mg_2P_2O_7$ phase is formed in the mix in addition to the 5.1.8 phase, thus improving the overall water resistance of the hardened material. Yu (1992) reported the development of a water-resistant magnesium oxychloride cement that consists of MgO grains evenly distributed in a 5.1.8 gel.

Hardened magnesium oxychloride cement is also susceptible to carbonation by the CO_2 in the air. In this process the 3.1.8 phase gradually converts into a basic magnesium chlorocarbonate hydrate phase (de Castellar *et al.*, 1996):

$$3Mg(OH)_2.MgCl_2.8H_2O +$$
$$2CO_2 \rightarrow Mg(OH)_2.2MgCO_3.MgCl_2.6H_2O + 4H_2O \quad (15.1)$$

Carbonation of the 5.1.8 phase may also occur, but it progresses much more slowly. The carbonation may be associated with crack formation in the material.

Magnesium oxychloride cement must not be combined with steel reinforcement or put in contact with steel surfaces, as chloride ions present in the pore solution and the low pH of the latter promote steel corrosion. However, magnesium oxychloride cement may be combined with glass fibers, cellulose fibers, wood chips, and a variety of other materials. Other drawbacks of this cement are poor dimensional stability and poor freeze–thaw resistance.

Because of its sensitivity to water, magnesium oxychloride cement is unsuitable for outdoor applications, or for applications in which exposure to water is likely. The binder is most commonly used for industrial flooring, mainly because of its good elastic properties. Examples of other uses of this binder are:

- lightweight internal partition walls;
- wood-cement boards;
- boards for fire protection;
- polishing bricks.

Magnesium oxysulfate cement is a variant of Sorel cement. It is obtained by mixing calcined magnesia with a magnesium sulfate solution. Alternatively magnesium oxide may be combined with a sulfuric acid solution. A variety of magnesium oxysulfate phases may be formed in the hardening process, of which $3Mg(OH)_2.MgSO_4.H_2O$ and $5Mg(OH)_2.MgSO_4.3H_2O$ are the most important. Most of the properties of the hardened material, including strength, are very similar to those of magnesium oxychloride cement. The cement is used mainly for making lightweight insulating boards.

Magnesium oxyacetate binder may be produced by combining calcined magnesia with magnesium acetate $[Mg(CH_3.COO)_2]$ (Harmuth *et al.*, 1998). The setting characteristics of the binder are similar to those of the magnesium oxysulfate cement. The binder develops considerable strength. The nature of the reaction products has not yet been studied.

15.2   GLASS CEMENTS

A glass cement may be obtained by grinding a glass of the system $CaO–Al_2O_3–SiO_2$ to a fineness of about 300–500 m^2/kg. The oxide composition of the glass should lie in the range 45–55% CaO, 22–40% Al_2O_3 and 12–26% SiO_2. Minor additions of TiO_2 and ZrO_2 may also be present, and may have a positive effect on the quality of the final product (MacDowell, 1991). A glass phase of this composition may be produced by heating an appropriate blend of lime or limestone, alumina, and quartz or amorphous SiO_2 to temperatures exceeding 1600 °C, followed by rapid quenching of the resulting melt in an excess of water.

If mixed with appropriate amounts of water such a glass powder undergoes hydration associated with setting and hardening, yielding strätlingite or a hydrogarnet phase $[C_3(A,F)S_{(3-x)}H_{2x}]$, or a combination of both, as products of hydration. Strätlingite (also called gehlenite hydrate, C_2ASH_8) is an AFm phase having aluminosilicate anions as an interlayer between $[Ca_2Al(OH)_6]^+$ sheets. Upon heating, the compound loses its crystalline water and converts to C_2ASH_6 at 100 °C, and to C_2ASH_4 at 138 °C; it breaks down to an amorphous material at 140–210 °C.

The composition of the hydrogarnet phase may vary within the compositional region bounded by C_3AH_6, C_3FH_6, C_3AS_3, and C_3FS_3. As the glass phase constituting glass cement does not contain more than traces of iron, the product formed in the hydration of this binder is a solid solution of C_3AH_6 and C_3AS_3, which is collectively called "hydrogrossular." The end-member of this series, C_3AS_3 (called "grossular"), has a cubic structure in which the silicon, aluminum, and calcium are in tetrahedral, octahedral, and distorted cubic coordination respectively. In hydrogrossular some or all of the silicons in the structure are omitted, and the charge is balanced by replacing each of the oxygen atoms to which it was attached by hydroxyl groups. In hydrated glass cement pastes the composition of the resulting hydrogarnet phase may vary, depending on the oxide composition of the original glass phase. A hydrogarnet composition corresponding to the formula $C_3AS_{0.3-0.7}H_{4.6-5.4}$ was determined in a hydrated glass cement paste from XRD spacing (Hommertgen and Odler, 1991). The thermal stability of hydrogarnet phases is better than that of gehlenite hydrate; a loss of crystalline water gets under way only at temperatures well above 200 °C.

The properties of glass cement depend greatly on the composition of the glass phase and thus on the quality of the hydrates formed. In general, the hydrogarnet phase predominates as the product of hydration in cements with SiO_2 contents below 15%, whereas at SiO_2 contents above 20% mainly strätlingite is formed (MacDowell 1986, 1991; Hommertgen and Odler, 1991; MacDowell *et al.*, 1991; Yan and Yang, 1997). A combination of both phases is present at SiO_2 contents between 15% and 20%.

Cements yielding mainly the hydrogarnet phase tend to exhibit a fast initial hydration, short setting times, and high short-term strengths, but only moderate final strengths. Cements yielding mainly gehlenite hydrate exhibit a longer setting time and a slower strength development, but relatively high final strengths (MacDowell 1986, 1991; Hommertgen and Odler, 1991). Particularly favorable results may be obtained in some glass cements in which both phases are formed simultaneously in the course of hydration. Table 15.1 summarizes the properties of three glass cements of different composition.

In addition to its composition, the properties of glass cement are also influenced by the way in which the original melt was cooled. In general, the reactivity of the cement increases with increasing cooling rate (Hommertgen and Odler, 1991).

The hydration of the glass cement is usually not complete, and non-hydrated residua of larger cement particles may remain embedded in the hydrate matrix even after long hydration times.

Just like the strength development, the release of the heat of hydration also depends on the composition of the cement. In cements yielding mainly the hydrogarnet phase an intensive exothermic maximum is apparent within the first hour of hydration. Such a maximum is markedly delayed and the overall amount of released heat is significantly reduced if strätlingite is formed (Hommertgen and Odler, 1991).

Table 15.1 Properties of glass cements.

	A	B	C
Composition			
CaO (%)	50	50	50
Al_2O_3 (%)	35	30	25
SiO_2 (%)	15	20	25
w/c	0.40	0.40	0.40
Hydrates formed	Hydrogarnet	Hydrogarnet + strätlingite	strätlingite
Combined H_2O (%)			
1 d	21	24	3
3 d	23	28	4
28 d	23	34	26
Compressive strength (MPa)			
1 d	13	38	0
3 d	15	43	0
28 d	20	62	43

Source: Hommertgen and Odler (1991)

The texture of the hardened material also depends on the quality of the formed hydrates. At equal initial water/cement ratios and degrees of hydration the total porosity and medium pore size are distinctly higher if the hydrogarnet phase, rather than strätlingite, is formed as the product of hydration (MacDowell, 1986; Hommertgen and Odler, 1991). The specific surface area (BET) of the hydrated material typically lies below 10 m^2/g.

It has been observed that in some glass cement pastes, especially those containing the hydrogarnet phase, the specific surface area declines and the medium pore size increases with prolonged curing under water. Along with this, the strength of the material declines as well, though only moderately. These phenomena were attributed to a gradual recrystallization of the existing hydrates associated with a coarsening of the texture of the hardened paste (Hommertgen and Odler, 1991).

An addition of gypsum alters the hydration reaction, as ettringite is formed as the main hydration product (see section 11.4). A simultaneous addition of gypsum and calcium hydroxide is deleterious to the hydration process.

The development of glass cement is still in an experimental stage. Some compositions that exhibit a very fast setting may be considered for highway repair works (MacDowell and Chowdhury, 1990). The cement also exhibits very good resistance against diluted acids and alkaline solutions regardless of the hydration products formed, and may be applied in corrosive environments (MacDowell, 1991). Owing to the relatively low pH of its pore solution, and the absence of calcium hydroxide among its hydration products, glass cement may also be used as the matrix in glass fiber reinforced composites (Hommertgen and Odler, 1991).

Another type of glass cement has been developed by Nakatsu *et al.* (1994). This binder contains a $CaO–Al_2O_3–SiO_2$ glass with a CaO/SiO_2 molar ratio of 1.0 and an Al_2O_3 content of 5 wt%. Such a glass may be produced by heating the appropriate blend of starting materials to 1750 °C and subsequent quenching. The binder is produced by grinding the glass to a specific surface area of 700 m^2/g (Blaine), and by adding to it limited amounts of a Portland clinker powder to act as a hydration accelerator. Upon mixing with water the cement sets and hardens, yielding a C-S-H phase as hydration product. This cement exhibits a low heat of hydration: 72, 100, and 150 J/g after 7, 28, and 90 days respectively.

Suzuki *et al.* (1998) produced a glass corresponding to the composition $12CaO.7Al_2O_3$ that also contained Na_2O or K_2O. This was produced by fusion at 1500 °C and quenching. After grinding and mixing with water the resulting cement yielded the phase C_2AH_8 as a product of hydration. The hydration reaction was intensified if Li_2O, instead of Na_2O or K_2O, was added to the system. At an addition of 3% of this oxide, lithium aluminate hydrate was formed as a product of hydration in addition to C_2AH_8.

15.3   ALKALI SILICATE BINDERS

A number of crystalline alkali silicates exist in the alkali oxide–SiO_2 systems:

- $Li_2O.SiO_2$; $Li_2O.2SiO_2$
- $2Na_2O.SiO_2$; $Na_2O.SiO_2$; $Na_2O.2SiO_2$
- $K_2O.SiO_2$; $K_2O.2SiO_2$; $K_2O.4SiO_2$

All of them exhibit a high solubility in water and a low melting temperature. In addition to their crystalline forms, alkali silicates may also exist in the form of alkali silicate glasses with variable SiO_2/alkali oxide ratios, which may be produced by a rapid cooling of the pertinent melts. The term "water glass" designates concentrated solutions of alkali silicates in water. They may be produced by dissolving solid alkali silicates or alkali silicate glasses in hot water. They may also be produced by dissolving amorphous SiO_2aq. in concentrated alkali hydroxide solutions.

In water glass solutions the SiO_2 is present mainly in the form of $(SiO_4)^{4-}$ monomers, but also exists in the form of larger units, such as $(Si_2O_7)^{6-}$ or $(Si_3O_{10})^{8-}$ chains and $(Si_3O_9)^{6-}$ or $(Si_4O^{12})^{8-}$ rings. The degree of polycondensation increases with declining Me_2O/SiO_2 ratio of the alkali silicate.

Commercial water glass products typically contain about 40% of solids, and appear as highly viscous liquids. They may be used as adhesives at elevated temperatures up to about 500 °C.

Compacts made from mixes of water glass with some powdered materials such as metakaolin, fly ash, or granulated blast furnace slag exhibit hardening upon storage in air at ambient or elevated temperatures. The hardening of the system is due mainly to a lengthening of the existing silicate chains by polycondensation (Ikeda, 1997). Metallic ions released from the fillers may also be involved in the process by forming links of the following type:

$$-O-\underset{\underset{O}{|}}{\overset{\overset{O}{\|}}{Si}}-O^{(-)} + {}^{(-)}O-\underset{\underset{O}{|}}{\overset{\overset{O}{\|}}{Si}}-O- + Me^{2+} \rightarrow -O-\underset{O}{\overset{\overset{O}{\|}}{Si}}-O-Me-O-\underset{O}{\overset{\overset{O}{\|}}{Si}}-O- \qquad (15.2)$$

The hardening reaction progresses rather slowly, and the ultimate strengths are low. In general, the strengths are higher in dry than in wet samples. Soaking in water and redrying increases the strength, probably by promoting an additional polycondensation of the existing silicate chains (Ikeda, 1997). The strength also increases with increasing amounts of ultrafine particles in the filler, probably by promoting the release of metallic ions.

If combined with alkali fluosilicates, water glass solutions exhibit setting and hardening based on a chemical reaction between the dissolved alkali silicate and the fluosilicate (Odler and Hennicke, 1991):

$$2Na_2O.nSiO_2 + Na_2SiF_6 \rightarrow (2+n)SiO_2 + 6NaF \qquad (15.3)$$

The products of reaction are an amorphous form of SiO_2 and crystalline sodium fluoride. Unlike most other cementitious systems, the setting and hardening is based not on a hydration process, but on a chemical reaction in which water is neither taken up nor released. The reaction progresses quite rapidly, and at ambient temperature is nearly completed within about two days. The setting time may be controlled by varying the SiO_2/Na_2O ratio of the water glass employed. In general, the setting time is extended as this ratio declines. Upon drying the hardened material exhibits a distinct shrinkage.

The texture of the hardened material consists of an amorphous SiO_2 matrix with embedded NaF crystals. In samples with too high a starting fluosilicate/water glass ratio some residual Na_2SiF_6 may also be present. The porosity of the material is located mainly in pores with radii of around 100 nm. The BET_{N2} specific surface area of the material dried at 120 °C lies at around 3m^2/g.

If water glass with a Na_2O/SiO_2 molar ratio of 1:3.3 is employed, the stoichiometric $Na_2O.3.3SiO_2/Na_2SiF_6$ weight ratio for reaction (15.2) equals 73.5:26.5%. This corresponds to 13.7 g of Na_2SiF_6 per 100 g of water glass with a 40% solids content, and to a water/solid ratio of 1.1:1. To prevent segregation and to obtain a homogeneous hardened material, a suspension of this kind must be rotated in a closed container until setting, unless it is combined with a suitable aggregate. Table 15.2 summarizes the properties of the hardened material. It is apparent that the obtained strengths are exceptionally favorable compared with other inorganic cementitious systems at equal porosities. Upon heating, the hardened material remains unchanged up to about 550 °C. The following is the sequence of reactions taking place at even higher temperatures:

> \>500 °C: gradual conversion of the primary formed amorphous SiO_2
> into quartz and tridymite
> \>750 °C: formation of crystalline sodium disilicate ($Na_2O.2SiO_2$)
> \>800 °C: evaporation of sodium fluoride
> ~850 °C: melting

In samples that contain residual fluosilicate this phase starts to dissociate to NaF and SiF_4 at around 550 °C.

Table 15.3 summarizes the properties of a series of mortars made with a sodium silicate binder and different aggregates. Even though the obtained strengths varied with the aggregates employed, all mortars exhibited a favorable flexural/compressive strength ratio (between 1:3.7 and 1:5.4). In all instances the sample exhibited a distinct shrinkage upon drying.

Table 15.2 Properties of hardened sodium silicate binder.

Density of original water glass – Na_2SiF_6 suspension 1.39 g/ml		
Density of hardened material dried at 100 °C		0.66 g/ml
Porosity of hardened material dried at 100 °C		73 vol.%
Compressive strength of hardened material dried at 100 °C ($w/s = 1.1$)		
	12 h	7.0 MPa
	1 d	14.8 MPa
	2 d	17.9 MPa
	28 d	26.9 MPa
Modulus of elasticity of hardened material dried at 100 °C ($w/s = 1.1$)		
	12 h	2.75 GPa
	1 d	4.13 GPa
	2 d	4.79 GPa
	28 d	6.10 GPa

Table 15.3 Properties of mortars made with sodium silicate cement.

Aggregate	Basalt 0/1 mm	Crushed brick 0/1 mm	Fireclay 0/1 mm	Corund 0/1 mm	Crushed glass 0/1 mm
Room temp. 4 d + drying 120°C					
Compressive strength (MPa)	12.6	19.6	15.3	11.6	23.5
Flexural strength (MPa)	3.25	4.43	4.13	2.16	6.29
Linear shrinkage (mm/m)	-4.1	-7.5	-9.5	-10.5	-8.3
Heating to 900°C					
Compressive strength (MPa)	23.4	26.4	19.0	40.4	–
Flexural strength (MPa)	4.59	6.24	4.49	7.08	–
Visual assessment	No change	Bending	Bending	No change	Melting

Source: Odler and Hennicke (1991)

Among test specimens that had been heated to 900 °C some exhibited no visible changes, some were moderately bent, and one even melted, depending on the aggregate employed. The strength was generally increased by exposure to elevated temperature. There are several fields of application, where the use of an alkali silicate based binder may be advantageous:

- Because of its hardening kinetics an alkali silicate binder may be used in applications where rapid strength development is required.
- Its thermal stability up to high temperatures makes this binder suitable for high-temperature applications. The upper limit seems to be a temperature of about 800–900 °C.
- In combination with lightweight fillers (such as expanded perlite), concretes or mortars may be produced with strength/dry density ratios more favorable than those that can be obtained with other inorganic binders, including Portland cement. (Typical compressive strength for dry density of 400 kg/m is 1.4–1.6 MPa.)
- Because of its resistance to all acids, except hydrofluoric acid, the binder (in combination with a suitable aggregate) may be used in applications in which contact with acid solutions is expected.

15.4 BELITE-ALUMINATE CEMENT

Belite-aluminate cement or aluminate-belite cement, also called porsal cement, contains β-C_2S (belite) and the calcium aluminate phases CA and $C_{12}A_7$ as its main constituents. It may be produced by firing a raw meal made from limestone, clay, and bauxite at temperatures not exceeding 1250–1300 °C (Zakharov, 1974). The reaction products must be formed through a solid–solid interaction of the starting materials. A sintering of the raw meal at higher temperatures in the presence of a melt must be avoided, to prevent the formation of the hydraulically non-reactive gehlenite phase (C_2AS). It is useful to add to the raw meal also limited amounts of calcium sulfate, to act as a mineralizer and also as a stabilizer of β-C_2S. Small amounts of CaF_2 may also be added, to intensify the clinker formation even further.

The phase composition of the porsal clinker may vary in the following range: β-C_2S, 55–75%; CA, 15–30%; $C_{12}A$, 0–10%; glass phase, 0–10%; and – in the presence of SO_3 – also $C_4A_3\bar{S}$, 0–15%. To obtain porsal cement, the clinker, which is easy to grind, is ground to cement fineness, without adding calcium sulfate to it, to act as set retarder.

The hydration products formed in the hydration of porsal cement include the C-S-H phase, C_4AH_{13} or C_2AH_8, and – if the original cement contains the $C_4A_3\bar{S}$ phase – also some ettringite. Strätlingite (gehlenite hydrate, C_2ASH_8) may also be formed in a reaction between the C-S-H and calcium aluminate hydrate phases, which are formed first (Raina *et al.*, 1978;

Viswanathan *et al.*, 1978). The hydrated paste contains only small amounts of portlandite. Only a very small fraction of the primary formed hexagonal calcium aluminate hydrates converts to C_3AH_6 and AH_3 in the latter course of hydration.

Porsal cement exhibits an acceptable strength development, comparable to that of Portland cement, and a high resistance against sulfate attack. It may be considered for applications similar to those of ordinary Portland cement.

Laxmi (1992) produced colored porsal cements. A blue product was obtained by adding to a low-iron raw meal 1.6 wt% Co_3O_4. A pink cement was produced by adding Fe_2O_3 (6.8 wt%) and ZrO (3.3 wt%).

15.5 FERRITE CEMENT

Ferrite cement contains the calcium aluminate ferrite phase as its main constituent. Its composition in the cement may vary between about C_4AF and C_6AF_2. The cement may be produced by burning a raw meal consisting of limestone, iron ore (such as magnetite), and bauxite to temperatures somewhat above 1300 °C.

In a hydration of the cement at ambient temperature calcium aluminate ferrite hydrate [$C_2(A,F)H_8$ or $CA_2[(Al,Fe)(OH)_5]_{2.3}H_2O$] is formed as the main product of hydration, together with some $C_3(A,F)H_6$ [or $Ca_3[(Al,Fe)(OH)_6]_2$]. These are analogues of the C_2AH_8 and C_3AH_6 phases (also formed in the hydration of CA or C_3A), in which some of the Al^{3+} is substituted by Fe^{3+}. With increasing temperature the amount of $C_3(A,F)H_6$ formed will increase at the expense of $C_2(A,F)H_8$. Some iron hydroxide gel may also be formed in the hydration. The hydration progresses rather slowly, attaining a degree of hydration of about 50% after 28 days. This rate may be increased by adding suitable additives to the mix, such as triethanolamin or sodium carbonate.

Ferrite cement is used only to a very limited extent. It currently finds use as a binder in the production of iron ore pellets.

15.6 SULFOFERRITE AND SULFOALUMINOFERRITE CEMENTS

In the system $CaO–Fe_2O_3–CaSO_4$ the sulfate-bearing phases $4CaO.3Fe_2O_3.SO_3$ (abbreviation $C_4F_3\bar{S}$) and $3CaO.Fe_2O_3.SO_3$ (abbreviation $C_3F\bar{S}$) exist, as well as monocalcium ferrite ($CaO.Fe_2O_3$, abbreviation CF) and dicalcium ferrite ($2CaO.Fe_2O_3$, abbreviation C_2F).

If a starting mix that contains the oxides CaO, Fe_2O_3, and SO_3 is burnt to high temperatures, initially CF is formed in a reaction between CaO and Fe_2O_3 at temperatures between 800 and 1100 °C. In the presence of $CaSO_4$

the primary formed CF reacts further, to yield $C_4F_3\bar{S}$ at 950–1200 °C. In mixes with high CaO contents – that is, with C/F > 2 – the CF reacts further with additional CaO at 1100–1200 °C to yield C_2F, which in turn may react with $CaSO_4$ to yield $C_3F\bar{S}$.

Out of the two calcium ferrite phases, C_2F hydrates appreciably faster than CF. In paste hydration about 65% of the former and 10% of the latter hydrate within 28 days at ambient temperature. Initially, low-basicity hydroferrites are formed as products of hydration, which contribute to a moderate strength of the hardened paste. However, after a few days of hydration these convert into cubic C_3FH_6 and amorphous $Fe(OH)_3$, and this conversion is associated with an almost complete loss of bonding properties.

In the hydration of C_3FS the products of hydration are a ferric AFt phase (ferric ettringite, $C_3F.3C\bar{S}.31H$), C_4FH_{13}, and iron hydroxide [$Fe(OH)_3$, abbreviation FH_3]. In the hydration of the low-basicity sulfoferrite C_4F_3S, initially a ferric AFm phase (ferric monosulfate, $C_3F.C\bar{S}.12H$) is formed, together with some ferric ettringite. However, the latter phase also subsequently converts to ferric monosulfate. Conversion of the formed hexagonal calcium hydroferrite into C_3FH_6 does not occur.

The hydration of both calcium sulfoferrites is much faster than that of calcium ferrites. Here the $C_3F\bar{S}$ phase hydrates faster than $C_4F_3\bar{S}$. In paste hydration both of them exhibit a favorable strength development. The hardening process is accompanied by a distinct expansion, which is greater in the hydration of C_3FS (15 mm/m compared with 7 mm/m for $C_4F_3\bar{S}$).

Sulfoferrite cements are modified Portland cements that – in addition to the phases common in ordinary Portland cement – also contain distinct amounts of calcium sulfoferrites (Osokin and Odler, 1992). To obtain such cements a special clinker that contains the sulfoferrite phases must be produced separately, and must subsequently be ground with ordinary Portland clinker and calcium sulfate.

A sulfoferrite-bearing clinker may be produced in a rotary kiln at burning temperatures of 1200–1350 °C. Its phase composition may vary – depending on the basicity of the raw meal and its gypsum content – in the following range: $C_4F_3\bar{S}$, 10–55%; $C_3F\bar{S}$, 0–60%; C_2S, 25–40%; calcium aluminoferrite phase and other constituents, 5–30%. In commercial sulfoferrite clinkers the composition of the sulfoferrite phases may vary as follows: $3CF.(0.75–1.5)C\bar{S}$ and $C_2F.(0.8–1.5)C\bar{S}$. About 5–15% of sulfoferrite clinker is combined with ordinary Portland clinker and calcium sulfate to obtain sulfoferrite cement.

In the combined hydration of sulfoferrite and Portland clinkers the same hydration products are formed as in the hydration of each of these constituents separately. However, the hardened paste contains very little calcium hydroxide, as most of it has been consumed in the formation of ferric ettringite.

The properties of sulfoferrite cement depend on the form of the sulfo-ferritic phase that predominates in the sulfoferritic clinker. The presence of the low-basicity $C_4F_3\bar{S}$ results in cements that exhibit favorable strength development and insignificant expansion. The hardened paste exhibits a distinct resistance against corrosion (seawater, sulfates), which increases with increasing C_4F_3S content in cement. By contrast, cements containing $C_3F\bar{S}$ exhibit greater expansion, lower strengths, and self-stressing proper-ties. It has been reported (Osokin *et al.* 1992) that a five- to tenfold increase of bending strength of sulfoferrite cements may be attained by a "mag-netic treatment" of the hydrating system.

Sulfoferritic cement has been produced industrially in the former Soviet Union.

Sulfo-aluminoferrite cement contains a sulfo-aluminoferrite clinker that may be produced by burning a raw mix composed of limestone, bauxite, clay, and gypsum at a burning temperature of about 1350 °C. The resulting clinker contains aluminoferrite (a $6CaO.Al_2O_3.2Fe_2O_3/6CaO.2Al_2O_3.Fe_2O_3$ solid solution) and sulfoaluminoferrite (a $6.3CaO.Al_2O_3.2Fe_2O_3.0.3SO_3/6.5CaO.2Al_2O_3.Fe_2O_3.0.5SO_3$ solid solution) phases (Krivoborodov and Samchenko, 1992), as well as dicalcium silicate. If interground with ordi-nary Portland clinker and gypsum an expansive cement is obtained (Osokin *et al.*, 1997).

15.7   ALKALI-ACTIVATED CEMENTS

A variety of powdered natural materials and industrial by-products rich in SiO_2 and Al_2O_3 may be activated with suitable alkaline activators to yield a cementitious system that sets and hardens if mixed with appropriate amounts of water.

All or a substantial fraction of SiO_2 and Al_2O_3 in the starting materials must be present in a non-crystalline, reactive form, as either a glass or an amorphous phase. The material may or may not also contain distinct amounts of CaO and lesser amounts of other oxides. Such starting mate-rials include granulated blast furnace slag and a variety of other industrial slags, class C and class F fly ash, metakaolin, and other types of calcinated clays.

The alkaline activators include sodium or potassium hydroxide, silicate, and carbonate.

The rate of setting and hardening of alkali-activated binders and the quality and quantity of the resulting reaction products depend on a variety of factors, including the quality of the starting aluminosilicate, the quality and amount of alkaline activator, and the temperature of reaction.

In all instances, at first the bonds between interlinked silicate and alu-minate tetrahedra of the starting material are broken down in the highly alkaline environment, and ionic species constituting the structure of the starting material enter the liquid phase. Subsequently, new reaction products are formed, causing setting and hardening. If a starting material is employed that contains – in addition to SiO_2 and Al_2O_3 – significant amounts of CaO (such as granulated blast furnace slag), a C-S-H phase with distinct amounts of incorporated Al^{3+} and with adsorbed alkalis is formed as the main product of reaction at ambient temperature. By contrast, in systems with little or no CaO (as in those based on metakaolinite) an amorphous zeolite-type reaction product predominates. Some crystalline phases may also be formed in the setting and hardening process. Alkali-activated slag cements are discussed in section 8.5.

Cementitious systems based on alkali-activated fly ashes are discussed in section 9.1.5.

15.7.1 Geopolymeric cements

The term "geopolymeric cements" or "geopolymers" is not clearly defined. In general, they are inorganic cementitious systems in which a three-dimensional zeolitic framework has been formed in a polycondensation reaction.

The starting materials include:

* an alumino-silicate precursor such as metakaolinite, which may be produced by calcining kaolinite to about 700 °C:

$$Al_2O_3.2SiO_2.2H_2O \rightarrow Al_2O_3.2SiO_2 + 2H_2O \qquad (15.4)$$

The precursor(s) may contain also variable, but limited, amounts of other oxides, mainly CaO, such as is the case in class F fly ashes.
* sodium or potassium polysilicate $(Na_2O.nSiO_2,\ K_2O.nSiO_2)$, and sodium or potassium hydroxide, which all are readily available in large volumes from the chemical industry. The role of the hydroxides is mainly to maintain a high pH of the system.

The products of the polymerization reaction that results in setting and hardening may be schematically expressed as follows:

$$Al_2O_3SiO_2 + Na_2O.nSiO_2 \rightarrow$$

$$\rightarrow [\text{-O-Si-O-} (\text{—Al}^{1-}\text{-O-Si-O-Si-O-})_n\text{—Al}^{1-}\text{-O—}].Na^{+}_{(n+1)} \qquad (15.5)$$

The structure of the resulting geopolymer consists of a rather randomly arranged three-dimensional network of corner-linked SiO_4 and AlO_4

tetrahedra. The Al/Si ratio may vary between about 1:1 and 1:4. To outbalance the negative charges brought in by the presence of AlO_4 tetrahedra a corresponding number of positive charges, in the form of monovalent or bivalent cations, must also be present in the structure. Such cations are located in spaces between the tetrahedra, together with some constitutional water.

Unlike naturally occurring aluminosilicates, which have a distinct crystalline structure, the structure of geopolymers is either amorphous or microcrystalline. Geopolymers formed at ambient temperature usually possess only a short-range order, not exceeding a few nanometers, whereas those produced at elevated temperatures may possess crystalline regions with dimensions in the order of a few micrometers.

The short-range order within the material may vary greatly even within a single geopolymer. By ^{29}Si (MAS) NMR one or a combination of resonances corresponding to five different local environment may be found: $Q^4(4Al)$, $Q^4(3Al)$, $Q^4(2Al)$, $Q^4(1Al)$, and Q^4. This implies that the SiO_4 tetrahedron may be linked within the geopolymeric structure with one, two, three, or even four neighboring AlO_4 tetrahedra, or just with four SiO_4 tetrahedra. The following peaks were found in a study of a low-calcium material cured for 14 days at 38 °C (Davidovits, 1993; Malek and Roy, 1997):

Peak assignment	Area (%)
Q^4 (4Al)	84.1
Q^4 (3Al)	11.0
Q^4 (2Al)	3.3
Q^4 (1Al)	3.2
Q^4	1.1

Aluminum is present in geopolymeric cements in its tetrahedrally coordinated form (Davidovits, 1993).

By selecting appropriate starting materials and by varying the conditions of processing and curing, it is possible to vary the properties of the produced geopolymeric systems over a wide range, and to tailor them to specific requirements. It is also possible to combine geopolymeric cements with ordinary Portland cement (Malek and Roy, 1997).

In general, geopolymeric cements exhibit rapid strength development. At room temperature a compressive strength of around 20 MPa may be achieved within 4 hours, and the 28 day strength may be in the range 70–100 MPa. The material also exhibits low shrinkage and good freeze–thaw resistance.

Upon heating, the hardened geopolymeric cement loses its constitutional water at about 150 °C, transforms to nepheline at about 900 °C, and melts at about 1300 °C (Palomo *et al.*, 1992).

In spite of their high alkali content, geopolymeric cements generally do not cause an alkali-aggregate reaction, as the alkali metal ions are tightly bonded within the structure of the material (Davidovits, 1993; Malek and Roy, 1997). Owing to their exceptional ability to bind foreign ions, geopolymeric cements seem also to be well suited to long-term containment of toxic and hazardous wastes.

Unlike Portland cement, geopolymeric cements are not dissolved by acidic solutions, and thus may also be applied in acid environments. Their corrosion resistance seems to be excellent.

Geopolymeric cements can be produced with a significantly lower consumption of energy, and thus with a lower CO_2 emission, than Portland cement. In addition, no CO_2 coming from the thermal dissociation of $CaCO_3$ is emitted in the production process. Thus it has been suggested that Portland cements should be replaced by geopolymer cements in an effort to lower industrial CO_2 emission and hence reduce global warming (Davidovits, 1993).

15.7.2   Commercially produced alkali-activated binders

Commercially produced alkali-activated binders have been introduced on the market in the United States under the name Pyrament and in France under the names Geopolymite and PZ-Geopoly.

Pyrament is a multicomponent cement with C_3S and C_2S as its major crystalline phases. The hydration is characterized by an initial formation of an alkali-rich phase responsible for rapid setting and early strength development. It is followed by a more conventional hydration process leading to the development of relatively high ultimate strength (Roy and Silsbee, 1992). The amount of portlandite formed in the hydration is strikingly low, if compared with ordinary Portland cement. An analysis of the liquid phase revealed the presence of high amounts of K^+ and a very high initial Al^{3+} concentration, which, however, dropped to very low levels within the first two hours of hydration (Roy and Silsbee, 1992).

15.8   ROMAN CEMENT

Roman cement was the predecessor of today's Portland cement. It was produced by calcining nodules of argillaceous limestone to temperatures below sintering: that is, to about 900–1000 °C. The binder obtained this way has been called inappropriately "Roman cement," even though it in no way resembles the binder in mortars used by the Romans.

The cement contains mainly the phases C_2S, CA, C_4AF, and free CaO, but no C_3S. It exhibits fast setting, owing to the hydration of CA, and rather low strengths. If produced from starting materials with a high MgO content it may also contain distinct amounts of free MgO, which may cause uneven volume changes during hardening (Grosvalds *et al.*, 1997). The latter can be avoided by wetting and storage of the fired material before milling.

Roman cement was invented in England and patented there in 1796. It was widely used in the first half of the nineteenth century. After about 1850 it was gradually replaced by Portland cement.

15.9 CEMENTITIOUS PHASES AND CEMENTS CONTAINING STRONTIUM AND BARIUM

Strontium and barium ions can partially or completely replace calcium ions in the calcium silicate and calcium aluminate phases.

Both distrontium and dibarium silicate are distinctly more reactive than dicalcium silicate, and it has been suggested that the rate of hydration of the latter clinker phase could be increased by a partial substitution of Ca^{2+} by Sr^{2+} or Ba^{2+} ions (Hanna *et al.*, 1970; Gawlicky 1992; Toreanu and Andronescu, 1992).

In the production of Portland clinker it appears possible to lower the energy consumption by adding by-products containing barium to the raw meal, and by simultaneously lowering its lime saturation factor. As a consequence of these measures the amount of dicalcium silicate in the clinker is increased at the expense of tricalcium silicate. Owing to the presence of Ba^{2+} ions in its crystalline lattice this phase becomes converted into its a' modification, and its reactivity is increased. At the same time Ba^{2+} also becomes incorporated into the aluminate phase and modifies its properties (Rajczyk and Mocun-Wczelik, 1992).

Also synthesized and studied have been sulfoaluminate phases in which calcium ions in the structure of tetracalcium trialuminate sulfate ($C_4A_3\bar{S}$) are partially substituted by Sr^{2+} or Ba^{2+} (Long *et al.*, 1992; Yan, 1992; Feng and Cheng, 1996). $3CaO.3Al_2O_3.SrSO_4$ yields in its hydration an AFt phase similar to ettringite, C_2AH_8, and $SrSO_4$. $3CaO.3Al_2O_3.BaSO_4$ yields the phases $BaSO_4$ and C_2AH_8, which at elevated temperature converts to C_3AH_6 (Yan, 1992).

In strontium aluminate and barium aluminate cements the main constituents are monostrontium or monobarium aluminate. Both these binders are discussed in section 10.9.

The use of cements containing barium and especially strontium is highly restricted by the limited availability of suitable raw materials. Cements containing barium may be considered for shielding against gamma or X-ray radiation. Less important is a possible use of cements containing Sr or Ba in refractory concretes.

15.10   BORATE CEMENT

Solutions of some alkaline borates such as potassium tetraborate ($K_2B_4O_7.4H_2O$) or sodium octaborate ($Na_2B_8O_{13}.4H_2O$) in combination with calcium chloride exhibit cementing properties based on the formation of crystalline calcium borates. The bonding of these systems is based on the development of a three-dimensional structure held together by hydrogen bonds.

Combinations of boric oxide with polyacrylic and polyphosphonic acid exhibit quick setting and hardening based on the formation of hydrogen-bonded complexes.

15.11   HYDRAULICALLY REACTIVE CALCIUM GERMANATES, STANNATES, AND PLUMBATES

Dicalcium and tricalcium germanate have evoked scientific interest as their structure is isotypic with that of dicalcium and tricalcium silicate. Both compounds possess a distinct hydraulic reactivity. Tricalcium germanate exhibits a very fast initial hydration without any induction period, yielding, besides calcium hydroxide, three or more crystalline hydrate phases as products of hydration.

The hydration rate of dicalcium germanate is significantly slower, but the hydration products are similar to those formed in the hydration of tricalcium germanate. The bonding characteristics of the formed C-G-H phases are better than those of the C-S-H phase formed in calcium silicate hydration. Calcium germanates have no practical use as inorganic cements, because of their prohibitive costs.

Calcium stannates and plumbates also exhibit hydraulic reactivity, but are of no practical interest. Calcium titanates are not hydraulically reactive.

15.12   BLENDS OF BASIC CEMENTS

Blends of two or more binders may yield products whose chemistry of hydration and cementing characteristics may or may not differ significantly from those of pure cements constituting the blend. The properties of the new binder will depend both on the quality of the original binders being blended together, and on their mutual ratio in the blend. The properties of the resulting combination of binders may be either beneficial or unwelcome. Table 15.4 shows the properties of blends of selected cements as published by Heble *et al.* (1997).

Table 15.4 Properties of blends of different cements.

Compositions of original cements

Ordinary Portland cement (OPC)
  $C_3S + + +$; $C_2S + + +$; $C_3A +$; $C_4AF +$; gypsum +
Calcium aluminate cement with 50% Al_2O_3 (CAC-A)
  $CA + + + +$; $CA_2 + +$; $C_{12}A_7 + +$
Calcium aluminate cement with 73% Al_2O_3 (CAC-B)
  $CA + + +$; $CA_2 ++$; $C_{12}A_7 + +$
Calcium sulfoaluminate-belite cement (CSAB)
  $C_4A_3S + + + +$; $C_2S + + +$; gypsum + +
Calcium sulfoaluminate clinker (CSAC)
  $C_4A_3S + +$; $C_2S + + +$; $C + +$

Phase composition of hydrated material and relative strength (pastes $w/c = 0.4$ hydrated 3 d)

Combination	Ratio	Hydrate phases formed	Compressive strength (%)
1. OPC + CAC-A	90:10	P(H), E(L)	101
	50:50	P(M), E(L), S(L), C_3AH_6(L)	74
	10:90	P(L), C_3AH_6(L)	56
2. OPC + CAC-B	90:10	P(H), CAH_{10}(L)	100
	50:50	P(M), MS(L), S(H), C_2H_8(M)	84
	10:90	P(L), MS(L), S(L), C_2AH_8(M), C_3AH_6(L)	70
3. OPC + CSAB	90:10	P(H), E(M), MS(L)	108
	50:50	P(L), E(M), MS(L)	44
	10:90	P(M), E(M), MS(L), S(L)	135
4. OPC + CSAC	90:10	P(H), E(H)	152
	50:50	P(H), E(H)	109
	10:90	P(H), E(M)	128
5. CAC-A + CSAB	90:10	CAH_{10}(M), C_2AH_8(M)	110
	50:50	P(L), E(M), MS(L), S(M)	118
	10:90	P(M), E(H), MS(L), S(L)	166
6. CAC-A + CSAC	90:10	P(M), E(L)	43
	50:50	P(M), MS(M)	21
	10:90	P(H), E(M), MS(L)	109

P = portlandite; E = ettringite; S = strätlingite; MS = monosulfate
H = high; M = medium; L = low

Note: In combinations 1–4 the strengths were compared with that of a paste made from pure OPC; in combinations 5–6 with the strength of a pure CAC-A paste.

Source: Heble *et al.* (1997)

REFERENCES

Branski, A. (1957) Barium and strontium cements (in German). *Zement-Kalk-Gips* **10**, 176–184.

Branski, A. (1961) Similarities and differences between calcium- strontium- and barium cements (in German). *Zement-Kalk-Gips* **14**, 17–26.

Branski, A. (1967) Strontium cements (in German). *Zement-Kalk-Gips* **20**, 96–101.

Cole, W.F., and Demediuk, T. (1955) X-ray, thermal and dehydration studies on magnesium oxychloride. *Australian Journal of Chemistry* **8**, 234.

Davidovits, J. (1989) *Geopolymers*. Univ. of Techn. Compiegne, France, Vol. 2, pp. 149–168.

Davidovits, J. (1991) Geopolymers: inorganic polymeric new materials. *Journal of Thermal Analysis* **37**, 1633–1656.

Davidovits, J. (1993) Polymeric cements to minimize carbon-dioxide greenhouse warming, in *Cement Based Materials: Presence, Future and Environmental Aspects* (eds M. Moukwa *et al.*), American Ceramic Society, pp. 165–182.

Davidovits, J. *et al.* (1990) Geopolymeric concretes for environmental protection. *Concrete International* **12**, 30–39.

de Castellar, M.D. *et al.* (1996) Crack in Sorel's cement polishing bricks as a result of magnesium oxychloride carbonation. *Cement and Concrete Research* **26**, 1199–1202.

Demediuk, T., Cole, W.F., and Henber, H.V. (1955) Studies on magnesium and calcium oxychloride. *Australian Journal of Chemistry* **8**, 215.

Deng, D., and Zhang, C.H. (1996) The effect of aluminate minerals on the phases in magnesium oxychloride cement. *Cement and Concrete Research* **26**, 1203–1211.

Feng, X., and Cheng, X. (1996) The structure and quantum chemistry studies of $3CaO.3Al_2O_3.SrSO_4$. *Cement and Concrete Research* **26**, 955–962.

Gawlicky M. (1992) Studies of barium addition on the Ca_2SiO_4 hydration, in *Proceedings 9th ICCC, New Delhi*, Vol. 4, pp. 449–453.

Grosvalds, I., Lagzdina, S., and Sedmalis, U. (1997) Dolomitic roman cement: a low temperature hydraulic binder, in *Proceedings 10th ICCC, Göteborg*, paper 2ii009.

Hanna, K.M., Cook, R.I., and Kantro, D. (1970) Hydration of dibarium silicate. *Journal of Applied Chemistry* **20**, 334–340.

Harmuth, H., Neuherz, A., and Schrempf, S. (1998) Investigation of a magnesia binder in the system $MgO-Mg(CH_3.COO)_2-H_2O$. *Cement and Concrete Research* **28**, 811–814.

Heble, A.S., Viswanathan, K., and Chatterjee, A.K. (1997) Blends of basic cements of different phase compositions: hydraulic behaviour study, in *Proceedings 10th ICCC, Göteborg*, paper 2ii039.

Hommertgen, C., and Odler, I. (1991) Glass fiber composites made with a cement based on a hydraulically active $CaO-SiO_2-Al_2O_3$ glass. *Materials Research Society Symposium Proceedings* **211**, 93–104.

Hommertgen, C., and Odler, I. (1992) Investigations on hydraulically reactive glasses in the system $CaO-Al_2O_3-SiO_2$. *Materials Research Society Symposium Proceedings* **245**, 165–172.

Ikeda, K. (1997) Preparation of fly ash monoliths consolidated with sodium silicate binder at ambient temperature. *Cement and Concrete Research* **27**, 657–663.

Krivoborodov, Yu.R., and Samchenko, S.V. (1992) Sulfate-bearing solid solutions of calcium aluminates and ferrites, in *Proceedings 9th ICCC, New Delhi*, Vol. 3, pp. 209–215.

Laxmi, S. *et al.* (1992) Development of alumino-belitic coloured clinkers, in *Proceedings 9th ICCC, New Delhi*, Vol. 2, pp. 118–124.

Long, S., Wu, Y., and Liao, G. (1992) Investigation on the structure and hydration of a new mineral phase $3SrO.3Al_2O_3.CaSO_4$, in *Proceedings 9th ICCC, New Delhi*, Vol.4, pp. 418–421.

MacDowell, J.F. (1986) Hydrogarnet – gehlenite hydrate cement from $CaO-Al_2O_3$-SiO_2 glasses, in *Proceedings 8th ICCC, Rio de Janeiro*, Vol. 4, pp. 423–428.

MacDowell, J.F. (1991) Strätlingite and hydrogarnet from calcium aluminosilicate glass cements. *Materials Research Society Symposium Proceedings* **179**, 159–179.

MacDowell, J.F., and Chowdhury, A. (1990) Quick setting high early strength glass based cements for rapid highway repair. *Transportation Research Board Meeting*, paper CB023.

MacDowell, J.F., and Sorrentino, F. (1990) Hydration mechanism of gehlenite glass cements. *Advances in Cement Research* **3**, 143–152.

MacDowell, J.F., Sorrentino, F., and Capelle, M. (1991) Hydrogarnet–gehlenite hydrate equilibrium in glass cements, in *Proceedings 13th International Conference on Cement Microscopy, Tampa*, pp. 131–146.

Malek, R.I.A., and Roy, D.M. (1997) Synthesis and characterization of new alkali-activated cement, in *Proceedings 10th ICCC, Göteborg*, paper 1i024.

Menetrier-Sorrentino, D., Barret, P., and Saquat, S. (1986) Investigation in the system $MgO-MgCl_2-H_2O$ and hydration of Sorel cement, in *Proceedings 8th ICCC, Rio de Janeiro*, Vol. 4, pp. 339–343.

Nakatsu, K. *et al.* New low-heat cement with $CaO-Al_2O_3-SiO_2$ glass. *Ceramic Transactions* **4**, 265.

Nishi, F., and Marumo, M. (1996) Study on $3CaO.GeO_2$ (= C_3G) hydrates by means of SEM (in Japanese). *Saitama Kogyo Kiyo* **5**, 42–46 [ref. CA 125/228889].

Odler, I., and Hennicke, U. (1991) Investigations on Na-silicate based binders. *Materials Research Society Symposium Proceedings* **179**, 283–293.

Osokin, A.P., and Odler, I. (1992) Properties of cements containing calcium ferrite sulfate (in German). *Zement-Kalk-Gips* **45**, 536–537.

Osokin, A.P., Krivoborodov, Yu.R., and Dyukova, N.F. (1992) Sulfoferrite cements, in *Proceedings 9th ICCC, New Delhi*, Vol.3, pp. 256–261.

Osokin, A.P., Krivoborodov, Yu.R., and Samchenko, S.V. (1997) Melt structure and properties of clinkers containing calcium sulphoferrites and sulphoalumino-ferrites, in *Proceedings 10th ICCC, Göteborg*, paper 1i014.

Palomo, A. *et al.* (1992) Physical, chemical and mechanical characterization of geopolymers, in *Proceedings 9th ICCC, New Delhi*, Vol. 5, pp. 505–511.

Raina, S.J., Viswanathan, V.N., and Chatterjee, A.K. (1978) Early hydration characteristics of Porsal cement (in German). *Zement-Kalk-Gips* **31**, 516–518.

Rajczyk, K., and Mocun-Wczelik, W. (1992) Studies of belite cement from barium containing by-products, in *Proceedings 9th ICCC, New Delhi*, Vol. 2, pp. 250–254.

Roy, D.M., and Silsbee, M.R. (1992) Alkali-activated materials: an overview. *Materials Research Society Symposium Proceedings* **245**, 153–164.

Sorell, C.A., and Armstrong, C.R. (1976) Reaction and equilibria in magnesium oxychloride cement. *Journal of the American Ceramic Society* **59**, 51–54.

Suzuki, K. *et al.* (1998) Very early hydration of Li₂O-CaO-Al₂O₃-SiO₂ glass (in Japanese). *Nippon Kagaku Kaishi* **1998**(6), pp. 438–441 [ref. CA 129/84893].

Toreanu, I., and Andronescu, E. (1992) Orthosilicate compositions in CaO-SrO-BaO-SiO₂ system and their water reactivity, in *Proceedings 9th ICCC, New Delhi,* Vol. 2, pp. 107–113.

Trettin, R. *et al.* (1995) The hydration of hydraulically active calcium germanates. *Advances in Cement Research* **7**, 117–123, 139–142.

Urwongse, L., and Sorell, C.A. (1980) The system MgO–MgCl₂–H₂O at 32 °C. *Journal of the American Ceramic Society* **63**, 501–504.

Viswanathan, V.N., Raina, S.J., and Chatterjee, A.K. (1978) An exploratory investigation of Porsal cement. *World Cement Technology* **9**, 109–118.

Wang, L. (1995) Study on fracture mechanism in expansion of the hardened magnesium oxychloride cement (in Chinese). *Guisuanyan Xuebao* **23**, 471–476 [ref. CA 125/17314].

Xia, S., Zhuang, X., and Liu, F. (1995) Study on the reaction process between CO₂ and magnesium oxychlorides by infrared spectroscopy (in Chinese). *Guangpu Shiyanshi* **12**, 16–20 [ref. CA 125/121725].

Yan, P. (1992) Investigation on the hydration of Sr- and Ba-bearing sulfoaluminates, in *Proceedings 9th ICCC, New Delhi,* Vol. 4, pp. 411–417.

Yan, P., and Yang, W. (1997) Cementitious behaviour of CaO-Al₂O₃-SiO₂ glass (in Chinese). *Guisuanyan Xuebao* **25**, 480–484 [ref. CA 128/234213].

Yu, H. (1992) Study of new type of water resistant magnesium oxychloride cement (in Chinese). *Guisuanyan Xuebao* **20**, 374–381 [ref. CA 120/329711].

Zakharov, L.A. (1974) Alumino-belite cement, in *Proceedings 5th ICCC, Moscow,* Supplementary paper: Section III.

Zhang, C., and Deng, D. (1994) Research on water resistance of magnesium oxychloride cement (in Chinese). *J. Wuhan Univ. Technol. Mater. Sci. Ed.* **9**, 51–59 [ref. CA 123/264386].

Zhang, C., and Deng, D. (1995) Effect of aluminate minerals on magnesium oxychloride cement (in Chinese). *Guisuanyan Xuebao* **23**, 211–218 [ref. CA 124/35785].

16 Low-energy cements

The term "low-energy cements" is used to designate cements that may be used to replace ordinary Portland cement, at least in some applications, yet can be produced with a reduced consumption of energy. It is believed that a more extended use of such cements would both reduce the production costs of the binder and lower undesired emissions, especially those of CO_2.

The thermal energy consumed in the production of Portland clinker may be subdivided into two categories: the chemical enthalpy of clinker formation, and thermal losses. In addition, significant amounts of electrical energy are consumed in the production process to power the equipment used. In contemporary cement plants the following values may be considered typical in the production of ordinary Portland cement:

- enthalpy of clinker formation: 1600–1800 MJ/t (380–430 Mcal/t)
- thermal losses: 1500–1700 MJ/t (360–410 Mcal/t)
- consumption of electricity: 100–110 kWh/t (corresponds to 360–400 MJ/t of electric energy).

The chemical enthalpy of clinker formation is the energy/enthalpy needed to convert the substances constituting the raw meal into Portland clinker in the absence of any heat losses and without changing the overall temperature of the system. Since the formation of clinker consists of a series of chemical and physical processes that are both endothermic and exothermic in nature, the overall chemical enthalpy will be derived from the difference between the energy liberated in the exothermic reactions and the energy consumed in the endothermic reactions taking place.

Table 16.1 gives the energy balance of a clinkering process in which a raw meal of typical composition is converted into an ordinary Portland cement clinker. From the table it is apparent that the reaction in which by far the largest fraction of energy is consumed is the conversion of the calcium carbonate constituting the raw meal into calcium oxide and carbon dioxide:

$$CaCO_3 \rightarrow CaO + CO_2 \tag{16.1}$$

Table 16.1 Energy balance of the clinkering process (OPC clinker of typical composition).

	MJ/kg	*kcal/kg*
Heating raw meal 20–450°C	– 0.78	– 170
Clay dehydration	– 0.13	– 30
Heating dehydrated material 450–900°C	– 0.82	
Dissociation of $CaCO_3$	– 1.99	– 25
Heating calcined material 900–1400°C	– 0.52	– 125
Heat of fusion	– 0.11	– 25
Heat of formation of clinker phases	+ 0.42	+ 100
Cooling clinker 1400–20°C	+ 1.51	+ 360
Cooling CO_2 900–20°C	+ 0.50	+ 120
Cooling steam 450–20°C including condensation	+ 0.08	+ 20
Total	– 1.76	– 420

This is because, of all oxide constituents of the raw meal, calcium oxide is the most abundant (constituting typically about 70–75 wt% of the material), and its decomposition to CaO + CO_2 is a highly endothermic reaction (1782 kJ or 426 kcal per 1 kg $CaCO_3$, equals 3182 kJ or 760 kcal per 1 kg CaO). Thus a reduction of the overall energy consumption could be achieved by reducing the $CaCO_3$ content in the raw meal.

The amount of CaO chemically bound within the individual phases of Portland clinker varies significantly, and along with it the amount of $CaCO_3$ required to synthesize one weight unit of the particular clinker mineral. The CaO content of pure compounds is as follows:

- tricalcium silicate (C_3S): 73.7%
- dicalcium silicate (C_2S): 65.6%
- tricalcium aluminate (C_3A): 62.2%
- tetracalcium aluminate ferrite (C_4AF): 46.1%

Thus it is apparent that the $CaCO_3$ content in the raw meal, and hence the energy consumption, may be reduced by increasing in the clinker the amount of phases with lower CaO contents at the expense of those in which the CaO content is higher. Consequently, high-C_2S Portland cement (see section 2.4) and high-iron Portland cement (see section 2.8) qualify as low-energy cements, whereas a cement high in C_3S does not.

A cement with an even lower enthalpy of clinker formation is belite cement (see sections 3.2 and 3.3), in which the content of tricalcium silicate is reduced to zero or near-zero values. However, the absence of the latter phase results in a poor strength development of this type of cement. It has been suggested that the reactivity of the C_2S phase could be increased by extremely fast cooling of the produced clinker; however, such a measure would also increase the thermal losses significantly, and would eliminate most of the gains achieved by the reduction of the enthalpy of clinker formation.

Another approach that might help to reduce the enthalpy of clinker formation is to use starting materials that contain CaO in decarbonated form, rather than in the form of $CaCO_3$, as constituents of the raw meal. The presence of $CaCO_3$ in the raw meal cannot usually be completely eliminated by such a measure, but the amount of this constituent, and hence the enthalpy of the clinkering process, may be distinctly reduced. Raw materials that may be considered for this purpose include blast furnace slag and a variety of other metallurgical slags, type C fly ashes with high CaO contents, and nephelin sludge. The use of such starting materials may reduce the $CaCO_3$ content in the raw meal by up to 30% at a constant lime saturation factor, but such a measure can only be applied if these materials are available.

In addition to reducing the enthalpy of clinker formation, energy savings may also be achieved by reducing the thermal losses in the clinker production process. These include the heat lost with the hot clinker, the heat lost with the exhaust gases and dust, and radiation losses. In recent decades great progress has been made in developing cement kilns equipped with highly effective raw meal preheaters and clinker coolers. Further reduction of heat losses by modifications of the available equipment is still theoretically possible, but it must be expected that further steps in this direction will be small, and will be associated with increasing investment costs.

Chemical and physical measures that may lower the thermal losses in the production of Portland clinker include steps aimed at improving the burnability of the raw meal. A measure that has attracted particular attention is the use of mineralizers and fluxes to lower the maximum burning temperature. Reductions of the maximum burning temperature by 100°C and even more are possible by such an approach, resulting in energy savings of 5% and more.

Another measure to reduce the overall energy consumption consists in combining Portland clinker – in addition to calcium sulfate – with a variety of latent-hydraulic or pozzolanic materials, thus reducing the amount of clinker needed to produce one weight unit of binder. The amount of clinker contained in these composite cements (see sections 7,8 and 9) is significantly lower than that consumed in plain Portland cement. A series of cements has been developed in which the energy requirements are significantly reduced as compared with Portland cement. These energy savings result from a reduced $CaCO_3$ content in the raw meal and lower burning temperatures. Such cements include sulfobelite cement (see section 4.2), belite-fluoroaluminate cement (see section 5.2), alinite cement (see section 6.2), aluminate-belite cement (see section 5.3), and related binders.

Finally, a separate category of low-energy cements are represented by the alkali-activated cements (see section 15.7), including alkali-activated slag cement (see section 8.5) and geopolymeric cement (see section 15.7.1).

17 Fast-setting cements

Pastes made from conventional cements preserve their plasticity for several hours before setting and hardening. Such a plastic stage is necessary so that it is possible to produce a concrete/mortar mix of the desired consistency, transport it to the place where it is to be applied, and compact it after placing. In some special applications, however, cements with very short setting times are required. Examples of such applications are various repair works (repairs of pavements, for example) and emergency measures (such as plugging of leaks to prevent leakage of water or other liquids). Obviously, the use of such rapid-setting mixes requires particularly fast mixing, placing, and compacting. Mixes that set almost instantaneously cannot be used in the traditional way, and must be applied by spraying. The use of fast-setting cements in spraying applications makes it possible to build up thick layers of material in a single pass even on vertical backgrounds, thus increasing the effectiveness of this technology.

17.1 CONTROL OF PORTLAND CEMENT SETTING BY THE USE OF CHEMICAL ADMIXTURES

Some highly alkaline chemicals, such as alkali metal hydroxides, carbonates, silicates, and aluminates, cause a significant shortening of the setting time of Portland cement. They act by increasing the pH of the liquid phase, and thus accelerating the hydration of the tricalcium aluminate present in the cement. If alkali silicates are used, soluble SiO_2 is immediately available to react with the calcium hydroxide formed in the simultaneous hydration of tricalcium silicate, and additional amounts of the C-S-H phase are also formed. At moderate dosages of the additives the resulting strengths are increased, but at high dosages the rapid setting caused by the added alkali silicate results in a more disordered structure and a lower strength (Larosa-Thompson *et al.*, 1997).

A very effective acceleration of Portland cement setting may also be achieved by adding to the cement a finely ground calcium aluminate glass with a CaO/Al_2O_3 molar ratio between about 12:7 and 3:1 (Hirose *et al.*,

1997). The rapid setting of such systems is caused by the formation of ettringite in a reaction between the glass constituents and the calcium sulfate present in the cement. The setting time of the mix shortens with increasing CaO/Al_2O_3 molar ratio of the glass used, whereas the short-term compressive strength declines.

Some anions exhibit a distinct accelerating effect on Portland cement hydration. They include halides, nitrate, nitrite, formiate, thiosulfate, and thiocyanate. Their activity depends not only on the dosage used, but also on the identity of the associated cation. Generally, divalent and trivalent cations enhance the effectiveness of these anions more than monovalent ions do. This category of additives affects mainly the hydration of the alite phase, thus accelerating the development of early strength. The setting time is also shortened, but less effectively than with admixtures that increase the pH of the liquid phase.

17.2   FAST-SETTING GYPSUM-FREE PORTLAND CEMENT

Conventional Portland cements contain limited amounts of calcium sulfate (in the form of anhydrite or gypsum), interground with Portland clinker, to control setting. Under these conditions the tricalcium aluminate of the clinker reacts with the sulfate and water, to yield ettringite (Aft phase, $C_6A\bar{S}_3H_{32}$). This phase precipitates at the surface of the cement grain as a thin layer, which does not adversely affect the rheology of the cement suspension. The cement paste sets after several hours, owing to the hydration of the tricalcium silicate present and the formation of the C-S-H phase.

In the presence of no or insufficient amounts of calcium sulfate, an AFm phase (C_4AH_{13} or $C_4A\bar{S}H_{12}$, or a solid solution of both) is formed, instead of the AFt phase. This phase is formed very rapidly after mixing with water, in the form of platelet-like crystals up to 10 μm in size, which precipitate throughout the paste and interfere with its free flow. Such a sudden stiffening of the paste (called quick or flash setting) is generally unwelcome, but cement compositions of this type may be employed as rapid-setting cements. To obtain a distinct conversion from the plastic to the solid state, the clinker used in such cements must have a high C_3A content and must be low in SO_3. The strength development of such cements lags behind those produced with an adequate amount of gypsum and exhibiting "normal" setting.

Rapid setting may also take place in Portland cements that contain calcium sulfate, if they simultaneously contain elevated amounts of K_2O. In this case setting is due to the formation of syngienite $[K_2Ca(SO_4)_2.H_2O]$ in the paste.

In systems that contain ground Portland clinker without interground calcium sulfate the rheology of the paste may be controlled by adding calcium lignosulfonate in combination with an alkali carbonate to the mix (see low-porosity cement, section 2.16.1). The mix exhibits good flow even at low water/cement ratios.

The setting time of these mixes may vary over a wide range, and depends greatly on the amount of the additives employed. In general, the setting time is extended with increasing amounts of the lignosulfonate, whereas the carbonate has little effect on the time of setting. Thus well-flowing cement pastes with a short setting time may be produced by combining ground Portland clinker (300–350 m^2/kg) with about 0.5–0.75% Na_2CO_3 and about 0.25% sodium lignosulfonate. Mixes of this composition exhibit a good initial flow, and tend to set rather suddenly within less than 30 min (Odler *et al.*, 1978).

It has also been reported that mixes with setting times between about 10 and 30 min may be obtained by combining ordinary Portland cement with about 0.5–2.0% K_2CO_3 and about 0.3–1.0% lignosulfonate (Sanitsky and Sabol, 1992).

17.3 FAST-SETTING CEMENTS CONTAINING THE PHASES $C_{11}A_7.CAF_2$ OR $C_{12}A_7$

Cements that contain the calcium fluoroaluminate phase ($11CaO.7Al_2O_3.CaF_2$) in combination with calcium sulfate exhibit a very fast setting and initial strength development, owing to the formation of the ettringite phase. In alite-fluoroaluminate cements (see section 5.2) the fluoroaluminate is combined with tricalcium silicate, whereas in belite-fluoroaluminate cements (see section 5.3) dicalcium silicate, but no tricalcium silicate, is present. Variable amounts of the ferrite phase may also be present in these cements, but usually no, or only reduced amounts of, tricalcium aluminate. After a fast setting and rapid initial strength development associated with ettringite formation, the strength increase slows down as a large fraction of the fluoroaluminate and calcium sulfate is consumed. In alite-fluoroaluminate cements and in calcium fluoroaluminate modified Portland cement a renewed fast strength development, associated with tricalcium silicate hydration, gets under way after several hours, whereas the strength development is much more sluggish in belite-fluoroaluminate cements.

The setting time of these cements is typically between several minutes and half an hour, but may be conveniently extended by the addition of suitable retarders, such as citric acid.

In addition to cements that contain the calcium fluoroaluminate phase ($11CaO.7Al_2O_3.CaF_2$), fast-setting binders may also be obtained by combining Portland cement with $C_{12}A_7$ ($12CaO.7Al_2O_3$) that has been produced separately (Kondo *et al.*, 1997). However, the handling of such cements is more difficult than that of conventional calcium fluoroaluminate cement, owing to the faster formation of ettringite prior to setting, associated with a rapid deterioration of the rheology of the fresh mix. After long-term storage in air the ettringite phase formed in the hydration of cements containing calcium fluoroaluminate in combination with belite may react with the CO_2 in the air, converting to gypsum, alumina gel, calcite and/or vaterite:

$$C_6A\bar{S}_3H_{32} + 3\bar{C} \rightarrow 3C\bar{C} + 3C\bar{S}H_2 + AH_3 + 23H \qquad (17.1)$$

This may result in an increase of capillary porosity and decline of strength in belite-fluoroaluminate cements (Knöfel and Wang, 1992). In contrast, alite-fluoroaluminate cements resist carbonation much better, owing to the presence of significant amounts of calcium hydroxide among the hydration products, which reacts with CO_2 yielding $CaCO_3$ and lowering the porosity of the cement paste.

For additional information on cements containing the $C_{11}A_7.CaF_2$ and $C_{12}A_7$ phases see Chapter 5.

17.4   FAST-SETTING CEMENTS CONTAINING THE PHASE TETRACALCIUM TRIALUMINATE SULFATE

Cements that contain the phase tetracalcium trialuminate sulfate ($C_4A_3\bar{S}$) in combination with calcium sulfate exhibit fast setting and fast early strength development. Such binders include sulfobelite cement (see section 4.2), sulfoalite cement (see section 4.3), calcium sulfoaluminate modified Portland cement (see section 4.4), and sulfoaluminate cement (see section 4.5). The setting and initial strength development are brought about by a rapid formation of ettringite in a reaction between $C_4A_3\bar{S}$ and calcium sulfate in these cements.

The strength development at later ages will depend mainly on the calcium silicate phases present in the particular cement. In those that contain tricalcium silicate, a renewed intensive strength development gets under way after a short period of slower strength growth due to a gradual slow-down of ettringite formation and the presence of an induction period in the hydration of C_3S. In sulfobelite cement, which contains only the C_2S phase but not the C_3S phase, the strength development after the initial phase of fast setting and strength increase is much more moderate, as C_2S hydrates much more slowly. The ferrite phase, which may be present in variable amounts in these cements, also contributes to strength development by forming an iron-rich AFt phase. Fast-setting cements based on the presence of the $C_4A_3\bar{S}$ phase must not contain calcium sulfate in amounts exceeding an optimum $C\bar{S}/C_4A_3\bar{S}$ ratio, as at higher calcium sulfate contents an undesired expansion of the hardened paste due to delayed ettringite formation may occur. Cements of such composition may be used, however, as expansive cements (see section 21.3.4).

Upon long-term exposure to air the ettringite phase may react with the CO_2 of air converting to gypsum, alumina gel, and calcite and/or vaterite, and this reaction will result in an increase of porosity and a decline of strength of the hardened cement paste (Knöfel and Wang, 1992). Sulfobelite cements are more susceptible in this respect than cements that also contain alite, in which CO_2 reacts preferentially with the calcium hydroxide formed in the hydration of this phase, lowering the porosity of the paste and protecting the stability of ettringite.

For additional information on cements containing the $C_4A_3\bar{S}$ phase see Chapter 4.

17.5   FAST-SETTING BLENDS OF PORTLAND CEMENT AND CALCIUM ALUMINATE CEMENT

Blends of ordinary Portland cement (OPC) and calcium aluminate cement (CAC) exhibit a very fast setting over a wide range of OPC/CAC ratios. Figure 17.1 shows a typical example of such behavior, but the exact setting time will also depend strongly on the characteristics of the individual cements employed, and on the water/solid ratio of the mix. After setting, the OPC + CAC blend exhibits rapid strength development, and measurable strength values may be achieved within less than 1 hour. The final strengths, however, are lower than those of either the Portland cement or the calcium aluminate cement alone. Figure 17.2 shows typical strength development of OPC + CAC blends as a function of their composition.

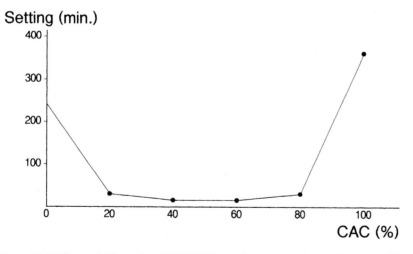

Figure 17.1 Effect of the ratio CAC/OPC on the setting time of the resulting cement paste.

Str. (MPa)

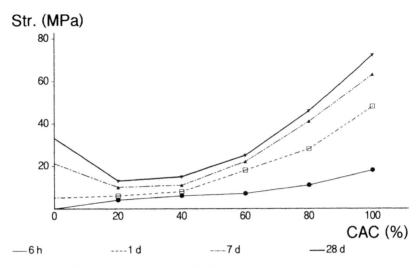

Figure 17.2 Effect of the ratio CAC/OPC on the compressive strength of concrete.

The reasons for the fast setting of OPC + CAC blends are not fully understood. It is assumed that in blends high in Portland cement the rapid setting is due to the formation of ettringite in a reaction between the CA phase of the calcium aluminate cement, calcium sulfate present in the Portland cement, and calcium hydroxide formed in the hydration of the tricalcium silicate phase present in Portland clinker:

$$CA + 3C\bar{S}H_2 + 2CH + 24H \rightarrow C_3A.3C\bar{S}.32H \tag{17.2}$$

A rapid setting of the tricalcium aluminate phase present in Portland clinker after depletion of calcium sulfate in reaction (17.2) may also be involved.

In the CAC-rich region setting is due to hydration of the calcium aluminate cement, which is accelerated in the high-pH environment produced by the presence of Portland cement in the mix.

In the later stages the hydration progresses similarly to that of pure Portland cement in OPC-rich mixes and also to that of pure calcium aluminate cement in CAC-rich mixes. In mixes of intermediate composition strätlingite (gehlenite hydrate, C_2SAH_8) may also be formed.

If OPC + CAC mixes are used as fast-setting cements, compositions in which Portland cement is the main constituent are preferred. Ratios of 75–90% of Portland and 10–25% of calcium aluminate cement may be considered typical.

To accelerate the setting process even further, it has been suggested that anhydrite ($CaSO_4$) and calcium chloride should be added to the OPC + CAC blend (Laxmi *et al.*, 1986). In blends of this kind the hydration process is characterized by an intensive initial heat peak caused by the dissolution of soluble species, and setting due to the formation of ettringite may occur within about 15 min. After a short induction period the hydration is accelerated again, and is characterized by the formation of additional ettringite, calcium monoaluminate hydrate (CAH_{10}), calcium chloroaluminate hydrate ($C_3A.CaCl_2.H_{11}$), strätlingite (C_2ASH_8), hydrogarnet (C_3AH_6), and AH_3. A hydration of alite also takes place.

The hydration process in OPC + CAC blends may also be accelerated by adding small amounts of a lithium salt, such as Li_2CO_3, to the mix. The lithium salt, which accelerates the hydration of the monocalcium aluminate phase very effectively, is also highly effective in mixes in which the calcium aluminate cement is the minor component.

For additional information on OPC + CAC blends see section 10.10.1.

17.6 FAST-SETTING MAGNESIUM PHOSPHATE CEMENT

Calcined magnesia (MgO) in combination with diammonium hydrogen phosphate [$(NH_4)_2HPO_4$] or some other water-soluble phosphates yields a fast-setting binder that sets within minutes and yields measurable strengths within less than half an hour. The main reaction product of the setting/hardening reaction is struvite ($NH_4MgPO_4.6H_2O$). If needed, the rate of the reaction may be reduced by suitable retarders, such as borax (sodium tetraborate, $Na_2B_4O_7.10H_2O$).

The hardened magnesium phosphate paste and concrete mixes made with this binder exhibit an excellent bond to old Portland cement based concrete, and are very suitable for repair works on existing concrete structures, such as patching of potholes on highway surfaces.

For additional information on magnesium phosphate cements see section 12.1.

17.7 FAST-SETTING GLASS CEMENT

Some glasses in the system $CaO—Al_2O_3—SiO_2$, if ground to a high fineness, yield cements called glass cements (see section 15.2), which exhibit very fast setting and hardening. The main product of hydration in such cements is the hydrogarnet phase. Fast-setting glass cements must have a low SiO_2 content, as at higher contents of this oxide strätlingite rather than hydrogarnet is formed in the hydration, and this results in an extension of the setting time. Glass cements yielding the hydrogarnet phase as hydration product exhibit a very short setting time and fast early strength development, but only very moderate strength growth later on.

For additional information on glass cements see section 15.2.

17.8   MISCELLANEOUS FAST-SETTING CEMENTS

Other fast-setting cements not mentioned so far include alkali-activated slag cement (see section 8.5), alkali-activated fly ash binder (see section 9.1), geopolymeric cement (see section 15.7.1), zinc phosphate cement (see section 12.4), magnesium oxychloride and oxysulfate cement (see section 15.1), and alkali silicate binder(see section 15.3).

REFERENCES

Chen, Y., and Odler, I. (1992) On the origin of Portland cement setting. *Cement and Concrete Research* **22**, 1130–1140.

Hirose, S. *et al.* (1997) Study on hydration and the properties of hardening accelerators based on calcium aluminate glass, in *Proceedings 10th ICCC, Göteborg*, paper 3iii022.

Knöfel, D., and Wang, J. (1992) Investigation on pore structure of quick cement, in *Proceedings 9th ICCC, New Delhi*, Vol. 4, pp. 370–375.

Kondo, N. *et al.* (1997) Relationship between ettringite formation and the development of strength for rapid hardening cement, in *Proceedings 10th ICCC, Göteborg*, paper 2ii017.

Larosa-Thomson, J. *et al.* (1997) Sodium silicate applications for cement and concrete, in *Proceedings 10th ICCC, Göteborg*, paper 3iii024.

Laxmi, S. *et al.* (1986) Development of a fast setting high strength cement composition, in *Proceedings 8th ICCC, Rio de Janeiro*, Vol. 4, pp. 357–363.

Odler, I., Duckstein, U., and Becker, Th. (1978) On the combined effect of water soluble lignosulfonates and carbonates on Portland cement and clinker pastes. I. Physical properties. *Cement and Concrete Research* **8**, 469–480.

Sanitsky, M.A., and Sabol, Ch.S. (1992) Gypsum-free rapid hardening and mixed Portland cement, in *Proceedings 9th ICCC, New Delhi*, Vol. 3, pp. 438–443.

18 Cements for high-performance concrete

The term "high-performance concrete" (abbreviation HPC) is not defined precisely. Such concretes are understood to have compressive strengths exceeding distinctly those of normal, ordinary Portland cement based concrete, typically in the range 75–100 MPa or above.

The following principles must be observed to produce concrete with strengths of this order of magnitude (Sarkar and Baalbaki, 1992; Kato et al., 1997; Lang and Geiseler, 1997; Peukert, 1997; Sota et al., 1997; Breitenbucher, 1998; Duval and Kadri, 1998; Steigenberger, 1998).

- The water/cement ratio must be very low. It lies typically in the range $w/c = 0.28$–0.35. This is important, as in this way one obtains a hardened paste of low porosity and high intrinsic bonding strength. The water/cement ratio is the main factor that determines the strength of the hardened concrete.
- The content of the cement in the concrete mix must be relatively high. Amounts of 400–700 kg/m^3 may be considered typical. A high cement content is desired to attain an acceptable rheology of the fresh mix even at very low water/cement ratios.
- Microsilica (silica fume) should be added to the mix to increase further the strength of the hardened concrete. By adding this supplementary material the strength of the concrete may be increased by up to 15–20% (Duval and Kadri, 1998; Steigenberger, 1998). The optimum strength is attained at a replacement level of microsilica of about 10–15% per weight of cement in most concrete mixes. At addition levels up to about 10% the rheology of the fresh mix is not affected unfavorably. Excessive amounts of microsilica, however, may affect both the strength and the rheology of the fresh mix adversely, and may require increased amounts of superplasticizer to achieve a given consistency of the mix. The slump loss of the fresh mix also increases with increasing microsilica additions, especially at very low water/cement ratios (Duval and Kadri, 1998). Figures 18.1 and 18.2 show the effect of microsilica addition on the slump loss and strength of concrete at equal consistencies of the mix. The action of microsilica is complex. It

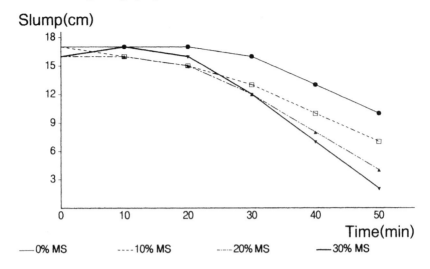

Figure 18.1 Effect of microsilica (MS) addition on the slump loss of Portland
cement based fresh concrete mixes [W/ (C + MS) = 0.25].
Source: Duval and Kadri (1998)

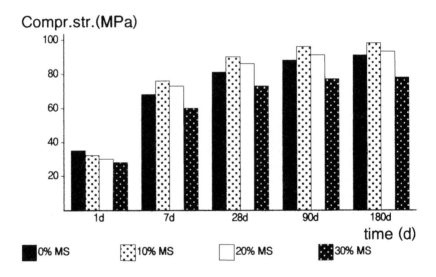

Figure 18.2 Effect of microsilica (MS) addition on compressive strength of
Portland cement based concrete at equal slump values [W/ (C + MS)
= 0.30].
Source: Duval and Kadri (1998)

reacts with calcium hydroxide released in the hydration of the cement,
yielding additional amounts of the C-S-H phase. At the same time the
presence of well-crystallized portlandite crystals, which act as points of
weakness within the structure of the hardened paste, is reduced or

eliminated. The added microsilica also causes a densification of the interfacial region, as it eliminates the presence of oriented portlandite crystals at the aggregate surface and improves the bond between the aggregate and the cement paste bond. A controversy exists about which of these effects is mainly responsible for the strength increase of the hardened concrete (Duval and Kadri, 1998). It has also been reported that some metakaolin products (at a 15% cement replacement level) may exhibit a more pronounced strength enhancement at early ages than microsilica (Curcio, 1998). This was explained by the high pozzolanic activity of these products, which is comparable to that of microsilica.

- A superplasticizer has to be added, to disperse the fine microsilica par-
  ticles and to improve the rheology of the fresh concrete mix. Sulfonated naphthalene or melamine based products have been found to be well suited for this purpose (Cheong *et al.*, 1997; Lang and Geiseler, 1997). To achieve the desired plasticity of the fresh mix, the amount of superplasticizer added must be relatively high; it will vary in different types of cement and at different additions of microsilica. In general, the required dosage of the superplasticizer increases with increasing C_3A content in the cement (Duval and Kadri, 1998) and with its increasing fineness (Sarkar and Baalbaki, 1992). It decreases with increasing C_2S content (Kato *et al.*, 1997). The effect of the super-plasticizer may be enhanced by adding only part of it to the mix at the time of mixing, introducing the rest several tens of minutes later (Sarkar and Baalbaki, 1992).

Portland cement is most widely used as the binder in high-performance concrete. However, not all types of Portland cement perform equally well. The properties of Portland cement produced in a given plant are not absolutely constant, but vary within a certain range. Even differences that are not as great as those common in existing plants may cause distinct variations in the rheology of the fresh concrete mix produced with such a cement, even though they do not cause unacceptable variations in the strength of the hardened material (Sarkar and Baalbaki, 1992). The use of a cement with an elevated C_2S content may be indicated in applications in which fast strength development is not required and a slower release of heat of hydration is desired. Such cement also exhibits higher strength development at later ages, compared with ordinary Portland cement (Kato *et al.*, 1997). Fine grinding of the cement increases the strength that is achieved, but also increases the demand for the superplasticizer (Sarkar, 1992; Nehdi *et al.*, 1998). Table 18.1 indicates the effect of cement composition on the strength of high-performance concrete.

Sarkar and Baalbaki (1992) and Peukert (1997) developed a special cement for high-performance concrete. To increase the strength of the concrete, the cement has a silicate modulus of 3.0 or more, a combined di- and

Table 18.1 Effect of cement composition on 28-day compressive strength of high-performance concrete.

		ASTM type I	*ASTM type II*		*ASTM type III*
Cement composition					
C_3S (%)		63	46		58
C_2S (%)		13	27		17
C_3A (%)		6	8		9
C_4AF (%)		11	9		5
Blaine (m^2/kg)		360	360		460
Microsilica (%)	0	10	0	10	10
Compressive strength (MPa)	60	74	75	95	91

W/(C + MS) + 0.30

Source: Sarkar and Baalbaki (1992)

tricalcium silicate content over 80%, and a correspondingly reduced C_3A and C_4AF content. The low C_3A content of the cement reduces its water requirement, which may be lowered even further by the use of anhydrite, rather than gypsum, as set regulator (Peukert, 1997). For very fast hardening, K_2CO_3 or Na_2CO_3 may be added to the mix. These alkaline activators have been found to be highly effective in this type of cement (Peukert, 1997). The development of still another modified Portland cement for high-performance applications was reported by Wang (1997) without giving any details about the composition.

The texture of cement pastes in high-performance concretes is characterized by high density and very low porosity. Owing to the pozzolanic reaction between the added microsilica and calcium hydroxide, the amount of crystalline portlandite in the hardened paste is reduced, and the formed C-S-H phase has a lower C/S ratio (Cheong *et al.*, 1997). The mature cement paste contains a significant fraction of non-hydrated cement.

The composition of the pore solution is altered in high-performance concrete as compared with concrete mixes with higher *w/c* and without added microsilica (Sota *et al.*, 1997). The Ca^{2+} content is higher and increases with time; the K^+ concentration is strongly reduced, and the Na^+ concentration is increased.

In addition to Portland cement, blended cements with added granulated blast furnace slag (up to 77 wt% replacement of Portland cement) or fly ash (up to 20 wt% replacement) may also be used for high-performance concrete (Sarkar and Baalbaki, 1992; Cheong *et al.*, 1997; Lang and Geiseler, 1997; Sheen and Hwang, 1997). However, when fly ash is added a strength retrogression at 28 days may be observed (Sarkar and Baalbaki, 1992).

Nehdi *et al.* (1998) studied the effect of a variety of microfillers (microsilica, finely ground silica, and limestone filler) replacing Portland cement, on the rheology of fresh high-performance concrete. Replacement levels of up to 20% were employed. For the same slump value, all of the fillers, except microsilica, required a reduced dosage of the superplasticizer. Both the flow resistance and the torque viscosity declined with increased fineness of the microfiller, microsilica being most effective in this respect.

In addition to compressive strength, the bending strength of high-performance concrete is also elevated, and increases proportionally with the compressive strength. Values exceeding 10 MPa are common. The brittleness of the material also increases with increasing strength. The creep of the hardened concrete declines significantly with the addition of microsilica. The resistance to corrosion and abrasion is improved.

The drying shrinkage of high-performance concrete is similar to that of normal concrete, but progresses much more slowly, owing to the reduced rate of drying.

High-performance concrete exhibits an excellent bond to old concrete and to the steel reinforcement, owing to the dense particle packing in the interfacial zone brought about by the presence of microsilica.

One important area where high-performance concrete may be effectively used is in the construction of high-rise buildings. If this kind of concrete is used, the use of reinforcement may be reduced and/or the dimensions of the structural members (columns, walls, etc.) can be lowered. The span of the flexural members may be increased. Owing to the high modulus of elasticity of high-performance concrete (45–50 GPa) structures may be designed to resist fatigue effects, which can cause structural failure with normal concretes.

High-performance concrete mixes with a rapid strength development may be used in the production of precast concrete elements, without the need for steam curing.

Because of its very dense microstructure, high-performance concrete may be effectively used in applications where it is exposed to chemical attack or abrasion. Very often additional protective measures, which are necessary if normal concrete is used, may be avoided.

If water-saturated high-performance concrete is exposed to an intensive fire it may explode, or spalling may occur. This is due to the high resistance of the dense concrete to steam movement, which results in a build-up of steam pressure within the material. Thus special measures, which may include the incorporation of suitable fibers into the mix, may be indicated if concretes with water/cement ratios below about 0.33 are employed.

REFERENCES

Breitenbucher, R. (1998) Development and application of high performance concrete. *Materials and Structures* **31**, 209–215.

Cheong, H.M., Lim, J.R., and Cho, D.W. (1997) Properties and microstructure of ultra high strength cement composites, in *Proceedings 10th ICCC, Göteborg*, paper 2ii037.

Curcio, F., DeAngelis, B.A., and Pagliolico, S. (1998) Metakaolin as a pozzolanic microfiller for high-performance mortars. *Cement and Concrete Research* **29**, 803–809.

Duval, R., and Kadri, E.H. (1998) Influence of silica fume on the workability and the compressive strength of high-performance concrete. *Cement and Concrete Research* **28**, 543–547.

Kato, H., Katumero, R., and Ushiyama, H. (1997) Properties of high strength concrete using belite-rich cement and silica fume (in Japanese). *Semento Konkurito Ronbunshu* **51**, 364–369 [ref. CA 128/220688].

Lang, E., and Geiseler, J. (1997) Utilization of high slag blastfurnace cement for high-performance concrete: influence on pore structure and permeability, in *Proceedings 10th ICCC, Göteborg*, paper 4iv001.

Nehdi M., Mindess, S., and Aitcin, P.C. (1998) Rheology of high-performance concrete: effect of ultrafine particles. *Cement and Concrete Research* **28**, 687–697.

Peukert, S. Cement for HPC concretes, in *Proceedings 10th ICCC, Göteborg*, paper 2ii034.

Sarkar, S.L., and Baalbaki, M. (1992) The influence of the type of cement on the properties and microstructure of high-performance concrete, in *Proceedings 9th ICCC, New Delhi*, Vol. 5, pp. 89–94.

Sheen, V.N., and Hwang, C.L. (1997) Hydration mechanism of high performance concrete, in *Proceedings 10th ICCC, Göteborg*, paper 2ii035.

Sota, J.D., Gaiccio, G.M., and Zerbino R.L. (1997) High performance concrete: study of mortar pore solution, in *Proceedings 10th ICCC, Göteborg*, paper 2ii036.

Steigenberger, J. (1998) High performance concrete for bridges: Composition and properties (in German). *Zement Beton (Vienna, Austria)*, **1998**(2), p. 13.

Wang, J.F. (1997) High performance cementitious binder, in *Proceedings 10th ICCC, Göteborg*, paper 2ii038.

19 Cements with reduced hydration heat evolution

The hydration of inorganic binders is an exothermic process, accompanied by the release of heat of hydration. As a consequence, the temperature of the cement paste (or mortar/concrete mix) increases in the course of hydration. At the same time the generated heat is also carried off into the environment.

The generation of heat of hydration may cause problems, if bulk concrete structures are being erected. Here – after placing the fresh mix – the temperature of the concrete may increase significantly, owing to the very low rate at which the generated heat is eliminated into the environment. Upon subsequent cooling, thermal gradients that develop within the hardened concrete generate tensile stresses, which may cause cracks within the material. To minimize such danger, measures must be taken to reduce the amount of heat liberated in the hydration. This may be done by selecting a binder with a low heat of hydration, and by reducing the amount of the binder within the mix as far as possible.

Another factor of importance is the rate at which the heat of hydration is released. In general, a faster rate of heat release will result in a higher maximum temperature of the concrete, and will increase the risks of crack formation. The rate at which the hydration heat is released, just like the total amount of heat generated, will depend on the phase composition of the binder. However, in addition to the composition of the binder, the rate of heat release will also depend on its fineness, and may be reduced by a coarser grinding of the cement.

Additional factors that affect the increase of temperature within the concrete body and the maximum temperature achieved include:

- the composition of the fresh concrete mix;
- the presence of admixtures that retard or accelerate the progress of hydration;
- the starting temperature of the mix;
- the ambient temperature;
- the geometry of the concrete body to be produced.

In some instances cooling of the concrete mix (for example by adding crushed ice, or by the use of pipes with circulating cold water, incorporated into the concrete mix) may be indicated, to prevent excessive temperature increase within the hardening concrete. If ice is added, it must be completely melted prior to setting, and the mix must be remixed. The addition of ice will increase the original water/cement ratio, and this must be taken into consideration.

Portland cement is a multi-phase binder in which the various phases possess different heats of hydration. Thus the overall heat of hydration of this binder will depend significantly on the phase composition of the clinker. Table 19.1 summarizes the heats of hydration of the individual phases.

Figure 19.1 shows the rate of heat evolution and the cumulative amount of heat evolved for a typical ordinary Portland cement. At ambient temperature the rate of heat evolution reaches its maximum within less than 24 hours, and declines in the subsequent course of hydration. Within this time about half of the total amount of hydration heat becomes released.

Figure 19.2 shows the typical amounts of heat liberated in different types of Portland cement (ASTM classification) after different hydration times. The Type II cement (moderately sulfate resistant) possesses a lower heat of hydration than type I cement (ordinary Portland cement, no limits for C_3A) because of a lower C_3A content (maximum 8%). The Type III cement (high early strength) possesses a high heat of hydration, especially after shorter hydration times, mainly because of finer grinding and a higher C_3A content (up to 15%). In the Type IV cement (low heat) the particularly low hydration heat values are mainly the consequence of a higher C_2S content (minimum 40%) and reduced C_3S (maximum 35%) and C_3A (maximum 7%) contents. Type V cement (sulfate resistant) also has a rather low heat of hydration, owing to limited C_3A and C_4AF contents (C_3A maximum 5%; C_3A + C_4AF maximum 20%). In concrete mixes made from Type IV cement the adiabatic temperature rise is particularly low, and the time at which the maximum temperature is reached is extended (Sone *et al.*, 1992). The reduced rate of hydration heat liberation is usually associated with a reduced rate of strength development. It has been observed, however, that at the same heat of hydration the compressive

Table 19.1 Heat of hydration of pure clinker minerals.

Starting phase	Reaction product	kJ/kg	kJ/mol
C_3S (+ H)	C–S–H + CH	520	118
β-C_2S (+ H)	C–S–H + CH	260	45
C_3A (+ CH + H)	C_4AH_{19}	1160	314
C_3A (+ H)	C_3AH_6	910	245
C_3A (+ $C\bar{S}H_2$ + H)	$C_4A\bar{S}H_{12}$ (AFm)	1140	309
C_3A (+ $C\bar{S}H_2$ + H)	$C_6A\bar{S}_3H_{32}$ (AFt)	1670	452
C_4AF (+ CH + H)	$C_3(A, F)H_6$	420	203

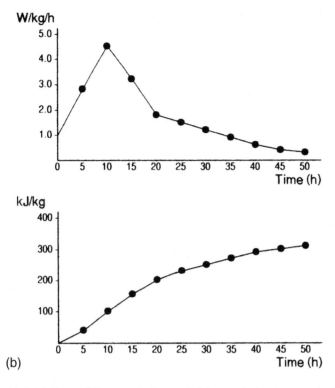

Figure 19.1 (a) Rate of heat evolution and (b) cumulative heat evolved in typical ordinary Portland cement.

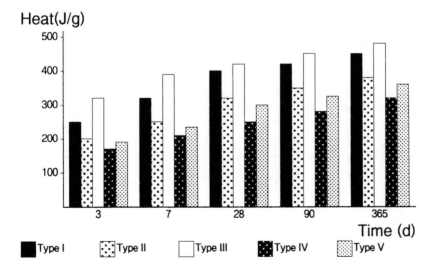

Figure 19.2 Cumulative evolution of hydration heat in different types of Portland cement.

strength of a low-heat Portland cement will increase with increasing C_2S content in the clinker (Yoshida and Igarashi, 1992).

The amount of heat released in the hydration of **Portland-slag cement** may vary over a wide range, and will depend on the amount of slag in the cement, on the reactivity of the slag employed, and on the fineness of grinding. At equal strengths the heat of hydration declines with increasing slag content in cement (Yoshida and Igarashi, 1992). Most cements of this type, especially those with a high slag content, qualify as low-heat cements. A "super low-heat cement" may be produced by keeping a high slag content in the cement and grinding the slag to a very high fineness (600 m^2/kg Blaine) (Tomisawa *et al.*, 1993).

The rate of hydration heat evolution in **Portland—fly ash cements** may vary greatly, mainly because of distinct variations in the reactivity of the individual fly ashes employed. As a general rule, however, the amount of heat released within the first days of hydration is lower than that of a control Portland cement. If, on the other hand, only the Portland clinker fraction is considered, the amount of released heat is usually higher, indicating an accelerating effect of the fly ash on the hydration of clinker constituents. A similar effect may also be observed in cements containing natural pozzolanas. Most blended cements that contain fly ash or natural pozzolana in sufficiently high amounts, in combination with Portland clinker and gypsum, qualify as low-heat cements.

Belite-rich Portland cement and **belite cement** possess a low heat of hydration due to the absence or presence of reduced amounts of tricalcium silicate among their constituents (Kuba *et al.*, 1997; Matsunaga *et al.*, 1997; Uchida, 1998). In general, the amount of released heat decreases with increasing C_2S content (Matsunaga *et al.*, 1997). At a constant C_2S content, the ratio of heat of hydration to strength in these cements also depends on the A/F content in the interstitial phase, and is at a minimum when the composition of the interstitial phase is close to C_6AF_2 (Ikabata and Takemura, 1997). The effects of fineness and SO_3 content on the heat evolution of belite cements are rather small (Nagaoka *et al.*, 1994).

Unlike belite cements, **sulfobelite cements** tend to exhibit a fast initial rate of hydration heat release, owing to a fast hydration of the $C_4A_3\bar{S}$ phase, and do not usually qualify as low-heat cements.

Calcium aluminate cement does not qualify as low-heat cement either, as its heat of hydration is high (500 k J/kg), and most of it is released within the first few days of hydration.

REFERENCES

Ikabata, T., and Takemura, H. (1997) Influence of the composition of the interstitial phase on the hydration properties of belite cements, in *Proceedings 10th ICCC, Göteborg*, paper 2ii007.

Kuba, S., Tajima, K., and Numata, S. (1997) Durability of low heat Portland cement and slag blends (in Japanese). *Semento Konkurito Ronbunshu* **51**, 648–653 [ref. CA 128/220662].

Matsunaga, A., Ito, T., and Chikamatsu, R. (1997) Study on strength and adiabatic temperature rise of high strength concrete using low heat Portland cement (in Japanese). *Semento Konkurito Ronbunshu* **51**, 370–375 [ref CA 128/220689].

Nagaoka, S., Mizikoshi, M., and Hotta, F. (1994) Effect of fineness and SO_3 content on strength development and adiabatic temperature rise of belite-rich cements (in Japanese). *Semento Konkurito Ronbunshu* **48**, 124–129 [ref. CA 123/151127].

Sone, T., Fujiyama, O., and Tanimura, M. (1992) Properties of concrete using low-heat cement containing a large amount of belite (in Japanese). *Semento Konkurito Ronbunshu* **46**, 392–397.

Terada, R., Tomita, R., and Tanaka, S. (1992) Some properties of concrete with belite type cement (in Japanese). *Semento Konkurito Ronbunshu* **47**, 142–147 [ref. CA 120/329775].

Tomisawa, T., Chikada, T., and Nagao, Y. (1993) Properties of super low heat cement incorporating large amounts of ground granulated blast furnace slag of high fineness. American Concrete Institute SP-132, pp. 1385–1399.

Uchida, K. (1998) High-performance concrete (in Japanese). *Kagaku Kogyu* **49**, 334–346 [ref. CA 128/298513].

Yoshida, K., and Igarashi, H. (1992) Strength development and heat of hydration of low heat cement, in *Proceedings 9th ICCC, New Delhi*, Vol. 3, pp. 16–22.

20 Binders with increased resistance to chemical attack

In many applications cementitious systems are exposed to chemical attack that may cause their gradual degradation, and may affect their service life adversely. Thus adequate measures must be taken to prevent or minimize this type of degradation and to ensure proper performance of the material.

Hardened cement pastes are inherently porous systems. Thus the chemical attack may not only take place at the surface of the cementitious body, but may also extend into the interior of the material as the corrosive agent migrates through the pore system into deeper regions. The corrosive action may be even greater if a corrosive liquid percolates through the cement paste. To reduce the extent of migration through the pore system, and thus to reduce the intensity of the chemical attack, it is important to keep the porosity of the cement paste as low as possible (van Eijk and Brouwers, 1998). This may be achieved by keeping the water/cement ratio of the starting mix low, and by protecting the material from exposure to the corrosive agent at the early stages of hydration.

A special situation occurs if concrete that is in contact with a corrosive salt solution is simultaneously exposed to drying-wetting cycles. Under these conditions the chemical corrosion may be enhanced by surface scaling caused by salt crystallization. Such an effect may be effectively minimized by introducing an air-entraining agent into the concrete mix (Yang et al., 1997). A concentration of the corrosive agent within the pore liquid and ultimately a salt crystallization within the pore space or at the concrete surface may also take place if water with the dissolved salt, after having migrated through the pore system, is evaporated at one of the concrete surfaces.

As the various hardened cement pastes react differently with different chemical agents, the selection of a proper binder is essential. In many instances damage of the cementitious material caused by chemical attack may be reduced or prevented completely by the use of a binder that is sufficiently resistant to the particular chemical agent.

In some instances, especially at high concentrations of the corrosive agent, it may be necessary to protect the concrete construction from direct contact with this agent. This may be done by the application of an appropriate coating or by other measures.

The following forms of corrosive action of chemical agents on hardened concrete may be distinguished:

- dissolution of the hydrated material in the corrosive liquid;
- a chemical reaction between the corrosive agent and constituents of the hydrated material – this is associated with the formation of a new phase, which causes expansion and ultimately a weakening and crack formation within the material;
- crystallization of the corrosive agents within regions of the pore system near the surface – this is associated with the generation of crystallization pressure, resulting in scaling of the concrete surface.

From among different inorganic binders, the resistance to chemical attack of Portland cement – the binder most widely used in concrete construction – has been studied the most extensively. It has been found that hardened Portland cement pastes perform well in many environments, but there are applications in which the use of other binders is more appropriate. In this chapter the action of the most common chemical agents on selected cementitious systems and the response of them to the chemical attack will be discussed.

20.1   WATER

Hardened cement pastes made from hydraulic binders consist of reaction products (hydrates) that have very low water solubility, and generally perform well in contact with water. However, upon prolonged exposure to flowing or percolating water a gradual degradation (dissolution) of the hardened cement paste may take place. Soft or deionized water is more dangerous in this respect than water in which distinct amounts of inorganic species are already dissolved.

In hardened Portland cement pastes alkali hydroxides are the species most loosely bound. They enter the liquid phase that is in contact with the paste surface first, followed by calcium hydroxide. In percolation experiments the alkali ion concentration in the liquid phase started to decline after reaching a maximum, whereas the calcium ion concentration increased to a value that remained constant for the rest of the experiment (Unsworth *et al.*, 1997).

The reactions taking place in a hardened Portland cement paste exposed to deionized water include dissolution of the existing portlandite and decomposition of the calcium silicate hydrate and hydrated aluminate (sulfate) phases, which are gradually deprived of calcium hydroxide. All these reactions take place simultaneously, but at different rates. The dissolution of portlandite progresses fastest, followed by the decomposition of ettringite, monosulfate, and the C-S-H phase. As this degradation process progresses, the C/S ratio in the C-S-H phase declines, and trivalent aluminum and iron ions, which are released in the decomposition of the AFt and AFm phases, become partially incorporated into the C-S-H structure, thus increasing the stability of this phase (Faucon *et al.*, 1996, 1998). Gradually a surface layer becomes established from which substantial amounts of calcium and lesser amounts of silicone have been leached out. Iron and magnesium are not dissolved in this process in significant amounts, and so their content in the surface layer increases. Magnesium precipitates as hydrotalcite, and iron becomes accommodated mainly as a constituent of iron-substituted hydrogarnet and an iron-containing low-calcium C-S-H phase (Faucon *et al.*, 1996). The process of degradation is initially governed by the laws of diffusion and later by the rate of surface dissolution (Faucon *et al.*, 1997). The final product is an amorphous material consisting mainly of hydrated silica with minor amounts of Ca, Fe, and Al.

The negative effect of leaching with water may be reduced by combining Portland cement with pulverized fly ash or silica fume (Unsworth *et al.*, 1997; van Eijk and Brouwers, 1998). In the presence of these additives the dissolved calcium hydroxide tends to react with the residual non-reacted fly ash, yielding additional amounts of C-S-H, and thus reducing the permeability of the paste. The optimum amounts of these additives to be interblended with Portland cement were found to be around 35 mass% of fly ash and about 8 mass% of silica fume (van Eijk and Brouwers, 1998). However, cement combined with 70 mass% of granulated blast furnace slag behaves not too differently from a cement that contains just Portland clinker alone (Faucon *et al.*, 1996).

20.2   ACIDS

The action of acids on hardened cements pastes is similar to that of plain water, but is significantly more intensive. Acids yielding products of limited solubility, such as H_2SO_4 or H_3PO4, are less aggressive than those whose reaction products are easily soluble, such as HNO_3 or HCl. In addition to the cement paste, carbonate aggregates, such as limestone or dolomite, are also attacked by the acid.

In the course of acid attack on hydrated Portland cement and related binders, including blended cements, the free calcium hydroxide is dissolved together with CaO bound within the C-S-H phase. In this way the CaO/SiO_2 molar ratio in this phase may be lowered below 1.0. In strong acids complete decomposition of the cement matrix may take place, leaving silica gel and gypsum as the final solid residua of the decomposition process (Ueda *et al.*, 1997).

A special case of acidic corrosion is that with carbonic acid. The partial pressure of CO_2 in air at sea level is about 32 Pa. Pure water in equilibrium with this CO_2 concentration contains 0.012 mmol/L of dissolved CO_2 (to a significant extent as $HCO_3^- + H^+$), and the acidity of the solution drops to pH = 5.6. In some natural waters the amount of dissolved CO_2 may be significantly higher and the pH value correspondingly lower. Water containing dissolved CO_2 may exhibit a distinct corrosion of concrete made with Portland cement. It not only dissolves free $Ca(OH)_2$ but may also decompose calcium carbonate present in the employed aggregate according to the equation

$$CaCO_3 + CO_2 + H_2O \rightarrow Ca^{2+} + 2HCO_3^- \tag{20.1}$$

The capacity of CO_2 dissolved in water to react with calcium hydroxide or calcium carbonate is limited, however, and declines greatly with increasing amount of Ca^{2+} already dissolved. Thus the term "aggressive CO_2" has been introduced, which is defined as the amount of dissolved CO_2 in one unit volume of water that can react with $CaCO_3$ according to reaction (20.1) at the given amount of Ca^{2+} already dissolved.

Portland cement (see section 0.00) is hardly suitable for use in acid environments. No significant differences have been found between the performances of ordinary and high-iron, low-C_3A Portland cements, but cements with a higher C_3S content, which liberate more calcium hydroxide in the course of hydration, seem to be more susceptible to acids (Matthew, 1992). Partial replacement of the clinker by pulverized fly ash does not noticeably improve the resistance of the cement to acids (Matthew, 1992). Portland slag cements are attacked by acids more severely than plain Portland cement (Madrid *et al.*, 1997). The rate of deterioration of concrete in acid environments becomes reduced as its quality, expressed in terms of 28-day compressive strength or water/cement ratio, is improved (Matthew, 1992).

Portland cement and related binders should not be used in applications where the concrete surface is permanently exposed to pH values lower than 7. Calcium aluminate cement (see section 10) resists acids somewhat better than Portland cement, and may be applied down to pH values as low as 4. Supersulfated cement (see section 8.4) also performs relatively well in diluted acid solutions down to pH = 4. Binders highly resistant to acid solutions include alkali silicate cement (see section 15.3), geopolymer cements (see section 15.7.1), and alkali-activated fly ash–slag cement (see section 9.1.5, and Lu and Li, 1997).

20.3 ALKALIS

Solutions of alkali hydroxides generally do not exhibit a corrosive action towards pastes of hardened Portland cement and related binders. Only at very high concentrations may a moderate corrosion become apparent, probably owing to degradation of the hydrated aluminate phases. Also highly resistant to alkaline solution are alkali-activated slag binders (see section 8.5). In contrast, hydrated calcium aluminate cement may be attacked by high-pH solutions.

20.4 SULFATES

Sulfates are regular constituents of most groundwaters, industrial waste-waters, and sewage waters, and may act corrosively on a variety of hard-ened cement pastes, including that of Portland cement, if present in sufficiently high concentrations. A degradation of concrete may also take place if sulfates, such as gypsum or anhydrite, are present in sufficiently high amounts in the aggregate used.

Chemical attack by solutions containing sulfates proceeds by an inward movement of a reaction front. In this way a surface region is produced whose thickness increases with time, and in which the reactions that take place first are those taking place in the greatest depth.

If a hardened ordinary Portland cement paste is in contact with a solution that contains SO_4^{2-} ions in combination with Na^+ or K^+, two principal reactions take place, a conversion of the monosulfate into ettringite and the formation of gypsum:

$$4CaO.Al_2O_3.SO_3.12H_2O + 2SO_4^{2-} + 2Ca^{2+} + 20H_2O$$
$$\rightarrow 6CaO.Al_2O_3.3SO_3.32H_2O \tag{20.2}$$

$$2Na^+ + SO_4^{2-} + Ca(OH)_2 + 2H_2O \rightarrow CaSO_4.2H_2O + 2Na^+ + 2OH^- \tag{20.3}$$

If residual, non-reacted C_3A is still present, this phase may also serve as a source of Al^{3+} for the formation of ettringite, whereas Al^{3+} incorporated in the C-S-H phase does not.

At a later stage of the process, after the available portlandite has been consumed, the sulfate solution also reacts with the C-S-H phase, yielding additional gypsum. In this reaction the CaO/SiO_2 ratio of the C-S-H phase becomes lowered and its mechanical strength reduced. If at this stage some residual non-reacted C_3A is still present, it may also react with SO_4^{2-} and Ca^{2+} to yield ettringite.

The formation of ettringite is associated with a generation of expansive forces, which may cause expansion of the material and spalling of surface layers. At the same time the decomposition of the C-S-H phase causes a loss of strength and of cohesion. Ultimately, a complete disintegration of the hardened cement paste may occur.

If concrete made with Portland cement is exposed to a calcium sulfate solution, there is initially a induction period in which the properties of the material change slowly, and virtually no damage to the concrete is visible. This is a period in which ettringite precipitates in the available pore space, without producing significant internal stresses within the material. Only after this introductory period does a disintegration of the material become apparent, which is caused by the formation of additional amounts of ettringite, producing internal stresses within the material. The length of the introductory period depends on the concentration of sulfates in the water, and is shortened as this concentration increases (Aköz *et al.*, 1995). If calcium sulfate, rather than alkali sulfate, acts as the corrosive agent, either present as a constituent of the aggregate or dissolved in water in contact with hardened concrete, a consumption of portlandite and a decalcification of the C-S-H phase does not take place. However, the ettringite formed may also cause an expansion and a degradation of concrete under these conditions.

If the sulfate ions are present in the form of magnesium sulfate, the existing calcium hydroxide is converted to brucite [magnesium hydroxide, $Mg(OH)_2$], and calcium sulfate is also formed:

$$Mg^{2+} + SO_4^{2-} + Ca(OH)_2 + 2H_2O \rightarrow CaSO_4.2H_2O + Mg(OH)_2$$
(20.4)

The C-S-H phase undergoes a gradual decalcification in which the C/S ratio gradually decreases. After reaching a sufficiently low C/S ratio, the residual low-lime C-S-H phase tends to convert to a near-amorphous M-S-H phase, most likely to a poorly crystallized serpentine (Gollop and Taylor, 1992–1996; Bonen, 1997), which does not exhibit cementing properties. Alternatively the decalcification of the C-S-H phase may progress further, until it is converted to amorphous hydrous silica. At the same time additional amounts of gypsum and brucite are also formed:

$$3Mg^{2+} + 3SO_4^{2-} + 3CaO.2SiO_2aq + 3 H_2O$$
$$\rightarrow 3CaSO_4.2H_2O + 3Mg(OH)_2 + 2SiO_2aq$$
(20.5)

In a partially corroded specimen the C/S ratio of the C-S-H phase is highest in the central core, and declines toward the sample surface as a consequence of progressing degradation. The degradation of the C-S-H phase in the presence of magnesium sulfate is significantly faster and more complete than with other sulfate solutions.

In parallel to the above reactions, the ettringite phase may also be formed in the $MgSO_4$ attack, but its amount is relatively low, as the hardened cement paste tends to disintegrate owing to the degradation of the C-S-H phase, before significant amounts of ettringite have been formed. The reasons why reactions (20.4) and (20.5) proceed to completion are the low solubility of magnesium hydroxide and the low pH of the solution in equilibrium with this phase. Only about 0.01 g/L of magnesium hydroxide dissolves in water, and the saturated solution has a pH of 10.5, which is too low to maintain the stability of the C-S-H phase. Thus, after all the free calcium hydroxide has been consumed [equation (3)], and the pH of the liquid phase has dropped below the stability range of C-S-H, this phase liberates calcium hydroxide to establish its equilibrium pH. This calcium hydroxide, however, is immediately also converted to magnesium hydroxide, as long as free magnesium sulfate is available.

The action of magnesium sulfate is associated with a migration of hydroxide ions towards the surface to produce insoluble brucite, and a migration of sulfate ions inwards to form gypsum. The calcium needed for gypsum formation is supplied first by the decomposition of calcium hydroxide and subsequently by decalcification of the C-S-H phase. Eventually, a double surface layer is formed consisting of an external layer of brucite, followed by a layer of gypsum.

Because of the simultaneous significant decomposition of the C-S-H phase, which accompanies the formation of gypsum and ettringite, the corrosive action of magnesium sulfate is greater than that of calcium or alkali sulfates (at equal SO_4^{2-} concentrations). Unlike attack by alkali sulfates, attack by magnesium sulfate is characterized mainly by a loss of strength and disintegration of the concrete under attack, rather than by an expansion.

The effectiveness of different forms of sulfate (at equal SO_4^{2-} concentrations) and the susceptibility of individual clinker phases and their blends to this ion are apparent from Table 20.1, which indicates the time it took to attain an expansion of the studied samples of 5 mm/m.

If hardened Portland cement paste or concrete is exposed to a solution of ammonium sulfate, this compound decomposes in the highly alkaline environment of the cement paste, and gaseous ammonia is liberated:

Table 20.1 Relative rate of expansion of clinker mineral blends in different sulfate solutions (equivalent SO_3 content = 1.2%).

Compounds	Time needed to expand 5 mm/m		
	1.8% MgSO₄	*Saturated CaSO₄*	*2.1% Na₂SO₄*
C_2S	28 days	Negligible expansion in 18 years	
C_3S	35 days	2.2 mm/m in 9 years	12 years
C_2S 80% + C_3A 20%	6 days	10 days	4 days
C_3S 80% + C_3A 20%	4 days	11 days	7 days
C_3S 80% + C_4AF 20%	16 days	1.5 mm/m in 3 years	400 days
Portland cement (mean value of 8 samples)	11 days	–	13 days

Mortars 1:10

Source: Eglinton (1998)

$$(NH_4)_2SO_4 + Ca(OH)_2 \rightarrow CaSO_4.2H_2O + 2NH_3 \tag{20.6}$$

The calcium sulfate formed reacts further with other constituents, producing ettringite and causing expansion. Thus the overall action of ammonium sulfate is a combination of an acidic and sulfate corrosion (Miletic and Ilic, 1997). As a result, ammonium sulfate is a particularly detrimental corrosive agent for Portland cement.

With respect to their resistance to sulfates, inorganic binders may be divided into three groups:

- binders that are susceptible to sulfate solutions, and expand and lose strength rapidly even at low sulfate concentrations;
- binders with an increased resistance to sulfates, which may, however, expand and decay at high concentration and/or at lengthy exposure to sulfates;
- binders that are resistant to sulfates even at high concentrations and exposure times.

Ultimately the resistance to sulfates will depend not only on the type of cement used, but also on the cation with which SO_4^{2-} is combined, on the concentration of the sulfate in solution, on the condition of exposure to the sulfate solution, and on the amount of the binder in the concrete mix and the water/cement ratio employed. In general, the sulfate resistance will increase with increasing amount of cement employed and with decreasing water/cement ratio.

20.4.1 Sulfate-resistant Portland cement

Sulfate-resistant Portland cement is characterized by a reduced C_3A content in clinker. Since sulfate attack on hydrated Portland cement involves mainly the C_3A phase, such a measure increases significantly the ability of the hardened concrete to resist the corrosive action of sulfates. The effectiveness of such a measure is apparent from Fig. 20.1, which shows the expansion of a series of cements made with different C_3A content (at an equal $C_3A + C_4AF$ content and a constant SO_3 content). Low-C_3A clinkers may be produced by reducing the Al_2O_3 content and/or by increasing the Fe_2O_3 content in the raw meal. In the latter case the amount of the ferrite phase in the clinker is increased at the expense of C_3A. Thus sulfate-resistant Portland cements are essentially high-iron, low-C_3A cements (see section 2.8) with an SO_3 content and specific surface area in the same range as common in ordinary Portland cement.

In the hydration of Portland cement at ambient temperature the existing C_3A reacts initially with added calcium sulfate to yield ettringite (AFt phase), as long as free calcium sulfate is available. After that, the ettringite

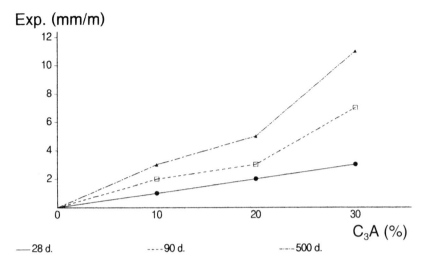

Figure 20.1 Effect of C$_3$A content in Portland cement on cement paste expansion
(C$_3$A + C$_4$AF = 30%).
Source: Odler 1991

already formed reacts with still unreacted C$_3$A and converts to monosulfate (AFm phase), which is a regular constituent of mature pastes made from ordinary Portland cement. If exposed to SO$_4^{2-}$ from an external source, monosulfate converts back to ettringite, and this is the reaction that causes expansion, spalling, and disintegration of the material. The ferrite phase also reacts with calcium sulfate interground with the clinker, but this reaction proceeds significantly more slowly. After free calcium sulfate has been consumed, the formed Fe-doped AFt phase does not tend to convert to the AFm phase, unlike that produced from C$_3$A. Instead, it mostly stays in the paste unchanged, and the fraction of it that converts to AFt is rather low.

It is not obvious why the extent of expansion due to delayed exposure to sulfates is reduced in pastes made from cements containing the ferrite phase rather than the tricalcium aluminate phase. It is assumed that this effect may be due to a reduced amount of newly formed ettringite, caused by the fact that a ferrite-doped AFt phase is the main aluminate/ferrite phase in these pastes (Lota *et al.*, 1995; Eglinton, 1998). It has also been suggested that the iron-doped AFt phase possesses a reduced expansivity due to an altered crystal morphology (Regourd *et al.*, 1980; Odler and Gasser, 1988).

Note, however, that even though in high-Fe$_2$O$_3$, low-Al$_2$O$_3$ cements the resistance to sulfate attack is significantly improved, it is not absolute. Cements of this type usually resist moderately high sulfate concentrations well, but at high concentrations they also tend to expand, though more slowly and to a lesser degree. As the ferrite phase may also participate in the expansion process, though to a lesser extent than C_3A, the amount of this phase in the clinker should also be limited. The US Standard Specification for Portland Cement (ASTM C150–951) specifies a C_3A content of 8.0% for Type II cement (moderately sulfate resistant) and a C_3A content of 5.0%, together with a C_3A + C_4AF content of 20%, for Type V cement (high sulfate resistant). Note also that "sulfate resistant" Portland cements do not exhibit any significant resistance to $MgSO_4$ attack, in which a degradation of the C-S-H phase, rather than an expansion due to AFt formation, plays the main role.

20.4.2 Fly ash–Portland cement and related binders

The resistance of Portland cement to sulfate attack may be increased significantly by substituting a fraction of the clinker (about 30%) with pulverized fly ash (Irassar and Batic, 1989; Mehta, 1992; Soroushian and Alhozaimi, 1992; Mangat and Khatib, 1995; Djuric *et al.*, 1996; Giergiczny, 1997; Krizan and Zivanovic, 1997; Miletic *et al.*, 1997). A similar effect may also be attained by the use of natural pozzolana (Sersale *et al.*, 1997) and silica fume introduced in amounts of about 10% (Aköz *et al.*, 1995; Giergiczny, 1997). Just as with Portland cement, blended cements are more sensitive to magnesium sulfate than to other sulfate forms, but at very high ash contents (70%) the cement may perform acceptably even in contact with magnesium sulfate solutions (Krizan and Zivanovic, 1997).

The effectiveness of different fly ashes as additives preventing sulfate-induced expansion can vary greatly. When the mineral composition of the fly ash is such that the formation of ettringite occurs before exposure to sulfate solution, the blended cement performs well. However, if at the time of exposure to sulfates phases vulnerable to sulfate attack (such as monosulfate) are present, the sulfate resistance of the binder is poor (Mehta 1986, 1992). It has also been reported that at lower replacement levels (10% and 20%) class F fly ashes perform better than class C ashes (Soroushian and Alhozaimi, 1992).

The beneficial effect of pozzolanic materials on sulfate resistance results from calcium hydroxide consumption and the formation of additional amounts of C-S-H, which results in a reduction of porosity and permeability. Soroushian and Alhozaimi (1992) reported a strong correlation between permeability and extent of expansion, irrespective of the type of fly ash or its content. A reduction of the C_3A content in the cement also contributes to improved performance (Irassar and Batic, 1989; Giergiczny, 1997). To obtain a proper performance of blended cements appropriate curing prior to exposure to sulfate solutions is essential (Mangat and Khatib, 1992).

The response of blended cements to sulfate attack may be distinctly improved by a carbonation of the surface of the material prior to exposure to sulfates (Sersale *et al.*, 1997). Improved resistance may also be achieved by adding finely dispersed calcium carbonate to the original mix. In its presence monocalcium carboaluminate ($C_3A.CC.H_{12}$) is formed instead of monosulfate, and this phase does not convert to ettringite in the course of sulfate attack (Piasta *et al.*, 1997).

20.4.3 Portland-slag cement

Portland-slag cements tend to exhibit improved resistance to sulfate attack compared with plain Portland cement at high slag contents (60% and more), whereas the resistance is scarcely affected or even made worse at low substitution levels (below 25%) (Mehta, 1992; Gollop and Taylor, 1992–1996; Giergiczny, 1997; Krizan and Zivanovic, 1997; Sersale *et al.*, 1997). The improved response to sulfate attack appears to be due mainly to a dilution effect, rather than to any special mechanism. An increase of the slag content at the expense of clinker results in a lower amount of C_3A in the system, which is the phase mainly responsible for the susceptibility of Portland cement to sulfate attack.

The mechanism of the sulfate attack on Portland-slag cement is broadly similar to that observed in pure Portland cement pastes. In an attack by alkali sulfates the dominant reactions are the partial decalcification of the C-S-H phase and the conversion of the existing monosulfate to ettringite. The amount of ettringite formed is lower than in the case of Portland cement, as monosulfate is the only phase that can serve as a direct source for ettringite formation, whereas neither Al^{3+} incorporated in the C-S-H phase nor that present in hydrotalcite are available for such a reaction, just like Al^{3+} in ettringite already formed independently of sulfate attack (Gollop and Taylor, 1992–1996).

The high resistance of cements containing high amounts of granulated blast furnace slag to alkali sulfate attack is mainly due to a lower amount of Al^{3+} readily available for ettringite formation: that is, that present in the form of monosulfate. For a given degree of substitution with slag, the susceptibility to sulfate attack increases with increasing Al_2O_3 content of the slag, as this form of Al_2O_3 may serve as an indirect source of Al^{3+} for the formation of ettringite (Mehta, 1992; Gollop and Taylor, 1992–1996), which may progress with a slower, but still distinct rate. The resistance of Portland-slag cement to sulfate attack may be further improved by the use of sulfate-resistant rather than ordinary Portland clinker as a constituent of this cement, and by keeping the amount of calcium sulfate high enough to prevent the conversion of primary formed ettringite to monosulfate (Gollop and Taylor, 1992–1996).

In an attack with $MgSO_4$ the dominant effect is the decomposition of the C-S-H phase and the formation of brucite, poorly crystallized serpentine (a magnesium silicate hydrate), and gypsum. The resulting damage is generally higher than that caused by alkali sulfate solutions of the same concentration. If exposed to magnesium sulfate, blended cements containing blast furnace slag usually perform more poorly than sulfate-resistant Portland cement.

Because in Portland-slag cement, unlike in Portland cement, the decomposition of the C-S-H phase is the main feature of sulfate attack rather than the formation of ettringite, disintegration and softening rather than expansion and cracking are the main visible characteristics of the process.

20.4.4 Supersulfated cement

Hydrated supersulfated cement is highly resistant to sulfate attack, and may be used in applications in which the concrete is exposed to highly concentrated sulfate solutions. The high sulfate resistance is due to the fact that most or all of the Al_2O_3 (except that in the residual, non-reacted slag) is present in the form of ettringite, or is bound within the formed C-S-H phase, and is not available for a reaction with sulfate ions. Such sulfate resistance is limited to alkali and calcium sulfate solutions, or to moderately acid solutions (down to about pH = 3.5), in which the sulfate ions are present in the form of ammonium sulfate or even as free sulfuric acid of low concentrations.

Hardened supersulfated cement is not resistant to magnesium sulfate attack, in which the concrete damage is due to a decomposition of the C-S-H phase by this agent.

20.4.5 Calcium aluminate cement

Concrete based on calcium aluminate cement performs rather well if exposed to sulfate solutions, especially if made with a low water/cement ratio and high cement content. However, cases of expansion and cracking have also occasionally been reported (Scrivener and Capmas, 1998). The reasons for the good sulfate resistance of this type of cement are not obvious. It is mostly attributed to a surface densification of the hardened material, resulting in a very low permeability of the formed surface layer, and/or to the absence of calcium hydroxide in the system (Scrivener and Capmas, 1998). Unlike Portland cement and related binders, magnesium sulfate solutions are less aggressive to calcium aluminate cement than alkali sulfate solutions. This is due mainly to the absence of the C-S-H phase in the hardened calcium aluminate cement pastes, which is particularly sensitive to the action of magnesium sulfate.

A very good durability in sulfate solution is achieved by combinations of calcium aluminate cement and granulated blast furnace slag (Osborne and Sing, 1995).

20.4.6 Miscellaneous sulfate-resistant binders

Cements that may be considered completely stable if exposed to sulfate solutions include geopolymer cement (see section 15.7.1), phosphate cements (see section 12), and alkali silicate cement(see section 15.3). A high resistance to sulfate attack is also exhibited by sulfobelite cement (see section 4.2), slag-lime binder (see section 8.2), and alkali-activated slag cement (see section 8.5).

20.4.7   Recommendations

Table 20.2 summarizes the recommendations of BS 8110 concerning the selection of an appropriate binder under differing severities of exposure to sulfates. The recommendations also include the minimum amount of cement in the concrete mix to be produced.

Table 20.2 Recommendations of BS 8110 for concrete exposed to sulfates.

| Severity of exposure conditions | *Sulfate content (SO₃)* | | | Type of cement recommended | Minimum cement content (kg/m³) |
| | *Soils* | | *Water* | | |
	Total SO₃ (%)	SO₃ in 1:2 soil/water extract (g/L)	g/L		
Moderate	0.2–0.3	1.0–1.9	0.3–1.2	Any	330
				OPC + 25–40% PFA	310
				OPC + 70–90% GGBFS	310
				SRPC or SSC	280
Severe	0.5–1.0	1.9–3.1	1.2–2.5	OPC + 25–40% PFA	380
				OPC + 70–90% GGBFS	380
				SRPC or SSC	330
Very severe	1.0–2.0	3.1–5.6	2.5–5.0	SRPC or SSC	370
Extreme	> 2.5	> 5.6	> 5.0	SRPC or SSC with protective coatings	370

OPC = ordinary Portland cement
SRPC = sulfate-resisting Portland cement
SSC = super-sulfated cement
PFA = pulverized fuel ash
GGBFS = ground granulated blast-furnace slag

20.5   CHLORIDES

Hardened concrete contains or may be exposed to chloride ions under different circumstances. Sources of Cl^- may be $CaCl_2$ used as accelerator of hardening (now widely prohibited), seawater, certain aggregates (especially in desert climates), and – most often – deicing salts used on roads.

In the presence of chlorides a calcium chloroaluminate hydrate, called Friedel's salt ($C_3A.CaCl_2.10H_2O$), may be formed in hydrated Portland cement and other cements containing the C_3A phase. However, in addition

Table 20.3 Compressive strengths of mortars made from different cements and stored 180 days in plain water or an NaCl solution (196 g/L).

Cement	Compressive strength		Strength (NaCl) /Strength (water)
	Water (MPa)	NaCl (MPa)	
OPC 100%	52.2	40.7	0.78
Blast furnace slag cement	51.4	50.4	0.98
OPC 92.5% + SF 7.5%	51.5	53.2	1.03
OPC 80% + GGBFS 20%	46.5	39.5	0.85
GGBFS 85% + WG 15%	108.2	111.7	1.02
GGBFS 95% + Na_2CO_3 5%	59.5	61.9	1.04

OPC = ordinary Portland cement
SF = silica fume
GGBFS = ground granulated blast-furnace slag
WG = water glass

Source: Deja (1997)

to Cl present in chemically bound form, a significant fraction of it also remains in the pore solution in the form of free Cl^- ions. The strength and durability of the hardened cement are not adversely affected by the presence of chlorides (see also Table 20.3).

Chloride ions present in the cement paste promote the corrosion of the reinforcement in steel-reinforced concrete, by causing a breakdown of the protective oxide film that develops at the steel surface at high pH. Under these conditions iron oxide (rust) is formed at the steel/concrete interface, and the expansion associated with it causes cracking and spalling of the surrounding concrete.

To prevent steel corrosion in reinforced concrete exposed to chlorides, measures must be taken to minimize the migration of chloride ions from the surface towards the steel reinforcement. Such measures include a sufficiently high cement content, a low water/cement ratio, and adequate curing of the concrete mix.

Under equal conditions the diffusion coefficients for Cl^- vary greatly in different cements (see Table 20.4). In general, blended cements that contain blast furnace slag, fly ash or silica fume exhibit a lower permeability to chloride ions and are more suitable for protecting the steel reinforcement from corrosion (Jensen and Pratt, 1989; Arya *et al.*, 1990; Schiessl and Raupack, 1992; Deja, 1997; El Sayed *et al.*, 1997; Wiens and Schiessl, 1997). An extremely low permeability was also found in alkali-activated slag cement activated with water glass (Deja, 1997).

The main reason for the decreased chloride-induced corrosion seen in structures made with blended cements is related to the refinement of the pore structure of the formed cement paste, resulting in a restricted mobility of the Cl^- ions (Schiessl and Raupack, 1992; Wiens and Schiessl, 1997).

Table 20.4 Diffusion coefficients of Cl⁻ in different cement pastes.

Cement	$10^{-13} m^2/s$
OPC 100%	38.8
Blast furnace slag cement	8.4
OPC 92.5% + SF 7.5%	6.4
OPC 80% + GGBFS 20%	14.2
GGBFS 85% + WG 15%	1.9
GGBFS 95% + WG 15%	19.6

OPC = ordinary Portland cement
SF = silica fume
GGBFS = ground granulated blast-furnace slag
WG = water glass

Source: Deja (1997)

Another factor involved is the binding of Cl⁻ ions by the existing hydrate phases. A fraction of Cl⁻ may be adsorbed at the surface of the C-S-H phase; however, this amount may probably still be involved in the corrosion process, though to a lesser extent. In mixes made from cements that contain C_3A in clinker, a distinct amount of Cl⁻ is bound in the form of Friedel's salt ($C_3A.CaCl_2.10H_2O$). In this way it becomes removed from the pore solution, and cannot participate in the corrosion process. In addition to C_3A, a fraction of Al^{3+} present in fly ashes or slags may also participate in the formation of Friedel's salt.

Sulfate-resistant Portland cement, unlike ordinary Portland cement, contains no or only reduced amounts of C_3A, and produces no or very little Friedel's salt, even in the presence of excessive amounts of Cl⁻. This cement is therefore particularly unsuitable for applications in which protection of the steel reinforcement from chloride-induced corrosion is required.

Galvanized steel is susceptible to chloride-induced corrosion in highly alkaline environments, as exist in mixes made with ordinary Portland, Portland-pozzolana, or Portland-slag cements. The corrosion may be reduced significantly by the use of supersulfate cement (Dass *et al.*, 1992).

In combined corrosion by solutions that contain simultaneously chlorides and sulfates, the surface deterioration and loss of strength are greater in mixes made with blended cements that contain blast furnace slag or silica fume, than in those made with plain Portland or Portland–fly ash cement. However, the corrosion of steel reinforcement is lower in the former cements (Al-Amoudi, 1993).

20.6 SEAWATER

Concrete exposed to seawater may be damaged by a chemical action of the dissolved salts, by crystallization pressure of these salts within the concrete body, and by mechanical attrition. The salts dissolved in seawater in largest amounts include chlorides, sulfates, sodium, and magnesium.

If Portland cement is used as the binder, the dominant chemical reaction is the formation of ettringite, for which the aluminate and calcium ions are supplied by the cement paste and the sulfate ions are supplied by the seawater. In the first few years this results in an increase of compressive strength, as a result of pore filling. However, thereafter the formation of additional amounts of ettringite causes internal stresses and a strength loss (Sakoda *et al.*, 1992). The concrete corrosion is most critical in and somewhat above the tidal zone, as in this region extensive salt crystallization is also involved in the corrosion process.

In steel-reinforced constructions the chloride ions present in seawater migrate towards the reinforcement, and may cause its corrosion. Rebar corrosion develops most rapidly in the splash zone, owing to a sufficient oxygen supply, high chloride penetration, and the presence of appropriate amounts of water (Sakoda *et al.*, 1992).

For adequate performance, concretes to be exposed to seawater must have a high cement content and a low water/cement ratio. In steel-reinforced structures proper cover of the steel is essential. Portland cement may be used in these applications, but the presence of pozzolanic constituents or blast furnace slag generally improves the corrosion resistance of the concrete (Eglinton, 1998). In a direct comparative study it was found that the penetration rate of chloride ions in concrete is smaller in mixes made with silica fume and blast furnace slag than in those made with ordinary or sulfate-resistant Portland cements (Sakoda *et al.*, 1992).

REFERENCES

Aköz, F. *et al.* (1995) Effects of sodium sulfate concentration on the sulfate resistance of mortars with and without silica fume. *Cement and Concrete Research* **25**, 1360–1368.

Al-Amoudi, O.S.B. (1993) Effect of chloride-sulfate ions on reinforcement corrosion and sulfate deterioration in blended cements. American Concrete Institute SP-132, pp. 1105–1123.

Arya, C., Buenfeld, N.R., and Newman, J.B. (1990) Factors affecting chloride binding in concrete. *Cement and Concrete Research* **20**, 291–300.

Bonen, D. (1997) The microstructure of concrete subjected to high magnesium and high magnesium sulfate brine attack, in *Proceedings 10th ICCC, Göteborg*, paper 41v022.

Bonen, D., and Cohen, M.P. (1992) Magnesium sulfate attack on Portland cement pastes. *Cement and Concrete Research* **22**, 169–180, 707–718.

Dass, K. *et al.* (1992) Corrosion susceptibility of galvanized and mild steel reinforcement in blended cement concrete, in *Proceedings 9th ICCC, New Delhi*, Vol. 5, pp. 302–308.

Deja, J. (1997) Chloride resistance of the pastes and mortars containing mineral additives, in *Proceedings 10th ICCC, Göteborg*, paper 4iv015.

Djuric, M. *et al.* (1996) Sulfate corrosion of Portland cement: pure and blended with 30% fly ash. *Cement and Concrete Research* **26**, 1295–1300.

Eglinton, M. (1998) Resistance of concrete to destructive agents, in *Lea's Chemistry of Cement and Concrete* (ed. P.C. Hawlett), Arnold, London, pp. 300–342.

El-Sayed, H.A. *et al.* (1997) Effect of silicate fume incorporation in concrete on corrosion protection of embedded reinforcement, in *Proceedings 10th ICCC, Göteborg*, paper 41v014.

Faucon, P. *et al.* (1996) Leaching of cement: study of the surface layer. *Cement and Concrete Research* **26**, 1707–1715.

Faucon, P. *et al.* (1997) Contribution of magnetic nuclear resonance techniques to the study of cement paste water degradation, in *Proceedings 10th ICCC, Göteborg*, paper 3v003.

Faucon, P. *et al.* (1998) Long-term behaviour of cement pastes used for nuclear waste disposal: review of physico-chemical mechanism of water degradation. *Cement and Concrete Research* **28**, 847–857.

Giergiczny, Z. (1997) Sulphate resistance of cements with mineral admixtures, in *Proceedings 10th ICCC, Göteborg*, paper 4iv019.

Gollop, R.S., and Taylor, H.F.W. (1992–1996) Microstructural and microanalytical studies on sulfate attack. *Cement and Concrete Research* **22**, 1027–1037; **24**, 1347–1358; **25**, 1580–1589; **26**, 1013–1028, 1029–1044.

Irassar, F., and Batic, O. (1989) Effect of low-calcium fly ash on sulphate resistance of OPC cement. *Cement and Concrete Research* **19**, 194–202.

Jensen, H.U., and Pratt, P.L. (1989) The binding of chloride ions in pozzolanic products fly ash cement blends. *Advances in Cement Research* **2**, 121–129.

Krizan, D., and Zivanovic, B. (1997) Resistance of fly ash blended cement mortars to sulphate attack, in *Proceedings 10th ICCC, Göteborg*, paper 4iv020.

Lota, J.S., Pratt, P.L., and Bensted, J. (1995) A discussion of the paper "Microstructural and microanalytical studies of sulfate attack" by R.S. Gollop and H.F.W. Taylor. *Cement and Concrete Research* **25**, 1811–1813.

Lu, C., and Li, R. (1997) Resistance to chemical attack of high performance FKJ concrete, in *Proceedings 10th ICCC, Göteborg*, paper 4iv 028.

Madrid, J. *et al.* (1997) Durability of ordinary Portland cement and ground granulated blastfurnace slag cement in acid medium, in *Proceedings 10th ICCC, Göteborg*, paper 4iv040.

Mangat, P.S., and Khatib, J.M. (1992) Influence of initial curing on sulphate resistance of blended cement concrete. *Cement and Concrete Research* **22**, 1089–1100.

Mangat, P.S., and Khatib, J.M. (1995) Influence of fly ash, silica fume and slag on sulfate resistance of concrete. ACI Materials Journal **92**, 542–552.

Matthew, J.D. (1992) The resistance of PFA concrete to acid ground waters, in *Proceedings 9th ICCC, New Delhi*, Vol. 5, pp. 355–362.

Mehta, P.K. (1986) Effect of fly ash composition on sulfate resistance of cement. Journal of the American Concrete Institute **83**, 994–1000.

Mehta, P.K. (1992) Sulfate attack on concrete: a critical review, in *Material Science of Concrete III* (ed. J. Skalny), American Ceramic Society, pp. 105–130.

Miletic, S.R., and Ilic, M.R. (1997) Effect of ammonium sulphate corrosion on the strength of concrete, in *Proceedings 10th ICCC, Göteborg*, paper 4iv023.

Miletic, S. *et al.* (1997) Fly ash: useful material for preventing sulfate corrosion. *Studies in Environmental Science* **71**, 355–364.

Odler, I., and Gasser, M. (1988) Mechanism of sulfate expansion in hydrated Portland cement. *Journal of the American Ceramic Society* **71**, 1015–1020.

Osborne, G.J., and Sing, B. (1995) The durability of concrete made with blends of high alumina cement and ground granulated blastfurnace slag. ACI SP-153, Vol. 2, pp. 885–909.

Piasta, W.C. *et al.* (1997) Influence of limestone powder filler on microstructure and mechanical properties of concrete under sulphate attack, in *Proceedings 10th ICCC, Göteborg*, paper 4iv018.

Regourd, M., Hornain, H., and Mortureux, B. (1980) Microstructure of concrete in aggressive environments, in *Durability of Building Materials and Components*, ASTM, Philadelphia, PA, pp. 253–268.

Sakoda, S., Takeda, N., and Sogo, S. (1992) Influence of various cement types on concrete durability in marine environments, in *Proceedings 9th ICCC, New Delhi*, Vol. 6, pp. 175–181.

Schiessl, P., and Raupack, M. (1992) Influence of the type of cement on the corrosion behaviour of steel in concrete, in *Proceedings 9th ICCC, New Delhi*, Vol. 5, pp. 296–301.

Scrivener, K.L., and Capmas, A. (1998) Calcium aluminate cements, in *Lea's Chemistry of Cement and Concrete* (ed. P.C. Hawlett), Arnold, London, pp. 709–778.

Sersale, R. *et al.* (1997) Sulphate attack of carbonated and uncarbonated Portland and blended cement mortars, in *Proceedings 10th ICCC, Göteborg*, paper 4iv017.

Soroushian, P., and Alhozaimi, A. (1992) Correlation between fly ash effects on permeability and sulfate resistance of concrete, in *Proceedings 9th ICCC, New Delhi*, Vol. 5, pp. 196–202.

Ueda, H., Kimachi, Y., and Kydo, T. (1997) Chemical resistance of the surface of cement paste and mortar to weak acid and acid rain (in Japanese). *Semento Konkurito Ronbunshu* **51**, 636–641 [ref. CA 128/220661].

Unsworth, H.P. *et al.* (1997) Microstructural and chemical changes during leaching of cementitious materials, in *Proceedings 10th ICCC, Göteborg*, paper 4iv012.

van Eijk, R.J., and Brouwers, H.J.H. (1998) Study on the relation between hydrated Portland cement composition and leaching resistance. *Cement and Concrete Research* **28**, 815–828.

Wiens, U., and Schiessl, P. (1997) Chloride binding of cement paste containing fly ash, in *Proceedings 10th ICCC, Göteborg*, paper 4iv016.

Yang, Q., Wu, X., and Huang, S. (1997) Concrete deterioration due to physical attack by salt crystallization, in *Proceedings 10th ICCC, Göteborg*, paper 4iv032.

21 Expansive cements

21.1 DEFINITION

Cement pastes generally undergo small but distinct volume changes upon hydration and in the course of curing. Volume changes of this kind are usually of little significance, but may become critical under some special conditions.

In the hydration of Portland cement (and most other cements) the anhydrous cement constituents react with the mixing water to form hydration products. Even though the total volume of solids increases under these conditions, the volume of the formed hydrates is less than the sum of the volumes of the original cement and water consumed in the hydration reaction:

$$V_{cement} + V_{water} > V_{hydrates}$$

In a typical ordinary Portland cement paste this **chemical shrinkage** amounts to 0.03–0.08 ml per gram of cement at a w/c of ~0.50 (Köster and Odler, 1986; Justines et al., 1997; Paulini et al., 1997; Takahashi et al., 1997). As long as the cement paste is still plastic the chemical shrinkage results in a corresponding decrease of its external volume. After setting, when the paste is not able to deform freely anymore, the **external shrinkage** (also called **atogenious shrinkage**) represents only a fraction of the total chemical shrinkage that has occurred (Takahashi et al., 1997). The rest of the chemical shrinkage (and the resultant volume change) manifests itself as a corresponding increase of the internal porosity of the hardened material.

External shrinkage of the cement paste may be avoided by moist curing. Under these conditions an uptake of water from the environment causes both a complete filling of the pore system and even a limited overall expansion of the paste. However, upon subsequent curing at normal humidity or in dry air the hardened paste gradually loses water and exhibits a **drying shrinkage** that outweighs the original expansion that took place under moist curing conditions. The resulting tensile stresses may cause cracks in the hardened concrete, especially under restrained conditions.

If the hydration of the paste takes place without moist curing the stresses in the set paste caused by chemical shrinkage combine with those caused by drying shrinkage, and the probability of crack formation increases correspondingly. Figure 21.1 shows schematically the volume changes taking place in an ordinary cement due to hydration and curing.

Expansive cements are inorganic binders that generate expansive (compressive) stresses in the hardened paste in the course of hydration, counteracting the tensile stresses generated by chemical shrinkage and drying shrinkage. Thus the generation of these expansive stresses may prevent the formation of cracks in concrete in the course of drying. Expansive cements may also be employed to produce a controlled expansion of the concrete mix, in which the extent of expansion is restricted to an acceptable degree by the presence of steel reinforcement.

If the given goal is just to prevent concrete cracking due to chemical and drying shrinkage, the magnitude of the internal compressive stresses generated by the expansive action of the expansive cement constituent in the course of initial water curing must reach a level high enough to prevent the generation of tensile stresses in the material upon subsequent drying. Usually a small residual compressive stress of the order of 0.2–0.7 MPa remains in the dried concrete. Cements that meet these requirements are called **shrinkage-compensated cements**. Their possible applications include the production of crack-free concrete structures and products, such as multistory car garages or water storage tanks. The setting and hardening properties of concrete mixes made with shrinkage-compensated cements

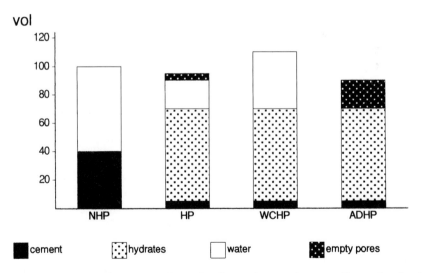

Figure 21.1 Schematic presentation of volume changes in an ordinary Portland cement paste due to hydration and curing. NHP, non-hydrated paste; HP, air-cured hydrated paste; WCHP, water-cured hydrated paste; AHDP, air-dried hydrated paste.

differ little from those made with ordinary cements, but the impermeability of the hardened concrete increases significantly.

If, however, the expansive stresses generated initially are higher than needed just to outbalance the tensile stresses caused by drying, then the hydrated material remains under expansive stresses even after it has dried completely. Such stresses may attain a magnitude of up to 8 MPa. To prevent excessive expansion and crack formation under these conditions the concrete body must be restrained or reinforced by a suitable reinforcement. Binders of this kind are called **self-stressing cements**. They may be employed in the production of prestressed concrete products and structures in which chemical reactions rather than mechanical measures are employed to get the reinforcement under stress. Possible applications include precast concrete products and very thin, strongly reinforced high-strength concrete products or structures. Figure 21.2 indicates schematically the development of volume and the orientation of stresses in pastes made from normal, stress-compensated and self-stressing cements.

Expansive cements consist of a cementitious component responsible for the cohesion and strength of the hydrated material, and an expansive component that produces expansive stresses in the hydrated material. In some expansive cements it is possible to produce directly a binder that contains both the cementitious and expansive components, but more often the two components are first produced separately, to be blended together or interground in a separate production step. The advantages of such an approach include the following:

- The expansive component may be interblended or interground with the cementitious component in different ratios, depending on the required magnitude of expansion of the finished cement.
- If the expansive cement is produced by interblending the expansive and

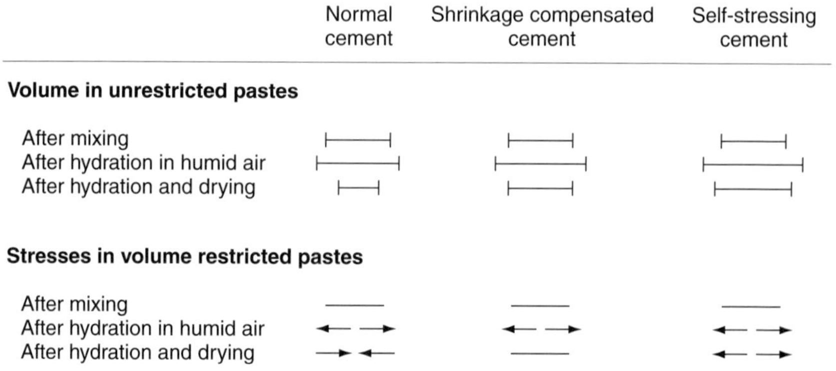

	Normal cement	Shrinkage compensated cement	Self-stressing cement
Volume in unrestricted pastes			
After mixing	⊢———⊣	⊢———⊣	⊢———⊣
After hydration in humid air	⊢————⊣	⊢—————⊣	⊢—————⊣
After hydration and drying	⊢——⊣	⊢————⊣	⊢——————⊣
Stresses in volume restricted pastes			
After mixing	———	———	———
After hydration in humid air	← →	← →	← →
After hydration and drying	→ ←	———	← →

Figure 21.2 Volume changes and stresses in OPC, shrinkage-compensated and self-stressing cement pastes due to hydration and drying.

cementitious components, the two of them may be ground to different finenesses, so that their respective rates of hydration can be at least partly controlled.

The expansive component may of course also be marketed as a separate product, to be added to the concrete mix at the job site.

In the most widely used expansive cements the cementitious component consists of ordinary Portland cement, but other binders such as calcium aluminate cement may also be employed. Most commonly, the expansive stresses in the course of hardening are created by the formation of ettringite ($C_3A.3C\bar{S}.32H$). This phase is formed in the reaction of a constituent containing aluminum oxide with calcium sulfate, calcium oxide or hydroxide and water. Less common than expansive cements based on ettringite formation are those based on the formation of calcium hydroxide [$Ca(OH)_2$] from calcium oxide (CaO), or magnesium hydroxide [$Mg(OH)_2$] from magnesium oxide (MgO). Other known expansive reactions that, however, are not employed in expansive cements are the alkali-silica reaction and the alkali-carbonate reaction. Both reactions are known to cause concrete failure if the starting mix contains a high-alkali cement in combination with an alkali-sensitive aggregate. Recently (Li *et al.*, 1996a, 1996b) an expansive reaction in the system $C_3A–C_3S–Na_2SO_4$ was reported, associated with the formation of a phase whose composition was described by the formula $4CaO.0.9Al_2O_3.1.1SO_3.0.5Na_2O.16H_2O$ (U phase).

21.2 MECHANISM OF EXPANSION

Just like the hydration of the cementitious component, the reaction of the expansive component with water is accompanied by a chemical shrinkage. Thus the resultant external expansion of the paste, which accompanies the hydration of an expansive cement, cannot be explained by an increase of the volume of the reactants formed in this chemical process. It is obvious that an expansion of the system would not take place if the hydration occurred in a simple through-solution reaction, in which the anhydrous starting component was readily dissolved and the hydration product precipitated randomly from the liquid phase. To produce an expansion a particular reaction mechanism seems to be required. Several hypothesis have been put forward to explain the expansion process:

- *Expansion due to a topochemical formation of reaction products.* In pastes in which the dissolution rate of the anhydrous starting component is relatively slow but the crystallization rate of the hydration product is fast, the reaction product may precipitate before the dissolved ions can migrate away from the surface of the anhydrous component. Under these conditions the crystallization of the resulting hydrate does not take place randomly from the liquid phase; the reaction product precipitates at the surface of the grains of the anhydrous starting material. The anhydrous phase becomes surrounded by a layer of hydration products. As the reaction progresses, the liquid phase migrates through the capillary and gel pores of the already formed hydrate to the grains of the residual anhydrous phase, and additional amounts of reaction product are formed at the grain/hydrate interface. Such a chemical process, which is bound to a certain location (in this instance to the surface of the grains of the anhydrous material), is called a **topochemical reaction**. The hydrate formed *in situ* has a substantially greater volume than the parent anhydrous phase and – if the process has taken place after setting of the cement paste – this leads to local stresses distributed randomly within the hardened material. This in turn will cause an increase of the external volume of the sample. A topochemical reaction may cause an expansion independently of the degree of crystallinity of the hydrate formed and without an uptake of water from the environment.

• *Expansion due to crystallization pressure.* Stresses within the cement paste and an increase of its external volume may be generated by an oriented, anisotropic growth of the products of hydration. Such a form of crystal growth may be produced if the reaction products are formed in a topochemical reaction at the interface between the anhydrous phase and the hydrate. Under such conditions the newly formed reaction products exert pressure on those already in existence and formed earlier. This may cause a situation in which the crystals of the reaction product grow in a direction perpendicular to the original solid/liquid interface. Such a form of crystal growth is associated with an increase of porosity. The external expansion may take place without an uptake of water from the environment.

• *Expansion due to swelling.* According to this theory expansion may occur when polar molecules of water are adsorbed at the surface of the hydration product, causing interparticle repulsion. A distinct expansion by this mechanism may take place only if the hydration product is present in a sufficiently dispersed (colloidal) form, with a large enough specific surface area and contact with an outside source of water.

• *Expansion due to osmotic pressure.* It is assumed that the hydration product forms a semi-permeable membrane around the grains of the anhydrous starting material, which hinders the free migration of ions between the liquid in immediate contact with the surface and the bulk solution. The resultant difference in the concentration on the two sides of the membrane generates differences in the osmotic pressure, and this in turn may generate internal stresses and ultimately an increase of the external volume of the paste.

To produce an external expansion of the cement paste the timing of the expansive reaction is also relevant. Prior to setting, as long as the paste is still plastic, no internal stresses needed to produce an expansion can be generated within the paste. An expansion can occur only if the paste has lost its plasticity because of setting and has attained a certain strength. However, it must still remain sufficiently extensible to accommodate the expansion taking place, without the formation of cracks. When the paste reaches the stage at which it loses this deformability because of progressing hydration, the expansive reaction must come also to its end, to prevent cracking of the hardened paste.

21.3 EXPANSIVE CEMENTS BASED ON ETTRINGITE FORMATION

In this class of cements the expansion is caused by the formation of the ettringite phase. Ettringite may be expressed by the formula $6CaO.Al_2O_3.3SO_3.32H_2O$ (abbreviation $C_6A\bar{S}_3H_{32}$ or $C_3A.3C\bar{S}.32H$) or – more accurately – as $[Ca_3Al(OH)_6.12H_2O]_2.(SO_4)_3.2H_2O$. The structure of this phase consists of hexagonal columns of the empirical formula $Ca_3Al(OH)_6.12H_2O$. The channels existing within these columns contain four sites per formula unit with six calcium atoms. Of these four sites, three are occupied by sulfate groups and one by two water molecules. Ettringite may be formed by crystallization from oversaturated solutions, which must contain at least 0.215 g $CaSO_4$, 0.043 g CaO, and 0.035 g Al_2O_3 per liter. It is produced in the form of acicular crystals, the size of which may vary depending on the calcium hydroxide concentration in the liquid phase. At higher $Ca(OH)_2$ concentrations ettringite tends to be formed as a fine-gained material consisting of crystals about 1 μm long and 0.25μm wide, whereas it precipitates in the form of long, thin needles if the concentration of free calcium hydroxide is low (Degenkolb and Knoefel, 1996). It easily loses some of its bound water if heated above about 60 °C.

In expansive cements based on ettringite formation, ettringite is formed in the reaction of a suitable aluminum-bearing phase with calcium sulfate, calcium hydroxide, and water. Suitable aluminate phases include CA, C_3A, and especially $C_4A_3\bar{S}$. The most common forms of calcium sulfate used in expansive cements are gypsum (β-$CaSO_4.2H_2O$) and anhydrite ($CaSO_4$-II). Calcium hydroxide is formed in the hydration of tricalcium and dicalcium silicate present in expansive cements, in which they act as cementing constituent. It may also be added to the system as a separate constituent or formed in the hydration of calcium oxide (CaO) introduced separately.

The process of expansion caused by ettringite formation has been studied using a variety of compounds as aluminum sources, but most extensively on cements containing the compound $C_4A_3\bar{S}$. It is known that the formation of ettringite does not produce expansion of the hardened cement paste under all conditions, and cementitious systems exist in which the formed ettringite contributes to the strength of the hardened paste, rather than causing an expansion. The formation of ettringite appears to cause expansion only under the following conditions:

- The ettringite must be formed after setting, at a stage when the paste has already attained a certain rigidity and the capability to transfer expansive stresses originated by ettringite formation. (Thus ettringite, regularly formed in ordinary Portland cement prior to setting, does not cause expansion.)
- Ettringite must be formed in a topochemical reaction to be able to produce the expansive stresses needed for the expansion to take place (see Fig. 21.3). Microscopical studies have revealed that expansion occurs only if ettringite is formed on the surface of the anhydrous aluminum phase and exhibits an oriented growth (Moldovan and Butucescu, 1980; Ogawa and Roy, 1981, 1982; Wang *et al.*, 1985; Herrick *et al.*, 1992). The cement does not expand if ettringite is formed in a through-

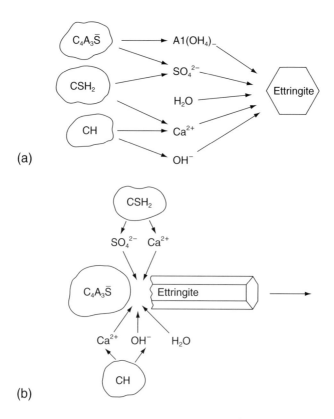

Figure 21.3 Mechanism of ettringite formation from $C_4A_3\bar{S}$: (a) through-solution mechanism (no expansion); (b) oriented crystal growth (expansion).

solution reaction in the intergrain space (Moldovan and Butucescu, 1980; Deng and Tang, 1994; Kasselouri *et al.*, 1995). In such instance it tends to contribute to the strength of the hardened paste, rather than generate internal stresses. To promote a topochemical formation of ettringite and to produce expansion, sufficiently high concentrations of $Ca(OH)_2$ must be present in the liquid phase (Mehta, 1973b; Aluno-Rosetti *et al.*, 1982; Clastres *et al.*, 1984; Deng and Tang, 1994; Degenkolb and Knoefel, 1996). At high calcium hydroxide concentrations aluminate ions cannot migrate too far from the phase serving as Al^{3+} source, and ettringite precipitates preferentially at its surface in a topochemical reaction. At low concentrations of calcium hydroxide, $Al(OH)_4^{4-}$ ions can migrate more freely, and ettringite can precipitate evenly throughout the liquid phase without generating expansive stresses. It has been reported that ettringite tends to be formed in a through-solution reaction at lime concentrations from 0.011 to 0.020 mol/L, and topochemically if this concentration is higher (Deng and Tang, 1994). The low concentrations of calcium hydroxide in the liquid phase explain the absence of expansion in some cements in which significant amounts of ettringite are formed in the hydration, such as supersulfated cement (see section 8.4) or sulfobelite cement (see section 4.2). In addition to its effect on the mechanism of ettringite formation, calcium hydroxide also slows down the rate at which ettringite is formed from calcium aluminates or sulfoaluminate (Mehta, 1973b).

• The expansion also depends on the mutual ratio of the aluminate and sulfate, and on their absolute amounts. It has been reported that an expansion occurs only at SO_3/Al_2O_3 molar ratios above 1.9 (Sudoh *et al.*, 1980). At a constant aluminate content the extent of expansion generally increases with increasing amounts of sulfate in the cement (Wang *et al.*, 1992a). To obtain a distinct expansion the amount of gypsum in the cement should range between 20 and 45 wt%.

• The expansion gets under way only after a critical amount of ettringite has been produced. This amount increases with increasing initial water/cement ratio of the mix. It has been postulated that the precipitation of ettringite at the surface of $C_4A_3\bar{S}$ grains causes an increase of the effective volume of these particles; an external expansion of the paste gets under way when the expanding particles come into contact with and exert pressure on each other (Bentur and Ish-Shalom, 1974, 1975; Ogawa and Roy, 1981, 1982).

• An uptake of water from the environment is not essential for the expansion to take place, but it does increase the extent of expansion (Mather, 1973; Odler and Gasser, 1988).

It has also been suggested that swelling of colloidal ettringite formed at high $Ca(OH)_2$ concentrations, rather than the formation of ettringite itself, is the main cause of expansion in expansive cements based on ettringite formation (Mehta 1973a, 1982; Mehta and Hu, 1978). In support of this hypothesis Mehta and Wang (1982) demonstrated a linear expansion of up to 4% in compacts of "colloidal" ettringite and C_3A-free Portland cement. However, from their measurement of expansion in compacts of ettringite, Aluno-Rosetti *et al.* (1982) concluded that the swelling of ettringite alone could not account for all the volume increase seen in expansive cements, and that a simultaneous development of crystallization pressure due to an anisotropic growth of ettringite must also be involved.

In conclusion, based on the available experimental evidence, it appears that a topochemical formation of ettringite and its oriented growth are essential for the expansion process to take place (Lafuma, 1929; Mather, 1973; Bentur and Ish-Shalom, 1974, 1975; Moldovan and Butucescu, 1980; Ogawa and Roy, 1981, 1982; Aluno-Rosetti *et al.*, 1982; Wang *et al.*, 1985; Odler and Gasser, 1988; Herrick *et al.*, 1992; Deng and Tang, 1994). A subsequent swelling of the formed ettringite may also be involved, and may enhance the extent of the overall expansion, but it does not seem to be essential for the expansion to take place.

To produce expansion in expansive cements a variety of aluminous phases may be considered, but calcium aluminate or sulfoaluminate phases are most commonly employed. Owing to differences in their reactivities (Mehta, 1973b; Odler and Colán-Subauste, 1999), not all of them perform equally well, to produce expansion without cracking. Tetracalcium trialuminate sulfate ($C_4A_3\bar{S}$) appears to be the most suitable, as most of this phase hydrates in the desirable time range of several hours to several days: that is, in a period in which the cement paste has already set but is not yet excessively rigid.

Monocalcium aluminate (CA) hydrates rather rapidly initially, but still a significant fraction of it reacts shortly after setting: that is, at a time that is favorable for the expansive process. The reaction ceases within a few days.

Unlike these two phases, tricalcium aluminate (C_3A) yields significant amounts of ettringite in the early stages of hydration: that is, at a time when the cement paste is still plastic, and when this reaction does not cause expansion. After setting, the formation of ettringite continues at a slow rate for an extended period of time, and is not terminated even after the concrete has lost its ability to expand without cracking.

Finally, the reaction of the phase $C_{12}A_7$ and its fluorine-substituted derivative $C_{11}A_7.CaF_2$ is too fast to be used in expansive cements. Moreover, the formation of ettringite in this instance takes place in a through-solution reaction, not associated with expansion.

Other aluminum-bearing materials that have been suggested for use in expansive cements include calcium aluminate hydrates existing in prehydrated high-alumina cement, industrial slags with a high Al_2O_3 content, and alunite [a natural mineral of the composition $KAl(SO_4)_2.(OH)_6$] in its original or precalcined form. However, cements based on these aluminum sources are rarely used.

21.3.1 Type K expansive cement

Type K expansive cement is the expansive cement used worldwide in the largest amounts. It contains calcium sulfoaluminate ($C_4A_3\bar{S}$) in combination with calcium sulfate as the expansive component and Portland clinker as the component producing the cementitious matrix. The expansive component is burnt separately in the form of a calcium sulfoaluminate clinker, which is subsequently interground with Portland clinker and additional calcium sulfate in the form of gypsum or anhydrite. Alternatively, the sulfoaluminate clinker may be ground separately and added to Portland cement in amounts that may vary depending in the intended use of the final expansive cement.

Under industrial conditions calcium sulfoaluminate clinker is produced by burning a raw meal containing limestone, bauxite, and anhydrite in a rotary kiln at a maximum temperature of about 1300 °C. The following is the sequence of reactions taking place in the course of burning (Su *et al.*, 1992):

- ~530 °C: dehydration and decomposition of bauxite
- ~900 °C: decomposition of $CaCO_3$
  formation of gehlenite (C_2AS) as a transitional phase
- >1000 °C: reaction of C_2AS with $CaSO_4$ to yield $C_4A_3\bar{S}$ and a-C_2S
  direct formation of $C_4A_3\bar{S}$ from CaO, Al_2O_3 and $CaSO_4$
- >1100 °C: direct formation of β-C_2S
- ~1250 °C: C_2AS and free CaO disappear
  formation of sulfospurrite [$Ca_5(SiO_4)_2(SO_4)$, abbreviation $C_5S_2\bar{S}$]
- >1280 °C: sulfospurrite decomposes back to α-C_2S and $CaSO_4$.

The produced clinker is polymineralic in nature, and contains – in addition to calcium sulfoaluminate – limited amounts of dicalcium silicate, anhydrite, and free lime. Sulfospurrite may also occasionally be present (Herrick *et al.*, 1992; Su *et al.*, 1992). The chemical composition of the sulfoaluminate phase may vary over a limited range, with a partial substitution of Ca^{2+} by Mg^{2+}, and Al^{3+} by Si^{4+}, Fe^{3+}, or Ti^{4+} (Su *et al.*, 1992). The clinker has a high porosity, with a liter weight of 650–850 g/L, and is easy to grind.

In the hydration of type K expansive cement the following reaction is responsible for its expansion:

$$C_4A_3\bar{S} + 8C\bar{S}H_2 + 6CH + 74H_2O \rightarrow 3C_6A\bar{S}_3H_{32} \qquad (21.1)$$

$C\bar{S}H_2$ (gypsum) is formed in the hydration of anhydrite present in the calcium sulfoaluminate clinker, and may also be directly interblended or interground into the cement. Calcium hydroxide (CH) is formed in the hydration of free lime present in the calcium sulfoaluminate clinker, and also in the hydration of alite and belite present in Portland clinker, both being constituents of the expansive cement.

Microscopic investigations on type K expansive cement pastes (Herrick *et al.*, 1992) indicate that ettringite is preferentially formed topochemically around the anhydrous sulfoaluminate grains. It exhibits an oriented growth. Most of the anhydrite originally present in the sulfoaluminate clinker dissolves and converts to gypsum before reacting with the sulfoaluminate. The free lime, originally present in the expansive component, appears to hydrate initially *in situ* to give a metastable cubic form of calcium hydroxide, which over several days converts into its stable hexagonal modification.

In concrete mixes the microstructure of the interfacial zone differs markedly from that of the plain bulk paste (Lu *et al.*, 1996). Here, ettringite is the main hydrate present, but its morphology varies, depending on the amount of gypsum in the original cement and the conditions of hydration. Three-dimensional restraint of the hydrating mix may improve the interfacial microstructure: the thickness of the interfacial zone decreases and the size of the formed crystalline hydration products becomes smaller.

In a stoichiometric mix corresponding to reaction (21.1), the hydration of $C_4A_3\bar{S}$ follows parabolic kinetics, corresponding to a diffusion-controlled mechanism (Hanic *et al.*, 1989). The half-time of the reaction was found to be 57 h. In paste hydration of a typical industrially produced expansive type K cement, about 60% of the existing $C_4A_3\bar{S}$ is hydrated within the first 24 hours, but small amounts of non-reacted residual $C_4A_3\bar{S}$ may still be detectable at up to 28 days (Herrick *et al.*, 1992). In the stage that follows the initial period of fast hydration, the rate of $C_4A_3\bar{S}$ hydration declines distinctly. At a constant degree of hydration the hydration rate increases with increasing gypsum content in the cement (Wang, 1992a).

The effect of alkalis on the hydration and expansion of type K expansive cement was studied by Wang *et al.* (1986). As expected, the Ca^{2+} concentration of the liquid phase was lowered and its pH value increased with increasing alkali content in cement. At the same time the formation of ettringite became accelerated and the extent of expansion increased. The ettringite that was formed even at pH = 13.3 was crystalline rather than amorphous.

It has also been reported that the initial rate of ettringite formation may be accelerated, and the extent of expansion increased, by adding silica fume to the mix (Lobo and Cohen, 1993). However, after a few days of hydration the rate at which additional ettringite is formed declines significantly as a consequence of the decreasing pH of the liquid phase.

For additional data on calcium sulfoaluminate-modified Portland cements see section 4.4.

21.3.2 Type M and related expansive cements

Type M expansive cement is a combination of ordinary Portland cement, calcium aluminate cement, and additional gypsum or calcium hemihydrate, typically in the ratio 66:20:14 (Mikhailov, 1960; Budnikov and Kravchenko, 1968). It may be produced by mixing these constituents in a plant. Alternatively, an expansive additive consisting of high-alumina cement, gypsum or hemihydrate, and hydrated lime may be added to the fresh concrete mix during mixing. By this approach the extent of expansion may be controlled by the amount of additive added to the mix.

In type M, just as in type K cement, the matrix of the hardened paste is produced by the hydration of the existing Portland cement. Ettringite, the phase responsible for expansion, is formed in a reaction of calcium monoaluminate (CA), the main constituent of high-alumina cement, with calcium sulfate and calcium hydroxide, formed in the hydration of Portland cement and also added to the system as a constituent of the expansive additive:

$$CA + 3C\bar{S}H_2 + 2CH + 24H \rightarrow C_3A.3C\bar{S}.32H \qquad (21.2)$$

The expansion characteristics of type M cement may be controlled by the proportions of its constituents and the temperature of curing. If allowed to hydrate at ambient temperature and at a low water/cement ratio, calcium aluminate monosulfate ($C_3A.C\bar{S}.12H$) may be formed initially, in addition to trisulfate (ettringite). Upon exposure of such a concrete to water, monosulfate also converts to trisulfate, and this process is associated with an expansion.

At higher water/cement ratios and at a higher temperature trisulfate is formed directly as the sole calcium sulfoaluminate. As this reaction progresses very rapidly, most of the trisulfate is formed prior to setting, without causing expansion. A limited expansion takes place only after setting, as the residual non-hydrated CA also undergoes hydration associated with ettringite formation.

To obtain satisfactory results, concrete mixes made with type M cement must be heat-cured at a temperature exceeding 70 °C. At these temperatures trisulfate is thermodynamically unstable; it cannot be formed in the hydration, and monosulfate is formed instead, without causing expansion. Upon subsequent curing in water at ambient temperature, the primary formed monosulfate converts to trisulfate, and this reaction is associated with a generation of expansive stresses in the concrete body. Type M expansive cement is suitable for the production of precast reinforced concrete products to be heat-cured in the course of production, but is hardly suitable for use at ambient temperature. Quick setting and unreliability are problems commonly encountered in the use of this type of cement.

To avoid the problems encountered with standard type M cement, it has been suggested that prehydrated calcium aluminate cement should be used as the source of Al^{3+} ions, rather than the non-hydrated binder (Fu *et al.*, 1995a, 1995b). The phase composition of such a prehydrated material will depend on the temperature at which the hydration has taken place. At temperatures around 5 °C the phase CAH_{10} and at around 20–25 °C the phase C_2AH_8 are formed as the main hydration products, yet both of them tend to convert over time to the more stable C_3AH_6 phase (+ AH_3). At even higher temperatures the C_3AH_6 phase is formed directly in the hydration of CA.

A comparison of expansive additives based on non-hydrated and hydrated calcium aluminates revealed great differences in their hydration kinetics and capability to generate expansion (Fu *et al.*, 1995a, 1995b): additives that contained CAH_{10} and C_2AH_8 exhibited a distinctly faster ettringite formation than non-hydrated high-alumina cement. The expansion was terminated within three hours of hydration. Thus it must be expected that the use of these hydroaluminates alone as aluminum-bearing components in formulations of expansive cements may cause only slight expansion, owing to the unfavorable timing in the formation of ettringite. The use of high-alumina cement hydrated at elevated temperature, in which the C_3AH_6 phase was the main aluminum-bearing constituent, resulted in reduced and delayed ettringite formation and expansion. The expansion of expansive cements that contain this phase as aluminum source may last up to 28 days, and may eventually cause crack formation in the hardened concrete. Thus C_3AH_6 appears not to be suitable for used as the aluminum source in shrinkage-compensated or expansive cement. It was concluded that optimum results with type M expansive cement may be obtained with an additive that contains as its aluminum-bearing constituent particles of non-hydrated high-alumina cement with a layer of CAH_{10} or C_2AH_8 on their surface.

21.3.3 Type S expansive cement

Type S expansive cement consists of Portland clinker with a particularly high content of the tricalcium aluminate phase in combination with elevated amounts of calcium sulfate (see also section 2.5). The formation of ettringite takes place according to the reaction

$$C_3A + 3C\bar{S}H_2 + 26H \rightarrow C_3A.3C\bar{S}.32H \tag{21.3}$$

A significant fraction of ettringite is formed rapidly immediately after mixing, but ettringite formed at this stage does not cause expansion. Later, the reaction becomes sluggish, and significant amounts of non-reacted C_3A may be present even after 7 days of hydration. Ettringite formed at this stage causes an expansion of the cement paste. The expansion takes place slowly, however, and may last for a prolonged period of time. If excessive, it may cause crack formation within the hardened paste. Such a tendency is enhanced by the fact that the cementitious matrix becomes gradually more rigid and unable to deform as the hydration progresses. Expansive cements of this type are used very rarely, because of difficulties in controlling the expansion process and in preventing crack formation.

21.3.4   Expansive sulfobelite cements

Sulfobelite cements contain belite (β-C_2S), calcium sulfoaluminate ($C_4A_3\bar{S}$), and calcium sulfate ($C\bar{S}$ or $C\bar{S}H_2$) as their main constituents. They may be produced in an expansive or non-expansive form by varying the proportions of these constituents. In general, the capacity of the cement to expand will increase with increasing content of sulfoaluminate and calcium sulfate. At the same time the content of free lime in the binder must be kept low. For additional data on sulfobelite cements see section 4.2.

21.3.5   Other expansive cements based on ettringite formation

Expansive calcium sulfoaluminate cement contains the phases $C_4A_3\bar{S}$, $C_5S_2\bar{S}$ (sulfospurrite), and $C\bar{S}$ (anhydrite) in variable ratios. Mixes of this cement made with relatively low water/cement ratios exhibit expansion due to ettringite formation, associated with an uptake of water from the environment. For additional data on calcium sulfoaluminate cements see section 4.5.

Expansive sulfoferrite cement consists of ordinary Portland clinker, calcium sulfate, and a separately produced sulfoferrite-bearing clinker, which contains the high-basicity calcium sulfoferrite phase C_3FS as its main constituent. The expansion in the hydration of this cement is due to the formation of a ferric AFt phase (ferric ettringite, $C_3F.3C\bar{S}.31H$). For more information on calcium sulfoferrite cements see section 15.6.

Wu and Wang (1980) developed an expansive cement on the basis of alunite. It consists of Portland clinker, anhydrite, fly ash or blast furnace slag, and natural alunite rock [with 20–50% alunite, $K_2SO_4.Al_2(SO_4)_3.4Al(OH)_3$]. The expansion of such a cement is due to the formation of ettringite from Al^{3+} and SO_4^{2-} ions from the alunite and calcium hydroxide liberated in the hydration of tricalcium silicate present in the clinker. Most of the ettringite is formed within the first few days of hydration, but its formation continues slowly even after 7 days. Owing to the rather good solubility of alunite, the ettringite phase is formed predominantly or exclusively in a through-solution reaction, and as a result the free expansion of the hardened cement paste is comparatively small. The ettringite precipitates in the free pore space of the paste and improves its impermeability and strength at later ages. The texture and properties of the paste are improved if the hydration takes place under restraint.

An expansive cement consisting of a combination of calcium aluminate cement and gypsum, called high-alumina/gypsum expansive cement, has been developed (Xue *et al.*, 1986; Stark and Chartschenko, 1997). At sufficiently high gypsum contents – that is, above about 40% – ettringite and hydroaluminate gel are formed as the sole products of hydration:

$$3CA + 3C\bar{S}H_2 + (26 + 2x)H \rightarrow C_3A.3C\bar{S}.32H + 2AH_x \qquad (21.4)$$

In the hardened cement paste the hydroaluminate gel acts as the main cementitious phase. The formed ettringite is mainly responsible for expansion, but also contributes to strength. Most of the hydration and expansion take place within the first 7 days. The extent of expansion may be increased and the progress of the reaction accelerated by increasing the temperature, by increasing the ratio CA/CA_2 in the calcium aluminate cement that is used, and by increasing the calcium sulfate content. If the calcium sulfate content in the cement is too low, calcium aluminate monosulfate may also be formed in the hydration process (Bayoux *et al.*, 1992):

$$3CA + C\bar{S}H_2 + (10 + 2x)H \rightarrow C_3A.C\bar{S}.12H + 2AH_x \qquad (21.5)$$

This reaction is not associated with expansion. The extent of expansion of a cement on the basis of calcium aluminate clinker and gypsum may be increased by adding to it opal, microsilica, and especially hydrated lime (Stark and Chartschenko, 1997).

Degenkolb and Knöfel (1997) developed a series of shrinkage-compensated mortars in which the cementitious component consisted of lime hydrate, or a combination of lime hydrate and granulated blast furnace slag, or a combination of the slag with gypsum and Portland cement. Calcium aluminate cement in combination with gypsum, or a combination of metakaolin and gypsum, or calcium sulfoaluminate together with gypsum, served as the components responsible for expansion.

Jiang and Roy (1995) produced an expansive cement by combining atmospheric fluidized bed combustion ash with alkali sulfate.

21.4 EXPANSIVE CEMENTS BASED ON CALCIUM OXIDE HYDRATION

In contact with water calcium oxide reacts to yield calcium hydroxide:

$$
\begin{array}{llll}
CaO & + & H_2O & \rightarrow & Ca(OH)_2 & \qquad (21.6) \\
56.08\ g & & 18.01\ g & & 74.09\ g \\
\underbrace{16.76\ ml \qquad\qquad 18.01\ ml} & & 33.08\ mm \\
\qquad\qquad 34.77\ ml
\end{array}
$$

In this reaction calcium oxide is first dissolved in water in the form of Ca^{2+} and OH^- ions, and calcium hydroxide precipitates from the liquid phase after it has become oversaturated with respect to the hydroxide. The reaction is associated with a chemical shrinkage: the volume of the resulting calcium hydroxide is smaller (by 4.86%) than the sum of the volumes of calcium oxide and water entering the reaction. At the same time the volume of the calcium hydroxide is greater by 97.4% than the volume of the original oxide.

If calcium oxide is formed in a thermal dissociation of calcium carbonate

$$CaCO_3 \rightarrow CaO + CO_2 \tag{21.7}$$

the texture of the reaction product and its reactivity will depend on the time and especially on the temperature at which the reaction was performed. In general, the specific surface area and internal porosity of the calcium oxide decline with increasing temperature and reaction time, and as a consequence the oxide reacts less readily with water to yield calcium hydroxide.

If allowed to react with water, calcium oxide produced by calcination of $CaCO_3$ at a low temperature (that is, below about 1200 °C) – also called soft-burnt free lime – dissolves rapidly, Ca^{2+} and OH^- ions migrate away from its surface, and calcium hydroxide precipitates randomly from the liquid phase in a through-solution reaction (Fukui, 1996). This process is not accompanied by an expansion if such a form of calcium oxide is present in a Portland cement paste undergoing hydration.

However, if the $CaCO_3$ is calcined at a high temperature (that is, above 1300–1400 °C) the dissolution of the produced calcium oxide – called hard-burnt free lime – is slowed down significantly. Under these conditions the hydroxide precipitates at the surface of the CaO grain (Fukui, 1996). The reaction continues as additional water migrates through the hydroxide layer toward the oxide/hydroxide interface, where additional amounts of hydroxide are formed at the expense of the oxide until the conversion of the oxide to hydroxide is complete. As the volume of the hydroxide is greater than that of the original oxide, this *in situ* reaction – commonly called a topochemical reaction – may generate expansive stresses in a cement paste that has already set and has already developed a semi-rigid structure (Deng *et al.*, 1995; Fukui, 1996). In addition to topochemical hydration, a swelling of the primary-formed calcium hydroxide may also contribute to the expansion (Ramachandran *et al.*, 1964).

Expansive cements based on calcium oxide are in essence Portland cements with a content of hard-burnt free lime in amounts high enough to produce the required expansion. In principle, such a cement may be produced either by burning the cement raw meal in a way that results in the desired free CaO content in clinker, or by adding to a cement with a "normal" (that is, low) free lime content additional amounts of calcium oxide in the form of hard-burnt lime. To produce a distinct expansion of a Portland cement paste the free CaO contents must lie typically above about 2–3 wt%.

In the hydration of a typical CaO-based expansive cement a significant fraction – that is, more than 50% – of the originally present hard-burnt free lime undergoes hydration prior to setting without causing expansion. The exact amount will depend on its thermal history and degree of dispersion. The expansion gets under way several hours after mixing, after the cement paste has attained a certain degree of rigidity, and is typically terminated within a few days. The extent of expansion increases proportionally with the amount of calcium hydroxide present, but the exact level of expansion will vary from one cement to another, and may also be influenced by the conditions of clinker burning and possibly also by any impurities that are present (Bensted, 1996).

The expansion is also promoted by the presence of additional $CaSO_4$ in the paste. This ingredient exhibits a retarding effect on the hydration of CaO, and thus increases the amount of it that remains in non-hydrated form at the time of setting.

Note that unlike calcium hydroxide formed in the hydration of calcium oxide, calcium hydroxide liberated in the hydration of dicalcium or tricalcium silicate under no circumstances generates expansion of a cement paste.

Expansive cements based on calcium oxide are rarely used because of difficulties in controlling the extent and the kinetics of expansion. It has been suggested that expansive additives based on free CaO should be used to prevent early age thermal cracking in hardening concrete (Ebensperger and Springenschmidt, 1992).

Cement compositions with a high content of hard-burnt CaO are suitable as demolition agents, owing to the relatively high stresses generated in the hydration/expansion process (Fukui, 1996).

21.5   EXPANSIVE CEMENTS BASED ON MAGNESIUM OXIDE HYDRATION

Magnesium oxide (MgO) hydrates in contact with water to yield magnesium hydroxide:

$$MgO \quad + \quad H_2O \quad \rightarrow \quad Mg(OH)_2 \qquad (21.8)$$

MgO	H_2O	$Mg(OH)_2$
40.31 g	18.01 g	58.32 g
11.48 ml	18.01 ml	24.71 mm

$$\underbrace{}$$

29.49 ml

Just as with CaO, this reaction is also associated with a chemical shrinkage and, at the same time, with an increase of the volume of solids.

If produced by a thermal dissociation of magnesium carbonate

$$MgCO_2 \rightarrow MgO + CO_2 \tag{21.9}$$

the reactivity of the oxide that is formed will depend on the temperature and time of calcination, but is generally significantly lower than that of CaO. The following degrees of hydration were found in experiments performed at 38 °C (Mehta, 1980):

Fineness	$< 45\,\mu m$			75–$150\,\mu m$		
Calcination temperature	900 °C	1000 °C	1100 °C	900 °C	1000 °C	1100 °C
Degree of hydration (%)						
1 d	74	6	0	60	1	0
3 d	89	81	6	67	27	0
28 d	100	92	87	90	77	67

It has been suggested that magnesium oxide should be used as an expansive additive in concrete mixes for mass structures to prevent cracking due to the temperature gradients that develop as the concrete cools from higher temperatures caused by heat of hydration to ambient temperature (Mehta *et al.*, 1980; Wang *et al.*, 1992b). To obtain a hydration rate that is adequate to prevent cracking in typical mass concrete structures, which may attain temperatures of 32–54 °C, the additive must be produced by calcination of magnesite ($MgCO_3$) at temperatures between 900 and 970 °C, as MgO made by calcination above 970 °C hydrates too slowly, and that calcined below 900 °C hydrates too rapidly to be useful to produce self-stress in such concrete. The MgO should be ground to a fineness of 300–1200 μm and introduced in amounts of around 5% (Mehta, 1980).

As an alternative solution it has been suggested that a clinker should be produced directly that contains free MgO, by burning a pertinent raw meal at 1380 °C (Wang and Lou, 1992). In laboratory experiments a cement with 5% MgO hydrating at 50 °C achieved complete conversion from MgO to $Mg(OH)_2$ within a year and produced a linear expansion of 10 mm/m.

At the temperatures existing within mass concrete the hydration of MgO gets under way relatively quickly, and expansive stresses develop continuously within half a year (Wang and Lou, 1992). They thus keep pace with the cooling shrinkage that is taking place in the mass concrete. At an MgO addition of about 5 wt% the ultimate linear expansion amounts to about 1 mm/m, and this has no harmful effect on the strength of the hardened concrete.

Unlike MgO-based binders, expansive cements (or expansive additives) based on ettringite formation or $Ca(OH)_2$ formation from CaO are of little benefit for controlling cooling shrinkage in mass concrete, because they would undergo complete hydration before the maximum temperature in the hardening concrete mix was reached. Consequently, the compressive stresses produced by these expansive reactions relax too soon to prevent crack formation in the concrete body in the course of cooling.

Unlike MgO calcined at low temperatures or present in clinker in limited amounts, free magnesium oxide (periclase) that is present in Portland cements made from raw meals with high $MgCO_3$ contents must be considered detrimental if present in amounts above 4–5 wt% MgO, as it tends to hydrate at ambient temperature for years, and ultimately may cause cracking in the concrete structure. For additional data on Portland cement that contains elevated amounts of free MgO see section 2.9.

21.6   MISCELLANEOUS EXPANSIVE CEMENTS

Wang *et al.* (1992b) developed a multicomponent expansive cement by grinding together ordinary Portland clinker with an expansive clinker, natural alunite, and gypsum. The expansive clinker had been produced by burning limestone, natural alunite, and gypsum at 1300 °C. It contained 45–50% C_3S and 35–40% CaO particles of size 15–35 μm, about half of them enclosed by C_3S. If the cement is mixed with water an expansion caused by the formation of $Ca(OH)_2$ occurs in the early stage of hydration – that is, between several hours and about 7 days – and is gradually overlapped by expansion caused by ettringite formation.

Ishida *et al.* (1997) developed an expansive agent containing free lime in combination with anhydrite and hauyne. Calcium hydroxide, ettringite, and in later stages also monosulfate, are formed in its hydration, causing expansion when added to a non-expansive cement. The additive is particularly effective in slag cements in which the performance of conventional expansive additives is rather limited.

REFERENCES

Aluno-Rosetti, V., Chioccio, G., and Paolini, A.E. (1982) Expansive properties of the mixture $C_4A\bar{S}H_{12}$ + $2C\bar{S}$. I. A hypothesis on expansive mechanism. *Cement and Concrete Research* 12, 577–585.

Bayoux, J., Testud, M., and Espinosa, B. (1992) Thermodynamic approach to understand the CaO-Al_2O_3-SO_3 system, in *Proceedings 9th ICCC, New Delhi*, Vol. 4, pp. 164–169.

Bensted, J. (1996) A discussion of the paper "Mechanism in hardened cement paste with hard -burnt free lime" by M. Deng *et al. Cement and Concrete Research*, **26**, 645–646.

Bentur, A., and Ish-Shalom, M. (1974, 1975) Properties of Type K expansive cement of pure components. *Cement and Concrete Research* **4**, 709–721; **5**, 139–145.

Budnikov, P.B., and Kravchenko, I.V. (1968) Expansive cements, in *Proceedings 5th ICCC, Tokyo*, Vol. 5, pp. 319–329.

Chartschenko, I., Rudert, V., and Wihler, H.D. (1996) Effects on the hydration process and the properties of expansive cement. *ZKG International* **49**, 432–443, 602–609.

Chatterji, S. (1995) Mechanism of expansion of concrete due to the presence of dead-burnt CaO and MgO. *Cement and Concrete Research* **25**, 51–56.

Clastres, P., Murat, M., and Bachiorrini, A. (1984) Hydration of expansive cements – correlation between the expansion and formation of hydrates. *Cement and Concrete Research* **14**, 199–206.

Degenkolb, M., and Knoefel, D. (1996) Expansive additives in lime- and slag-based mortars, in *Proceedings 18th International Conference on Cement Microscopy*, pp. 304–316.

Degenkolb, M., and Knöfel, D. (1997) Investigations on shrinkage compensated and expansive repair mortars and general behaviour to typical influences, in *Proceedings 10th ICCC, Göteborg*, paper 2ii074.

Deng, M., and Teng, M. (1994) Formation and expansion of ettringite crystals. *Cement and Concrete Research* **24**, 119–126.

Deng, M. *et al.* (1995) Mechanism of expansion in hardened cement pastes with hard-burnt free lime. *Cement and Concrete Research* **25**, 440–448.

Ebensperger, L.E., and Springenschmidt, R. (1992) Effectiveness of expansive additives avoiding early age thermal cracking, in *Proceedings 9th ICCC, New Delhi*, Vol. 5, pp. 578–584.

Fu, Y. *et al.* (1995a) Effect of chemical admixtures on the expansion of shrinkage-compensating cement containing a prehydrated alumina cement-based expansive additive. *Cement and Concrete Research* **25**, 29–38.

Fu, Y., Ding, J., and Beaudoin, J.J. (1995b) Expansion characteristics of a compounded expansive additive and pre-hydrated high-alumina cement based expansive additive. *Cement and Concrete Research* **25**, 1295–1304.

Fukui, H. (1996) Static demolition by calcium oxide (in Japanese). *Kayaku Gakkaishi* **57**, 62–70 [ref. CA 125/93869].

Hanic, F., Kapralik, J., and Gabrisova, A. (1989) Mechanism of hydration reactions in the system $C_4A_3\bar{S}$–$C\bar{S}$–CaO–H_2O referred to hydration of sulfoaluminate cements. *Cement and Concrete Research* **19**, 671–682.

Herrick, J., Scrivener, K.L., and Pratt, P.L. (1992) The development of microstructure in calcium sulfoaluminate expansive cement. *Materials Research Society Symposium Proceedings* **245**, 277–282.

Ishida, A. *et al.* (1997) Expansive behaviour of slag cement with expansive additive, in *Proceedings 10th ICCC, Göteborg*, paper 3iii017.

Jiang, W., and Roy, D.M. (1995) Expansive cement produced from AFBC ash by alkali sulfate activation approach. American Concrete Institute, SP-153, pp. 193–212.

Justines, H. *et al.* (1997) Influence of measuring method on bleeding chemical shrinkage values of cement pastes, in *Proceedings 10th ICCC, Göteborg*, paper 2ii069.

Kasselouri, V. *et al.* (1995) A study on the hydration products of a non-expensive sulfoaluminate cement. *Cement and Concrete Research* **25**, 1726–1736.

Köster, H., and Odler, I. (1986) Investigation on the structure of fully hydrated Portland cement and tricalcium silicate pastes. I. Bound water, chemical shrinkage and density of hydrates. *Cement and Concrete Research* **16**, 207–214.

Lafuma, H. (1929) Theory of expansion of cement (in French). *Revue des Materiaux de Construction et de Travaux Publics* **243**, 441–444.

Li, G., Le Bescop, P., and Moranville, M. (1996a) The U-phase formation in cement-based systems containing high amounts of Na₂SO₄. *Cement and Concrete Research* **26**, 27–34.

Li, G., Le Bescop, P., and Moranville, M. (1996b) Expansion mechanism associated with the secondary formation of the U-phase in cement-based systems containing high amounts of Na₂SO₄. *Cement and Concrete Research* **26**, 195–202.

Lobo, C., and Cohen, M.D. (1993) Hydration of type K expansive cement paste and the effect of silica fume. *Cement and Concrete Research* **23**, 104–115.

Lu, Y., Su, M., and Wang, Y.(1996) Microstructural study of the interfacial zone between expansive sulfoaluminate cement paste and limestone aggregate. *Cement and Concrete Research* **26**, 805–812.

Mather, B. (1973) A discussion of the paper "Mechanism of expansion associated with ettringite formation" by P.K. Mehta. *Cement and Concrete Research* **3**, 651–652.

Mehta, P.K. (1973a) Mechanism of expansion associated with ettringite formation. *Cement and Concrete Research* **3**, 1–6.

Mehta, P.K. (1973b) Effect of lime on hydration of pastes containing gypsum and calcium aluminates or calcium sulfoaluminates. *Journal of the American Ceramic Society* **56**, 315–319.

Mehta, P.K. (1980) Magnesium oxide additive for producing selfstress in mass concrete, in *Proceedings 7th ICCC, Paris*, Vol. 3, pp. V-6–9.

Mehta, P.K. (1982) Expansion of ettringite by water absorption. *Cement and Concrete Research* **12**, 121–122.

Mehta, P.K., and Hu, F. (1978) Further evidence for expansion of ettringite by water absorption. *Journal of the American Ceramic Society* **61**, 179–181.

Mehta, P.K., and Wang, S. (1982) Expansion of ettringite by water absorption. *Cement and Concrete Research* **12**, 121–122.

Mehta, P.K., Pirtz, D., and Komandant, G.J. (1980) Magnesium oxide additive for producing self-stress in mass concrete, in *Proceedings 7th ICCC, Paris*, Vol. 3, pp. V-6 to V-9.

Mikhailov, V.V. (1960) Stressing cement and the mechanism self-stressing concrete regulation, in *Proceedings 4th ICCC, Washington*, Vol. 2, pp. 927–955.

Moldovan, V., and Butucescu, N. (1980) Expansion mechanism of expansive cement, in *Proceedings 7th ICCC, Paris*, Vol. 3, pp. V-1–5.

Odler, I., and Colán-Subauste, J. (1999) Investigations on cement expansion associated with ettringite formation. *Cement and Concrete Research* **29**, 731–735.

Odler, I., and Gasser, M. (1988) Mechanism of sulfate expansion in hydrated Portland cement. *Journal of the American Ceramic Society* **71**, 1015–1020.

Ogawa, K., and Roy, D.M. (1981, 1982) $C_4A_3\bar{S}$ hydration, ettringite formation and the expansive mechanism. *Cement and Concrete Research* **11**, 741–750; **12**, 101–109.

Paulini, P. (1997) Chemical shrinkage as indicator for hydraulic bond strength, in *Proceedings 10th ICCC, Göteborg*, paper 2ii072.

Ramachandran, V.S., Sereda, P.J., and Feldman, R.F. (1964) Mechanism of hydration of calcium oxide. *Nature (London)* **201**, 288–289.

Stark, J., and Chartschenko, I. (1997) Theoretical fundamentals of the application of expansive cement concretes, in *Proceedings 10th ICCC, Göteborg*, paper 2ii010.

Su, M. *et al.* (1992) Research on the chemical composition and microstructure of sulfoaluminate cement, in *Proceedings 9th ICCC, New Delhi*, Vol. 2, pp. 94–100.

Sudoh, G., Ohta, T., and Harada, H. (1989) High strength cement in the CaO-Al_2O_3-SiO_2-SO_3 system and its application, in *Proceedings 7th ICCC, Paris*, Vol. 3, pp. V-153 to V-157.

Takahashi, T., Nakata, H., and Yashida K. (1997) Autogenous shrinkage of cement paste during hydration, in *Proceedings 10th ICCC, Göteborg*, paper 2ii070.

Wang, S. *et al.* (1985) Experiments on the mechanism of ettringite expansion, in *Proceedings 1985 Beijing International Symposium on Cement and Concrete*, Vol. 3, pp. 43–55.

Wang, S. *et al.* (1986) Effect of alkali on the expansive properties of sulfoaluminate cement pastes, in *Proceedings 8th ICCC, Rio de Janeiro*, Vol. 4, pp. 301–305.

Wang, Y., and Lou, Z. (1992) Retarded expansion for cooling shrinkage compensation in mass concrete, in *Proceedings 9th ICCC, New Delhi*, Vol. 5, pp. 129–135.

Wang, Y. *et al.* (1992a) A quantitative study of paste microstructure and hydration characteristics of sulfoaluminate cement, in *Proceedings 9th ICCC, New Delhi*, Vol. 4, pp. 454–460.

Wang, Y. *et al.* (1992b) The composition, properties and application of compound expansive agent, in *Proceedings 9th ICCC, New Delhi*, Vol. 5, pp. 564–570.

Wu, C.-W., and Wang, Y.-S. (1980) On alunite expansive cement and concrete, in *Proceedings 7th ICCC, Paris*, Vol. 3, pp. V-27 to V-32.

Xue, J. *et al.* (1986) Study on high self-stress aluminate cement, in *Proceedings 8th ICCC, Rio de Janeiro*, Vol. 4, pp. 306–311.

22 Cements for concrete mixes made with alkali-susceptible aggregates

A variety of rocks may cause expansion and crack formation in concrete if used as aggregate in combination with Portland cement that contains amounts of alkalis considered acceptable for normal use. Such **alkali-aggregate reactivity** occurs in two forms: as alkali-silica or alkali-silicate reactivity, and as alkali-carbonate reactivity, of which the former is much more common.

In Portland cement the Na^+ and K^+ ions are present mainly in the form of sulfates, and to a lesser extent are incorporated in the clinker phases. In the course of hydration the anions to which the alkali ions were bound enter the hydrate phases, and equivalent amounts of OH^- ions are formed. These enter the liquid phase together with the alkali ions, thus increasing the alkalinity of the pore solution to pH = 12.5–13.5.

In the **alkali-silica/silicate reaction** (abbreviation ASR) the OH^- ions present in the pore solution attack the reactive amorphous silica or crystalline (or semi-crystalline) silicate, which is a constituent of the aggregate, break the existing Si–O–Si bonds, and fragment the three-dimensional silica/silicate framework into separate silicate anions of varying sizes. The negative charges created in this way are balanced by the Na^+ and K^+ cations. At a sufficiently high degree of fragmentation the damaged silicate network becomes sufficiently deformable, imbibes water molecules, and expands. This in turn causes expansion and ultimately cracking of the concrete body (Hobbs, 1987, 1988; Swamy, 1992; Shayan *et al.*, 1996; West, 1996).

Rocks that may be susceptible to an alkali-silica/silicate reaction if used as concrete aggregates contain reactive SiO_2 in a variety of forms, such as opal, tridimite, cristobalite, or siliceous volcanic glass. Quartz may also participate in an alkali-silicate reaction, but only if sufficiently strained or microcrystalline. The susceptibility of different forms of SiO_2 to undergo alkali-silica/silicate reaction may vary greatly, and along with it the time taken for the resulting internal stresses to cause expansion and cracking of the concrete. If all other variables remain constant, in most aggregates – especially those that contain a highly reactive SiO_2 form (such as opal) – there exists a "pessimum" amount of the reactive constituent, at which the expansion is highest; it is lowered if the amount of this constituent is either reduced or increased from this pessimum. This reflects the fact that a maximum swelling of the alkali-silica gel formed in the ASR reaction occurs at an optimum ratio of alkali oxide to reactive SiO_2.

Three factors are required to cause damage to concrete by an alkali-silica/silicate reaction:

- The aggregate must contain a critical amount of a reactive form of SiO_2.
- The pore solution must contain sufficient amounts of an alkali hydroxide.
- Sufficient moisture must be available.

Thus by eliminating at least one of these factors the expansion and cracking of concrete may be prevented.

One approach to preventing expansion due to ASR consists in lowering the alkali content in the concrete mix to sufficiently low concentrations. It is generally accepted that an alkali-silica/silicate reaction in concrete made with Portland cement will not occur if the content of equivalent Na_2O (defined as $Na_2O^e = Na_2O + 0.66K_2O$) in the mix does not exceed 4 or even 3 kg/m^3. Such low alkali concentrations are usually not achievable with ordinary Portland cement, but may be achieved if a low-alkali Portland cement (see section 2.10) is used instead.

The ASR expansion may also be reduced or even prevented completely by replacing a fraction of the Portland cement in the concrete mix with fly ash or some natural pozzolanas, provided that they themselves do not introduce too much additional alkali (Hobbs, 1986, 1994; Nixon *et al.*, 1986; Nagataki *et al.*, 1991; Blackwell *et al.*, 1992; Shayan *et al.*, 1996). Alternatively, this may be done by the use of Portland–fly ash cement (see section 0.00) or Portland-pozzolana cement (see section 0.00), instead of ordinary Portland cement. By this measure the amount of alkali in the pore solution is reduced and the amount of reactive SiO_2 is increased, thus altering the ratio of alkali to reactive SiO_2 to a value at which the tendency of the concrete mix to expand is reduced or eliminated. A similar effect may also be obtained by the use of Portland-slag cement (see section 8.3) in the production of the concrete mix (Bakker, 1981; Chatterji, 1984; Hogan, 1985; Hobbs, 1986, 1987).

The presence of pozzolanic additives or blast furnace slag decreases expansion in most cases, but in instances where the alkali content in the mineral addition is too high the reduction in expansion may be insufficient, or the overall expansion may even be increased. The alkali cations present in fly ash or in the slag are almost completely incorporated in the glass phase, and enter the pore solution of concrete only in the course of hydration (Duchesne and Berube, 1994). Generally they tend to be less effective in raising the OH^- concentration in the pore solution than are equal amounts of alkalis released from Portland cement. In calculating the equivalent amount of Na_2O for mixes containing mineral additions it has been recommended that one-sixth of the total alkali content of the fly ash or one-half of that of the slag should be added to the amount contributed by the Portland cement component (Hobbs, 1986). In long-term experiments it was shown that fly ashes introduced to concrete mixes made with alkali-reactive aggregates were effective in preventing expansion with alkali contents in the mix as high as 7.0 kg Na_2O/m^3 (Shayan *et al.*, 1996).

It has been also recognized that ASR expansion may be prevented by adding appropriate amounts of a lithium salt to the concrete mix (McCoy and Caldwell, 1951; Stark *et al.*, 1993). Alternatively, the lithium compound may be introduced directly to the cement (Gajda, 1996). From among different lithium compounds lithium hydroxide monohydrate ($LiOH.H_2O$) was found to be the most suitable for this purpose. The optimum dosage of the lithium additive ranges between 0.8 and 1.2% $LiOH.H_2O$.

An **alkali-carbonate reaction** may takes place if an argillaceous (illitic) dolomitic limestone is used as concrete aggregate. Here expansion takes place as the consequence of swelling of the illite constituent, following a de-dolomitization of the dolomitic constituent by alkali hydroxide present in the pore solution (Deng and Tang, 1993; Tang *et al.*, 1994; Tong and Tang, 1995):

$$CaMg(CO_3)_2 + 2NaOH \rightarrow CaCO_3 + Mg(OH)_2 + Na_2CO_3 \qquad (22.1)$$

Unlike the alkali-silica/silicate reaction a "pessimum" concentration of the expansive constituent does not exist, and the expansion increases continuously with increasing amounts of dolomitic limestone (Tong and Tang, 1995). The use of low-alkali Portland cement (see section 2.10) as binder may also effectively prevent this form of alkali-induced expansion.

REFERENCES

Blackwell, B.Q. *et al.* (1992) The use of fly ash to suppress deleterious expansion due to AAR in concrete containing greywacke aggregate. *Proceedings 9th International Conference on AAR in Concrete, London*, pp. 102–109.

Deng, M., and Tang, N. (1993) Mechanism of dedolomitization and expansion of dolomitic rocks. *Cement and Concrete Research* **23**, 1397–1408.

Duchesne, J., and Berube, M.A. (1994) Available alkalis from supplementary cementing materials. *ACI Materials Journal* **91**, 289–299.

Gajda, J. (1996) Development of lithium bearing cement to inhibit alkali silica reactivity in hardened concrete. *World Cement* **27**, 58–62.

Hobbs, D.W. (1986) Deleterious expansion of concrete due to alkali-silica reaction: influence of pfa and slag. *Magazine of Concrete Research* 38, 191–2015.

Hobbs, D.W. (1988) *Alkali-Silicate Reaction in Concrete*, Thomas Telford Ltd, London, UK.

Hobbs, D.W. (1994) The effectiveness of PFA in reducing the risk of cracking due to ASR in concretes containing cristobalite. *Magazine of Concrete Research* **46**, 167–175.

McCoy, W.J., and Caldwell, A.G. (1951) New approach to inhibit alkali-aggregate expansion. *Journal of the American Concrete Institute* **22**, 693–706.

Nagataki, S., Ohga, S., and Inoue, T. (1991) Evaluation of fly ash for controlling alkali-aggregate reaction. *Proceedings 2nd International Conference on Durability of Concrete, Montreal*, pp. 955–972.

Nixon, P.J. *et al.* (1986) The effect of pfa with a high total alkali content on pore solution composition and alkali silica reaction. *Magazine of Concrete Research* **38**, 30–35.

Shayan A. (ed.) (1996) *Alkali-Aggregate Reaction in Concrete.* CSIRO, Melbourne, Australia.

Shayan, A., Diggins, R., and Ivanusec, I. (1996) Effectiveness of fly ash in preventing deleterious expansion due to alkali-aggregate reaction in normal and steam-cured concrete. *Cement and Concrete Research* **26**, 153–164.

Stark, D.C. *et al.* (1993) Eliminating or minimizing alkali-silica reactivity. SHRP (Strategic Highway Research Program) C 343, National Research Council, Washington, DC.

Swamy, R.N. (ed.) (1992) *The Alkali-Silica Reaction in Concrete.* Blackie & Son, Glasgow.

Tang, M. *et al.* (1994) Studies on alkali-carbonate reaction. *ACI Materials Journal* **91**, 26–29.

Tong, L., and Tang, M. (1995) Correlation between reaction and expansion of alkali-carbonate reaction. *Cement and Concrete Research* **25**, 470–476.

West, G. (1996) *Alkali-Aggregate Reaction in Concrete Roads and Bridges.* Thomas Telford Ltd, London.

23 Cements for elevated and high-temperature applications

Hardened cement pastes may lose some and eventually all of their chemically bound water and undergo phase transformations if exposed to elevated temperatures. These processes tend to be associated with changes in the physico-mechanical properties of the hardened material, and limit the temperature range within which the cement may be used. Eventually, each system undergoes melting and loss of cementing properties.

The changes that accompany the thermal degradation of the cementitious system include a decline of strength and changes in the external volume of material. These result from the thermal expansion of the material in the course of heating, from the shrinkage that is associated with the loss of free and chemically bound water, and from volume changes accompanying the phase transformations that take place under heating conditions. As a result of these volume changes cracks may be formed in the material, which may ultimately cause disintegration of the concrete body.

The temperature range in which the hardened paste exhibits acceptable performance varies greatly in different cements, and thus it is essential to select the proper binder in high-temperature applications.

In addition to the binder, it is also important to select the proper aggregate in the production of high-temperature mortar or concrete, as chemical changes taking place in this constituent of the mix may also limit the temperature range in which the hardened material can be applied.

Concrete mixes in which quartz is used as aggregate may not be employed at temperatures above 573 °C, as at this temperature quartz undergoes a phase transition from β-quartz to α-quartz, and this process is associated with a volume increase of 0.8%. Such an increase of the volume of the aggregate is associated with crack formation and strength loss.

Calcite (calcium carbonate, $CaCO_3$), the main constituent of limestone, starts to decompose to calcium oxide and carbon dioxide at around 700 °C. The decomposition temperature is even lower in dolomite [calcium-magnesium carbonate, $CaMg(CO_3)_2$], in which the decomposition starts at about 500 °C. In limestones that are partially "dolomitized" the actual decomposition temperature will depend on the Ca/Mg ratio in the crystalline lattice, and will decline with increasing MgO content in the rock. This must be taken into consideration if carbonate-based rocks are used as aggregate in concrete mixes.

Basalt rocks may be used for temperatures up to about 1000 °C and even higher, depending on their oxide composition.

For temperatures that are even higher, crystallized blast furnace slag, granular Portland clinker, calcined fireclays, or corundum may be considered as suitable aggregates.

23.1 CEMENT PASTES AND CONCRETES MADE WITH PORTLAND CEMENT AND RELATED BINDERS

The hydrate phases present in hydrated Portland cement undergo decomposition, associated with loss of chemically bound water, if exposed to elevated temperatures. Partial decomposition may get under way even at temperatures below 100 °C, but about half the total amount of combined water is lost in the temperature range between about 100 °C and 200–300 °C. Very small amounts of chemically bound water may still be present in the material even after exposure to a temperature of 1000 °C.

X-ray diffraction studies performed on hydrated Portland cement pastes reveal that the peaks belonging to the AFt phase (ettringite) disappear after a sufficiently long exposure to temperatures of around 100 °C, as the crystalline lattice collapses, owing to the loss of a significant fraction of its molecular water. A gradual decomposition of the C-S-H phase also gets under way at around 100 °C, but in the initial stage this does not affect adversely the intrinsic strength properties of this phase. Between about 400 and 500 °C a decomposition of the existing portlandite (calcium hydroxide) and its conversion to calcium oxide takes place (Tanaka *et al.*, 1982; Piasta *et al.*, 1984; Zürz *et al.*, 1986; Handoo *et al.*, 1997). At 500–600 °C the thermal decomposition of the C-S-H phase is almost complete, and it starts to be converted to β-dicalcium silicate, the final product of C-S-H thermal decomposition. At around 800–900 °C small amounts of C_3A and/or $C_2(A,F)$ may also be formed, depending on the amounts of Al_2O_3 and Fe_2O_3 in the original cement (Zürz *et al.*, 1986). Figure 23.1 shows the relative changes in the combined water content, the free $CaO/Ca(OH)_2$ content, and the compressive strength of a Portland cement as a function of temperature to which it has been exposed. Gypsum-free Portland cement (see section 2.16) resists elevated temperatures better than ordinary Portland cement (Skvara and Sevcik, 1999). The refractoriness of the binder may be increased even further by adding silica fume (12%) to the mix.

The thermal decomposition of hydrated composite cements is similar but not identical to that of plain Portland cement pastes (Zürz *et al.*, 1986). Here also the amount of combined water declines sharply with increasing temperature, and may drop to near-zero values upon exposure to 1000 °C (Fig. 23.2). The original amount of free calcium hydroxide in

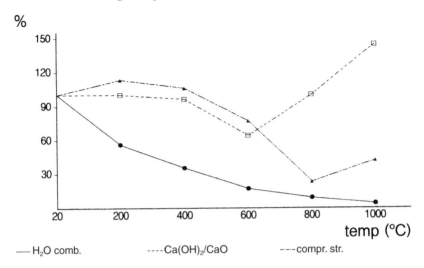

Figure 23.1 Combined H$_2$O, free CaO/Ca(OH)$_2$ and compressive strength of a mature ordinary Portland cement paste exposed to increasing temperatures (in per cent of the value existing at 20 °C).
Source: Zürz *et al.* (1986)

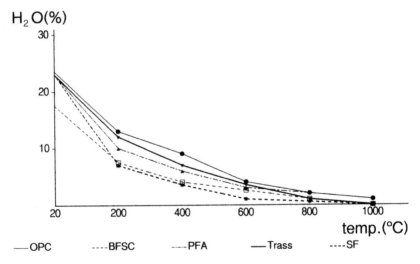

Figure 23.2 Content of bound water in cement pastes made with different cements and exposed to different temperatures. OPC, ordinary Portland cement; BFSC, blast furnace slag cement with 50% of granulated blast furnace slag; PFA, fly ash cement with 25% of pulverised fly ash; Trass = trass cement with 25% of trass; SF = Portland cement with 10% of added silica fume.
Source: Zürz *et al.* (1986)

pastes made with blended cements is lower, and the amount of calcium oxide formed in the decomposition process declines even further or may even drop to zero values, owing to solid-state reactions that may take place in the material at high temperatures (Fig. 23.3).

As to the phase transformations, in blast furnace slag cement pastes increasing amounts of melilite [a solid solution of gehlenite (C_2AS) and akermanite (C_2MS_2)] are formed as the amount of slag in the original cement increases, at the expense of β-dicalcium silicate, which is formed in Portland cement pastes under similar conditions. In pastes made from cements that contain distinct amounts of fly ash, β-dicalcium silicate is also the main final decomposition product. In parallel, quartz, which is a common constituent of fly ashes, tends to react at high temperatures with calcium oxide formed in the decomposition of portlandite, yielding additional amounts of dicalcium silicate and reducing the amount of free lime in the material. In pastes made from cements that contain natural pozzolanas, gehlenite (C_2AS) may also be formed as a product of thermal decomposition, whereas wollastonite (CS) may be formed if silica fume has been added to the original mix.

After exposure to elevated temperatures, the strength of the hardened Portland cement and blended cement pastes initially increases moderately up to about 200–400 °C (Zürz *et al.*, 1986; Handoo *et al.*, 1997), in spite of

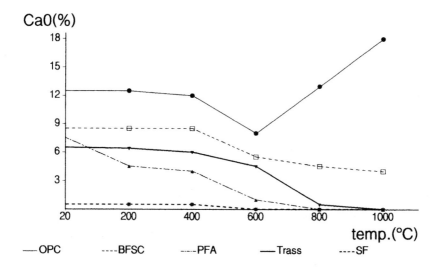

Figure 23.3 Effect of temperature on free CaO/Ca(OH)$_2$ content of hydrated cement pastes made with different cements. OPC, ordinary Portland cement; BFSC, blast furnace slag cement with 50% of granulated blast furnace slag; PFA, fly ash cement with 25% of pulverized fly ash; Trass, trass cement with 25% of trass; SF, Portland cement with 10% of added silica fume.
Source: Zürz *et al.* (1986)

compr.str. (MPa)

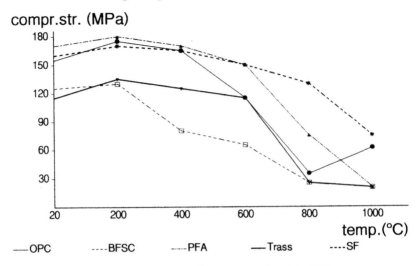

—OPC   ----BFSC   ----PFA   ——Trass   ----SF

Figure 23.4 Compressive strengths of pastes made with different cements and
exposed to increasing temperatures. OPC, ordinary Portland cement;
BFSC = blast furnace slag cement with 50% of granulated blast furnace slag; PFA, fly ash cement with 25% of pulverized fly ash; Trass,
trass cement with 25% of trass; SF, Portland cement with 10% of
added silica fume.

Source: Zürz *et al.* (1986)

a distinct loss of combined water and a corresponding increase of porosity. After reaching a maximum, the strength declines with increasing temperature of exposure. After exposure to temperatures of 800–1000 °C the
residual strength ranges typically between about 15% and 35% of the original strength of the paste. Figure 23.4 shows the effect of temperature on
the residual strength of pastes made from different types of cement.

Upon heating, a neat cement paste initially expands, owing to thermal
expansion. However, as the temperature increases, this expansion is over-
taken by shrinkage associated with the escape of free water that is present
in the pore system of the paste, and water liberated in the decomposition
of the hydrates that are present. The shrinkage caused by these phenomena is much larger than the normal thermal expansion, thus causing an
overall shrinkage of the paste. At high temperatures, after the paste has
lost all or most of its chemically bound water, some renewed thermal
expansion of the residual material may be observed. Figure 23.5 shows the
shrinkage of a series of cements taking place upon heating.

The shrinkage associated with the decomposition of the hydrated paste
may cause cracking within the material. The cracking may be even greater
in concrete, rather than plain cement mixes, as here the contraction of the
hardened cement paste is countered by the thermal expansion of the
aggregate particles.

In thermally decomposed cement pastes that contain distinct amounts

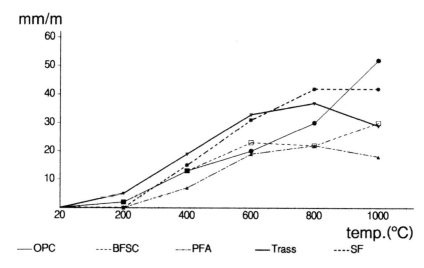

Figure 23.5 Shrinkage of cement pastes made with different cements after exposure to different temperatures. OPC, ordinary Portland cement; BFSC = blast furnace slag cement with 50% of granulated blast furnace slag; PFA, fly ash cement with 25% of pulverized fly ash; Trass, trass cement with 25% of trass; SF, Portland cement with 10% of added silica fume.

Source: Zürz *et al.* (1986)

of residual calcium oxide, this constituent rehydrates readily to calcium hydroxide, if in contact with water or exposed to humid air. This reaction is associated with expansion, which may cause further damage to concrete that has been exposed to fire, even in situations where the concrete structure has withstood the exposure to fire without disintegration or visible signs of destruction. Because of this, concrete structures made with blended cements, in which the amount of free lime present after exposure to fire is significantly reduced, are generally less damaged after being exposed to fire than those made with plain Portland cement.

If intended for applications at elevated temperatures, concrete mixes made with Portland cement may be exposed safely to temperatures of up to about 400 °C: that is, up to a temperature at which portlandite remains preserved in the hardened paste. Mixes made with blended cements may be used at even higher temperatures, provided that:

- the amount of free lime in the hardened paste after exposure to the targeted temperature remains low;
- the aggregate that is used is stable up to the targeted temperature;
- the decline of strength and increase of porosity caused by the exposure to the targeted temperature can be tolerated.

23.2   REFRACTORY CONCRETES BASED ON CALCIUM ALUMINATE CEMENT

The hydrate phases constituting hydrated calcium aluminate cement decompose gradually upon heating, losing virtually all their combined water by 800–1000 °C (see also section 10.8). This thermal decomposition is associated with a gradual loss of strength of the material. Starting at around 700–800 °C the decomposition of hydrate phases is superseded by solid-state reactions in which newly formed water-free phases gradually replace the original hydrates. This development of "ceramic bonding," replacing "hydraulic bonding," is associated with a renewed increase of strength of the material. In concrete/mortar mixes, solid-state reactions between the constituent of the hardened paste and the aggregate may also take place. The quality of compounds formed in these solid-state reactions depends on the compositions of the original binder and aggregate, and typically includes such high-melting-temperature phases as CA_2, CA_6, and MA (spinel). If the temperature is increased even further, the material eventually undergoes melting. The decline of strength taking place upon heating and the strength minimum that develops typically at temperatures between about 900 and 1100 °C are usually not too great, and do not interfere with the required high-temperature performance of the binder in most applications.

The maximum service temperature of the cement depends on the temperature of first melt formation, and this value increases with increasing A/C ratio of the cement. Thus to increase the refractory nature of the produced castable refractories, cements with up to 80% Al_2O_3 have been developed. Cements for high-temperature applications should also contain no or negligible amounts of Fe_2O_3, to increase their refractoriness even further. In the presence of a suitable aggregate the service temperature may be extended even beyond the fusion point of the pure cement (by up to 2000 °C), by the formation of higher melting-point eutectics. Table 23.1

Table 23.1 Heat-resistant and refractory concrete mixes.

Cement type	Al_2O_3 (%)	Aggregate	Maxiumum service temperature (°C)
Gray CAC	40	Basalt	700–800
Gray CAC	40	Emery	1000
Gray CAC	40	Chamotte	1300
Brown CAC	50–55	Brown fused alumina	1550
White CAC	70	Brown fused alumina	1650
White CAC	70	White fused alumina	1800
White CAC	80	White fused alumina	1850
White CAC	80	Tabular alumina	1900

Source: Scrivener and Capmas (1998)

summarizes the approximate service temperature limits of a series of cement/aggregate combinations.

Mixes whose service temperature is limited to 900–1000 °C are usually referred to as heat-resistant rather than refractory concretes. They are based on a low-Al_2O_3 cement in combination with basalt, granite, or similar aggregates. One typical area of application of such concretes is floors in industries where hot metal spillage may occur.

Conventional dense refractory castables contain typically 15–25% of calcium aluminate cement, and their refractoriness is governed mainly by the Al_2O_3 content in cement and the quality of the aggregate. In contrast, low-cement castables (LCC) contain only about 5–8% of calcium aluminate cement, plus silica fume and reactive alumina. These measures lead to a significant improvement of the refractoriness of the material. In recent developments the cement content has been pushed even lower, to 3–5% (ultra-low cement castables, ULCC), and silica fume has been replaced by ultrafine refractory alumina. High-performance refractory concretes have in many cases replaced refractory bricks, and their use is expanding.

In insulating refractory concretes the conventional aggregates are replaced by lightweight heat-resistant aggregates such as expanded vermiculite, expanded perlite, expanded chamotte, or pumice. These concretes are widely used to limit the heat losses of industrial kilns and high-temperature installations.

In addition to high temperature, the refractory concrete to be produced must also be resistant to possible chemical corrosion. Just as in applications at ordinary temperatures, porosity – apart from the chemical nature of the solid phases – is the main factor that controls the chemical resistance of refractory concrete. The same general principles as outlined for the production of a good non-refractory, low-temperature concrete for civil engineering applications must also be maintained in the production of refractory concrete.

One corrosive agent to which refractory concrete may be exposed during service is carbon monoxide (CO). To ensure an acceptable resistance of hydrated calcium aluminate cement to this agent, the cement must not contain iron oxide, which may undergo chemical reduction to bivalent iron under these conditions. In situations in which hydrogen may come into contact with concrete based on calcium aluminate cement, the silica content of the binder may be critical, and thus should be kept low.

To improve the performance of calcium aluminate cement in high-temperature concrete formulations, it has been suggested that it should be combined with water glass (Krivenko *et al.*, 1997). A binder of this type yields colloidal gibbsite (AH_3) and hexagonal calcium aluminate hydrates (CAH_{10} and C_2AH_8) as products of hydration at ambient temperature. Upon heating, these phases lose their chemically bound water, and this process is followed by the formation of gehlenite (C_2AS) at 800–1000 °C.

The residual strength increases with increasing silicate modulus of the water glass (SiO_2/Na_2O), and ranges between 60% and 100% of the initial strength.

23.3   PHOSPHATE-BONDED REFRACTORY CONCRETES

Phosphate-bonded refractory concretes are combinations of a phosphate binder and a refractory aggregate. The hardening of the system takes place as the result of a chemical reaction between these two constituents. To control the setting behavior small amounts of a "setter" (such as magnesium chromate) may also be added to the mix.

Based on the temperature at which the setting and hardening takes place, one has to distinguish between air-setting and heat-setting types of refractory concrete. In air-setting concretes the setting/hardening process takes place at ambient temperature, and mixes of this type may also be employed for other than high-temperature applications. In heat-setting mixes an elevated temperature is needed to achieve setting, and the use of such materials is limited to high-temperature applications.

The mixes may be supplied either as dry or wet binders. Dry binders are preferred, because of their more convenient handling and longer shelf life.

The binders most commonly used in heat-setting refractory concrete mixes are orthophosphoric acid (H_3PO_4) and mono-aluminum phosphate [MAP, $Al(H_2PO_4)_3$]. The most widely used aggregate is fused alumina, but a variety of other suitable aggregates (such as fireclay, or calcined magnesia) may also be used.

The chemical reactions taking place if a mix of phosphoric acid and a high-Al_2O_3 aggregate are exposed to elevated temperatures may be summarized as follows:

- room temperature: formation of mono-aluminum phosphate [$Al(H_2PO_4)_3$]
- around 250 °C: polymerization of $Al(H_2PO_4)_3$ and formation of aluminum pyrophosphate ($Al_4(P_2O_7)_3$)
- 500 °C: conversion of aluminum pyrophosphate to aluminum metaphosphate [$Al(PO_3)_3$]
- >500 °C: gradual formation of crystalline $AlPO_4$.

Other tertiary phosphates, such as $CrPO_4$ or $Mg_3(PO_4)_2$, may be formed instead of or in addition to aluminum phosphate, if aggregates other than Al_2O_3 are used as constituents of the mix.

The properties of the final product will depend on the type of binder and aggregate employed, on their mutual proportion, on the fineness of the aggregate, and on other factors. In general, the strength of the material will increase with increasing P_2O_5 content. Orthophosphoric acid has been found to be more effective than MAP (Sharma and

Chaturvedi, 1997). Under the most favorable conditions, using fused alumina as aggregate, a maximum service temperature of 1600 °C may be achieved.

Among air-setting phosphate-based refractory cements, magnesium phosphate cement is the most important (see also section 12.1). Its setting is based on a reaction between diammonium hydrogen phosphate $[(NH_4)_2HPO_4]$ and magnesium oxide (MgO). At ambient temperature struvite $(NH_4MgPO_4.6H_2O)$ is formed as the product of reaction. Upon heating above about 200 °C struvite loses water, and the decomposition product may react with additional MgO. These chemical changes may be described as follows:

$$NH_4MgPO_4.6H_2O \overset{(-H_2O)}{\longrightarrow} NH_4MgPO_4.H_2O \overset{(-H_2O, -NH_3)}{\longrightarrow}$$
struvite   dittmarite

$$MgHPO_4 \overset{(-H_2O)}{\longrightarrow} Mg_2P_2O_7 \overset{(+MgO)}{\longrightarrow} Mg_3(PO_4)_2$$
amorphous   pyrophosphate   orthophosphate

The reaction is associated with a moderate shrinkage, but the strength of the hardened material is preserved at least up to 1000 °C.

23.4 ALKALI SILICATE BINDERS

In alkali silicate binders the setting and hardening is based on a reaction between a water glass (a solution of an alkali silicate in water) and alkali fluosilicate (see also section 15.3). The reaction product is an amorphous form of SiO_2 with embedded crystals of alkali fluoride. The amorphous SiO_2 remains preserved until about 550 °C and converts gradually to quartz and tridimite at higher temperatures. At 750 °C crystalline sodium disilicate starts to be formed.

Concrete/mortar mixes made with this type of binder may be used up to about 700–900 °C, depending on the aggregate employed (Odler and Hennicke, 1991).

REFERENCES

Handoo S. *et al.* (1997) Effect of temperature on the physico-chemical and mineralogical characteristics of hardened concrete, in *Proceedings 10th ICCC, Göteborg*, paper 41v067.

Krivenko, P.V., Pushkaryeva, E., and Shapetko, S.V. (1997) Alkali aluminate binder, in *Proceedings 10th ICCC, Göteborg*, paper 2ii032.

Odler, I., and Hennicke, U. (1991) Investigations on Na-silicate binders. *Materials Research Society Symposium Proceedings* **179**, 283–283.

Piasta, J., Sevicz, Z., and Rudzinsky, L. (1984) Changes in the structure of hardened cement paste due to high temperature. *Materials and Construction* **17**(100), 191–296.

Scrivener, K.L., and Capmas, A. (1998) Calcium aluminate cements, in *Lea's Chemistry of Cement and Concrete* (ed. P.C. Hewlett), 4th edn, Arnold Publishers, London, pp. 709–778.

Sharma, S.C., and Chaturvedi, S. (1997) Effect of addition of phosphate bearing materials on strength development of the high temperature refractory concrete, in *Proceedings 10th ICCC Göteborg*, paper 4iv057.

Skvara, F., and Sevcik, V. (1999) Influence of high temperature on gypsum-free Portland cement materials. *Cement and Concrete Research* **29**, 713–717.

Tanaka, H., Totany, Y., and Saito, Y. (1982) Properties of hardened cement pastes after heating and their rehydration. *Rev. General Meeting Ceramic Association of Japan* **36**, 34–35.

Zürz, A., Odler, I., and Abdul-Maula, S. (1986) Thermal decomposition of hydrated cement pastes, in *Proceedings 8th ICCC, Rio de Janeiro*, Vol. 5, pp. 176–180.

24 Colored cements

Colored cements are manufactured so that concrete surfaces of a desired color can be produced. Cements of this type are used mainly for decorative purposes.

Portland cement, the binder most commonly used in civil engineering applications, normally has a gray color, which is caused by the presence of trivalent iron in its structure. Portland clinkers made from iron-free raw meals are white, and are used in the production of white cement (see section 2.7). These are also used mainly for decorative purposes.

There are several possible approaches in producing colored cements. A clinker of a desired color may be produced by adding small amounts of appropriate chromophoric ions to the raw meal, prior to burning. A green color of the clinker may be obtained by adding trivalent chromium (Cr_2O_3) to the raw meal. Cobalt ions (CoO or CO_3O_4) yield a blue color in the resultant clinker. A pink color may be obtained by a combination of zirconia (ZrO_2) and trivalent iron (Fe_2O_3). A violet color is obtained by adding a manganese compound (MnO_2) to the raw meal.

The raw meal is typically of Portland cement type, but other types of cement, such as an aluminobelite cement (Laxmi *et al.*, 1992), may also be produced in a colored form by adding appropriate chromophores. To eliminate the effect of iron on the resultant color of the produced cement the raw meal must be low in Fe_2O_3, unless the presence of iron ions is explicitly desired. In the course of burning, the added chromophores are incorporated into the crystalline lattices of the clinker phases, or yield separate compounds in reactions with other raw meal constituents. For example, cobalt oxide reacts with SiO_2 or Al_2O_3 to yield cobalt silicate (CO_2SiO_4) or cobalt aluminate $CoO.Al_2O_3$). As some of the chromophores may be toxic (compounds of chromium, for example) appropriate precautions have to be taken in the production process to prevent harm to human health. The properties of colored cements do not differ significantly from those of their non-colored counterparts. However, very strict control of the production process is necessary, to minimize possible variations in the color of the finished cement.

A colored cement may also be produced by blending an ordinary or white cement with appropriate amounts – up to 10% – of a suitable pigment. The pigment:

- must not be chemically attacked by the cement;
- must not detrimentally affect the setting and strength development of the cement;
- must have good color durability when exposed to light and weather conditions.

Even distribution and good dispersion of the pigment within the cement are essential to produce an even color of the concrete surface. This may be achieved by a sufficiently joint grinding or mixing of the cement and pigment.

In the production of colored cements based on Portland cement it is also possible to grind together the original clinker, gypsum, and pigment. To improve the distribution of the pigment even further it has been recommended that small amounts of a supersoftener and an organosilicone modifier (polyphenyl siloxane) should also be added to the cement/pigment blend (Kholoshin and Vavrenyuk, 1992). To eliminate the interference of the gray color of ordinary Portland clinker, the use of white cement/clinker is preferred. Such a measure improves the subjective perception of the color that is produced, and may also reduce the amount of pigment needed in the production of the colored cement.

Table 24.1 lists a range of pigments that may be used in combination with white or ordinary Portland cement in the production of colored cements. Even though Portland cement is the binder most widely used for

Table 24.1 Pigments for colored cements.

Color	Pigment	Composition
White	Titanium dioxide	TiO_2
Yellow	Yellow iron oxide Yellow iron hydroxide	Fe_2O_3 $Fe(OH)_3$
Red	Red iron oxide	Fe_2O_3
Green	Chromium oxide green Phthalocyanine CI Pigment green 7, 41 or 43	Cr_2O_3 Polyhalogenated Cu
Blue	Cu phthalocyanine CI Pigment blue 15	
Brown	Brown iron oxide	Fe_2O_3
Black	Carbon black Black iron oxide	Amorphous C Fe_2O_3

Source: BS 1014

this purpose other types of cement may also be employed, such as calcium aluminate cement.

A "white" cement has also been produced by jointly grinding an electric arc furnace slag with gypsum. Such a binder may also be combined with suitable pigments to produce colored cements, but these are only very rarely used.

Finally, it is also possible to introduce the pigment directly into the fresh mortar/concrete mix. Extensive mixing is essential to produce an even distribution of the pigment within the mix. The pigment must be alkali resistant, and must maintain its color unchanged under exposure to sunshine and other external factors.

REFERENCES

Akatsu, A.M. *et al.* (1970) Effect of Cr_2O_3 and P_2O_5 on the strength and color of Portland cement clinker, in *Cement Association of Japan, Review of the 24th General Meeting*, pp. 20–23.

Bensted, J. (1998) Special cements, in *Lea's Chemistry of Cement and Concrete* (ed. P.C. Hawlett), Arnold, London, pp.779–835.

Kholoshin, E.P., and Vavrenyuk, S.V. (1992) Resource-saving colored cement: mineral composition, in *Proceedings 9th ICCC, New Delhi*, Vol. 3, pp. 444–448.

Laxmi, S. *et al.* (1992) Development of alumino-belitic coloured cement, in *Proceedings 9th ICCC, New Delhi*, Vol. 2, pp. 118–124.

25 Cements for fiber-reinforced cementitious composites

A typical feature of hardened inorganic cement pastes is their low tensile/compressive strength ratio. To improve this ratio and to lower the brittleness of the material, cementitious systems may be reinforced with suitable fibers randomly distributed within the cement matrix (Fig. 25.1).

If such fiber-reinforced cementitious material is exposed to an external tensile stress, a deformation of the matrix takes place, which – as long as a good bond exists between the matrix and the fiber surface – is also transferred to the fibers embedded in the matrix. Thus both the matrix and the fibers are put under stresses, whose magnitude is proportional to the deformation, and – at an equal deformation – proportional to the modulus of elasticity of the particular material (Fig. 25.2). Consequently, fibers with a high modulus of elasticity – if introduced in sufficiently high amounts – have the capacity to reduce distinctly the stresses existing in the matrix at a given degree of deformation and thus increase the maximum load that the material can sustain without failure. In general, the tensile

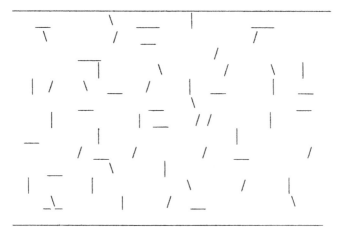

Figure 25.1 Schematic presentation of the distribution of fibers in a fiber-reinforced cementitious composite material.

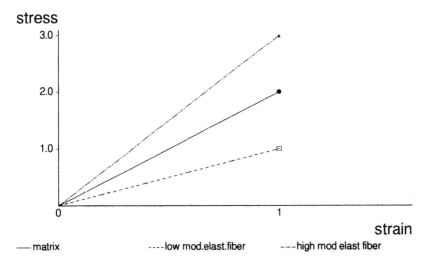

Figure 25.2 Stress–strain curves of fibers with different moduli of elasticity embedded in a cementitious matrix (schematic presentation, arbitrary values).

strength of the composite material will be determined by the strength properties of the matrix and the fibers, the amount and the geometry of the fibers, and the magnitude of the fiber/matrix bond.

The failure of fiber-reinforced composite material usually starts with excessive deformation and crack formation within the cement matrix. In parallel, failure of the fiber reinforcement also gets under way. It may result either in a gradual pull-out of the fibers from the matrix as the width of the crack gradually increases, or in a tear of the fibers themselves at or in the vicinity of the crack.

The mode of failure will determine the work of fracture: that is, the work needed to produce a complete failure of the material, also called the fracture energy. If the failure of the composite material is associated with failure of the fiber/matrix bond and subsequent pull-out of the fibers, the overall fracture energy will be quite high, and will generally increase with increasing magnitude of the friction between the two surfaces up to a point at which the friction exceeds the tensile strength of the fiber itself. Above this point the failure of the material changes from a non-brittle pull-out mode to a brittle tear-apart mode and – as a consequence – the overall work of fracture decreases sharply. Thus there is an optimum fiber/matrix bond strength at which the fracture energy of the composite material attains its maximum value and the material's behaviour is at its least brittle. Similarly, the material also fails in a brittle mode if the magnitude of the bond between the fiber surface and matrix exceeds the tensile strength of the fibers.

Figure 25.3 shows schematically the stress-strain curve of a fiber-reinforced cementitious material with a non-brittle mode of failure. Point

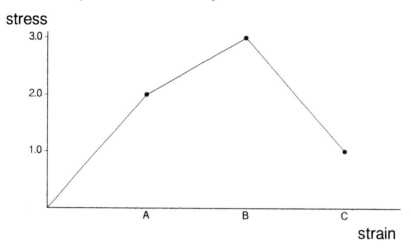

Figure 25.3 Schematic presentation of the stress–strain curve of a fiber-reinforced
cementitious composite material. A, first crack strength; A–B, propagation of cracks and gradual fiber pull-out; B, maximum strength
(modulus of rupture); B–C : continuation of crack formation and
fiber pull-out; C, failure of the composite material.

A indicates the strain value at which crack formation starts within the
cementitious matrix. Above this point the cohesion of the material is preferentially maintained by the presence of the fibers, which bridge the
formed cracks. As the strain of the material increases further, the existing
cracks widen and propagate, and the fibers undergo a gradual pull-out
from the matrix until complete failure. The stress to which the fibers are
exposed increases up to a maximum value (point B), which corresponds to
the maximum strength of the composite material, and which used also to
be called the modulus of rupture. At still higher strain values the fiber pull-
out continues, but the stress value declines, until separation of the material
takes place. The magnitude of the area under the stress-strain curve indicates the fracture energy of the material, and is a measure of its brittleness:
the lower the fracture energy, the more brittle is the material.

Table 25.1 summarizes the properties of the fibers most commonly used
in fiber-reinforced cementitious composites and compares them with those
of ordinary Portland cement paste.

Asbestos is an almost ideal fibrous material for fiber-reinforced composites, and has been widely used in combination with Portland cement as a
constituent of asbestos cement. Its use has been discontinued in most
countries, after it was discovered that it represents a serious health hazard.

Steel fibers guarantee good properties of the fiber-reinforced composite
material, but must be used only in combination with binders that protect
the fibers from corrosion. This requirement is generally fulfilled in cements
in which the pore solution of the hardened paste has a pH higher than

Table 25.1 Properties of fibers used in fiber-reinforced cementitious composite materials.

Fiber	Diameter (µm)	Density (kg/L)	Young's modulus (GPa)	Tensile strength (MPa)
Cement paste	–	2.4	–	3–7
Asbestos	0.02–0.4	3.2	85–140	570–1000
Steel	–	7.8	200	3000–4000
E-glass	10–20	2.4–2.6	70–80	1100–3900
AR-glass	10–20	2.7	75	3700
Graphite PAN-based	7–8	1.6–1.7	250–400	3000–4000
Graphite pitch-based	14–18	1.7–1.8	30–200	600–2000
Polyethylene	20–200	0.95	0.3	700
Polyacrylnitril	18–20	1.8	17	900
Cellulose	10–50	1.5	–	800

about 11. Portland cement is one of the binders that meets this requirement, and is most widely used in combination with steel fibers (Bentur *et al.*, 1985b, 1985c; Bentur, 1989; Marchese and Marchese, 1992). Both the compressive and the flexural strength of steel fiber reinforced concrete made with Portland cement may be improved by adding microsilica to the mix (Soroushian and Bayazi, 1988). Portland slag cement (Onabolu and Pratt, 1988) and Portland pozzolana cement (Marchese and Marchese, 1992) have also been found to be suitable binders for steel fiber reinforced materials. Steel fibers may corrode in regions close to the surface if the pH drops below the critical value as a result of carbonation. Chloride-based hardening accelerators, which tend to cause corrosion of steel, must not be used in cementitious mixes containing steel fibers. Compressive strengths in excess of 500 MPa and flexural strengths in excess or 80 MPa were obtained if chopped steel wool was combined with Portland cement and silica fume or crystalline silica, and the mix was first cured hydrothermally followed by curing at 400 °C for 24 hours (Wise *et al.*, 1988). Steel fibers must not be used in combination with gypsum binders, because the pH of the pore solution is too low. The use of calcium aluminate cement is possible, but should be discouraged, mainly because of an increase of porosity and a gradual decline of strength in the course of conversion.

Glass fibers seem to be the main candidate for replacing asbestos in fiber-reinforced composite materials. However, it has been found that fibers made from ordinary, E-type glass undergo corrosion if combined with ordinary Portland cement. As a consequence, the tensile strength of the material decreases over time (Proctor *et al.*, 1982; Bentur, 1989; Serbin, 1992). "Alkali resistant" fibers (called AR fibers) made from glass with a high ZrO_2 content have been developed to overcome this handicap. Such fibers resist alkaline corrosion significantly better, but it was found that, even with AR glass, glass fiber composites undergo strength loss and embrittlement, although the damage to the glass surface is significantly reduced (Bentur *et al.*, 1985a; Bentur and Diamond, 1987b; Bentur, 1989). It was also found that calcium hydroxide, liberated in the hydration of Portland cement, tends to precipitate preferentially at the glass fiber surface by heterogeneous nucleation. Such precipitation may increase the strength of the bond between the fiber surface and the cementitious matrix, or between two neighboring fibers, to values at which the composite material attains brittle properties. To improve the performance of glass fiber—Portland cement composites, microsilica may be added to the binder (Bartos and Zhu, 1996; Bentur and Diamond, 1987a, 1987b). This additive reacts with calcium hydroxide to yield the C-S-H phase, and thus greatly reduces the undesired presence of free portlandite at the fiber surface. It was observed, however, that although this additive inhibits notching of the glass surface, it does not reduce the pH of the pore solution sufficiently to inhibit hydroxylation and dissolution of the AR glass completely. Thus even in composites made with added microsilica, the flexural strength starts to decrease after reaching a maximum, and eventually drops to values similar to that of the unreinforced matrix. Other binders that have been used in combination with glass fibers include high-alumina cement (Majumdar *et al.*, 1981a, 1981b; Proctor and Litherand, 1985; Hu, 1991; Glukhovskiy *et al.*, 1992; Serbin, 1992; see section 10), belite cement (Serbin, 1992; see section 3.2), supersulfated cement (Majumdar *et al.*, 1981a; see section 8.4), sulfoaluminate cement (Hu, 1991; Wang *et al.*, 1997), gypsum (Serbin, 1992), magnesium-phosphate cement (Pera and Ambroise, 1998; see section 12.1), and glass cement (Hommertgen and Odler, 1991; see section 15.2).

In a separate development, Uchida *et al.* (1988) developed a special, Portland-type, low-alkali cement, which does not release calcium hydroxide in its hydration, to be used in combination with glass fibers.

There are two different types of **graphite** fibers, which differ distinctly in their properties (see Table 25.1). Graphite fibers may be used in combination with Portland cement without any danger of corrosion (Ali *et al.*, 1972; Ohama *et al.*, 1985; Akihama *et al.*, 1988; Banthia *et al.*, 1992; Fu and Chung, 1996; Fu *et al.*, 1996a, 1996b). A serious handicap of graphite fibers is their poor bond to the cement matrix. To improve this property and to increase the tensile strength of the composite material, microsilica, methylcellulose, a butadiene-styrene latex, or a combination of these may be added to the fresh mix (Fu *et al.*, 1996a). Even more effective is treatment of the fibers with ozone (Fu *et al.*, 1996b; Fu and Chung, 1998).

Synthetic organic fibers, such as those made from polyethylene, polypropylene, or polyacrylamide (except polyester fibers), generally exhibit a sufficiently high resistance to the high pH values existing in Portland cement pastes, and may be used in combination with this binder (Hannant, 1998). However, they exhibit a rather poor bond to the cementitious matrix, which does not improve distinctly with added microsilica (Dyczek and Petri, 1992). The nature of the bond is adhesional and mechanical (Rice *et al.*, 1988). The fracture process is usually characterized by pull-out of the fibers. The bond between the cement matrix and the fibers is improved if fibrillated rather than straight fibers are used (such as polypropylene) (Rice *et al.*, 1988). Synthetic organic fibers tend to be abraded by cement particles during pull-out. This results in an increase of the pull-out force and absorption of energy, compared with the stationary frictional bond strength (Wang *et al.*, 1988). This effect contributes significantly to the high energy absorption capability of such composite materials.

Synthetic organic fibers generally possess a low Young's modulus, and thus barely contribute to an increase of the "first crack" strength and only modestly to the increase of the modulus of rupture. To improve this handicap, synthetic organic fibers based on polyacryl nitrile (PAN fibers) have been developed, which have a significantly higher Young's modulus than conventional synthetic organic fibers. A significant increase of the modulus of rupture may be achieved by the use of these fibers (Odler, 1988).

Cellulose fibers of plant origin are being used in combination with Portland cement. The long-term durability of such materials must be questioned, however, given the high pH of the pore solution of this binder.

REFERENCES

Akihama, S., Suenaga, T., and Nagakawa, M. (1988) Carbon fiber reinforced concrete. *Concrete International* **10**, 40–47.

Ali, M.A., Majumdar, A.J., and Rayment, D.L. (1972) Carbon fiber reinforcement in cement. *Cement and Concrete Research* **2**, 201–202.

Banthia, N., Seng, J., and Pigeon, M. (1992) Strengthening and toughening mechanism in carbon fiber reinforced cement, in *Proceedings 9th ICCC, New Delhi*, Vol. 5, pp.492–504.

Bartos, P., and Zhu, W. (1996) Effect of microsilica and acrylic polymer treatment on the ageing of GRC. *Cement and Concrete Composites* **18**, 31–39.

Bentur, A. (1988) Interfaces in fibre reinforced cements. *Materials Research Society Symposium Proceedings* **114**, 133–144.

Bentur, A. (1989) Fiber reinforced cementitious materials, in *Materials Science of Concrete I.* (ed. J.P. Skalny), American Ceramic Society, pp. 223–284.

Bentur, A., and Diamond, S. (1987) Direct incorporation of silica fume into glass fibre strands as a means for developing GRCF composites of improved durability. *International Journal of Cement Composites and Lightweight Concrete* **9**, 214–222.

Bentur, A., and Diamond, S. (1987) Aging and microstructure of glass fiber cement composites reinforced with different types of glass fibers. *Durability of Building Materials* **3**, 201–226.

Bentur, A., Ben-Bassat, M., and Schneider, D. (1985a) Durability of glass-fiber-reinforced cements with different alkali-resistant glass fibers. *Journal of the American Ceramic Society* **68**, 203–208.

Bentur, A., Diamond, S., and Mindess, S. (1985b) Cracking process in steel fiber reinforced cement paste. *Cement and Concrete Research* **15**, 331–342.

Bentur, A., Mindess, S., and Diamond, S. (1985c) Pull-out process in steel fiber reinforced cement. *International Journal of Cement Composites and Lightweight Concrete* **7**, 29–37.

Dyczek, J.R.L., and Petri, M.A. (1992) FRC reinforced with short polypropylene fibers fracture process, in *Proceedings 9th ICCC, New Delhi*, Vol. 5, pp. 485–491.

Fu, X., and Chung, D.D.L. (1996) Submicron carbon filament cement matrix composites for electromagnetic interference shielding. *Cement and Concrete Research* **26**, 1467–1472.

Fu, X., Lu, W., and Chung, D.D.L. (1996a) Improving the bond strength between carbon fiber and cement by fiber surface treatment and polymer addition to cement mix. *Cement and Concrete Research* **26**, 1007–1012.

Fu, X., Lu, W., and Chung, D.D.L. (1996b) Improving the tensile properties of carbon fiber reinforced cement by ozone treatment of fibers. *Cement and Concrete Research* **26**, 1485–1488.

Fu, X., Lu, W., and Chung, D.D.L. (1998) Improving the strain-sensing ability of carbon fiber-reinforced cement by ozone treatment of the fibers. *Cement and Concrete Research* **28**, 183–186.

Glukhovskiy, V.V., Mikhailishina, N.Z., and Serbin, V.P. (1992) High-strength and durable composites with non-metallic reinforcement, in *Proceedings 9th ICCC, New Delhi*, Vol.5, pp. 474–478.

Hannant, D.J. (1998) Durability of polypropylene fibers in Portland cement-based composites: eighteen years of data. *Cement and Concrete Research* **28**, 1809–1817.

Hommertgen, C., and Odler, I. (1991) Glass fiber composites made with a cement based on a hydraulically active CaO—SiO_2—Al_2O_3 glass. *Materials Research Society Symposium Proceedings* **211**, 93–104.

Hu, G. (1991) The low pH value cement in GRC, in *Cement and Concrete Science & Technology* (ed. S.N. Ghosh), ABI Books Ltd, New Delhi, pp. 237–252.

Majumdar, A.J., West, J., and Larner, L. (1977) Properties of glass fibres in cement environment. *Journal of Materials Science* **5**, 927–936.

Majumdar, A.J., Sing, B., and Evans, T.J. (1981a) Glass reinforced supersulfated cement. *Composites* **12**, 177–183.

Majumdar, A.J., Sing, B., and Ali, M.A. (1981b) Properties of high alumina cement reinforced with alkali resistant glass fibers. *Journal of Materials Science* **16**, 2597–2609.

Marchese, B., and Marchese, G. (1992) Constitution of mortars and pull-out of steel fibers, in *Proceedings 9th ICCC, New Delhi*, Vol. 5, pp. 461–467.

Odler, I. (1988) Structure and mechanical properties of Portland cement—polyacrylnitril fiber composites. *Materials Research Society Symposium Proceedings* **114**, 153–158.

Ohama, Y., Amano, M., and Endo, M. (1985) Properties of carbon fiber reinforced cement with silica fume. *Concrete International* **10**, 58–62.

Onabolu, O.A., and Pratt, P.L. (1988) The effect of blast furnace slag on the microstructure of the cement paste/steel interface. *Materials Research Society Symposium Proceedings* **114**, 255–261.

Pera, J., and Ambroise, J. (1998) Fiber-reinforced magnesia-phosphate cement composites for rapid repair. *Cement and Concrete Composites* **20**, 31–39.

Proctor, A., and Litherand, K.L. (1985) Improving the strength retention of GRC by matrix and fibre modification, in *Proceedings 5th International Congress on GRC, Darmstadt*, pp. 45–52.

Proctor, A., Oakley, D.R., and Litherand, K.L. (1982) Development in the assessment and performance of GRC over 10 years. *Composites* **13**, 173–179.

Rice, E.K., Vondran, G.L., and Kundbargi, H. (1988) Bonding of fibrillated polypropylene fibers to cementitious materials. *Materials Research Society Symposium Proceedings* **114**, 145–152.

Serbin, V.P., Glukovskiy, V.V., and Serbina, R.V. (1992) Composites: cement and glass fibres, in *Proceedings 9th ICCC, New Delhi*, Vol. 5, pp. 479–484.

Soroushian, P., and Bayasi, Z. (1988) Silica fume effects on the pull-out behaviour of randomly oriented steel fibres from concrete. *Materials Research Society Symposium Proceedings* **114**, 187–196.

Uchida, U. *et al.* (1988) Development of low-alkaline cement for new GRC, in *Proceedings MRS International Meeting on Advanced Materials, Tokyo*, Vol. 13, pp. 169–178.

Wang, Y. *et al.* (1997) Sulfoaluminate cement with low alkalinity, in *Proceedings 10th ICCC, Göteborg*, paper 4iv030.

Wang, Y., Li, V.C., and Backer, S. (1988) Analysis of synthetic fiber pull-out from a cement matrix. *Materials Research Society Symposium Proceedings* **114**, 159–165.

Wise, S. *et al.* (1988) Chopped steel fiber reinforced chemically bonded ceramic (CRC) composites. *Materials Research Society Symposium Proceedings* **114**, 197–203.

26 Binders to be used in combination with wood

Wood chips, wood fibers, wood flakes, and similar wood-based starting materials, in combination with suitable inorganic binders, are widely used in the production of reinforced cementitious composites, such as wood chip cement boards (also called cement-bound wood chip boards).

In using wood in combination with inorganic binders one must bear in mind that wood from every species of tree contains constituents that – after mixing with an inorganic binder and water – may enter the liquid phase, and may or may not interfere with the hydration process. The quality and amount of these soluble constituents differ in different wood species, and may also vary geographically and seasonally.

Wood-reinforced cementitious materials may be compacted to different degrees, depending on the binder/wood chip ratio and on the degree of pressure applied to the starting mix in the course of setting and hardening. In most instances wood-based cementitious composite materials exhibit lower bulk densities and better thermal insulating properties than materials made with inorganic fillers. Owing to swelling and contraction of the wood (and to a lesser extent also of the hardened cement paste) under fluctuations of humidity, and upon wetting and drying, the volume stability of wood-based cementitious products is inferior to that of materials made with inorganic fillers (Fan *et al.*, 1999). This effect may be enhanced by carbonation and chemical degradation of the wood chips (Fan *et al.*, 1999). The flammability of cement-bound particle boards is significantly reduced compared with wood-based products bound with organic resins, but their bulk density tends to be higher.

A binder widely used in combination with wood is **hemihydrate gypsum** (see section 11.2.1). The setting and hardening of this binder, which are barely affected by water-soluble constituents present in the wood, are rapid, which is advantageous in industrial mass production of wood-chip-based boards. Owing to the partial solubility of hardened gypsum in water, the use of gypsum-based wood-chip boards is limited to indoor applications with a limited exposure to water. Gypsum-based particle boards possess a higher thickness swelling, higher water absorption, and inferior mechanical properties compared with products based on Portland cement.

All these properties may be improved by adding Portland cement (about 10% per weight of gypsum) to the starting mix (Deng *et al.*, 1998).

Magnesium oxychloride cement (see section 15.1) is also well suited to be combined with wood. Such a combination is most commonly used for industrial floorings. The main advantage of this material in such applications is its favorable elastic properties.

Wood chips or wood fibers may also be combined with **Portland cement** to produce wood chip–cement plates (also called wood fiber reinforced cement composites or wood-cement composites) (Lin *et al.*, 1994; Sorushia *et al.*, 1994; Simatupang *et al.*, 1995). In the conventional production of these products, blends of wood chips or flakes with cement and limited amounts of water are kept under compression at elevated temperature until the material attains the needed cohesion and sufficient strength. Flakes and chips from only a limited number of wood species are suitable as a reinforcing material, as in many other species the water-soluble oligosaccharides present in the wood may interfere with the hydration of the cement. It may also be sensible to store the wood chips for several months before use, to lower the content of undesired constituents by spontaneous biodegradation reactions taking place in the material.

To accelerate the process of hardening it has been suggested that carbon dioxide should be injected into the spreadable mix of wood, cement, and water, or that ammonium, sodium or potassium carbonate should be added (Simatupang *et al.*, 1995). Under these conditions the cement sets in a very short time, owing to the formation of calcium carbonate. Effective acceleration of the hardening process may also be achieved by the use of a fast-setting cement produced by combining Portland cement with limited amounts of calcium aluminate cement (see section 10.10.1).

To improve the strength properties it has been suggested that silica fume should be added to the starting mix (Lin *et al.*, 1994).

Wetting-drying and freezing-thawing tests have confirmed the acceptable performance of Portland cement-bound wood fiber reinforced materials (Parviz *et al.*, 1994). In using such products in combination with other construction materials one has to bear in mind, however, the relatively large longitudinal changes of the wood-based board in the course of wetting and drying. The resistance of wood fiber–cement composites to moisture changes may be improved, and the gradual loss of strength due to temperature cycling may be reduced, by pretreatment of the wood fibers with an aqueous acrylic emulsion or aqueous alkylalkoxysilane (Blankenhorn *et al.*, 1999).

A combination of Portland cement and clay has also been suggested for use as a binder in the production of wood chip based lightweight concrete (Bouguerra *et al.*, 1998)

As an alternative to Portland cement, **alkali-activated blended cements** (see section 8.5 and 9.1.5) have also been suggested as a binder in wood-based materials (Lin *et al.*, 1994). The hardening of these binders is little affected by the quality of the wood employed.

REFERENCES

Blankenhorn, P.R. *et al.* (1999) Temperature and moisture effects on selected properties of wood fiber–cement composites. *Cement and Concrete Research* **29**, 737–741.

Bouguerra, A. *et al.* (1998) Effect of microstructure on the mechanical and thermal properties of lightweight concrete prepared from clay, cement and wood aggregates. *Cement and Concrete Research* **28**, 1179–1190.

Deng, Y., Furuno, T., and Uehara, T. (1998) Improvement on the properties of gypsum particle board by adding cement. *Journal of Wood Science* **44**, 98–102.

Fan, M.Z. *et al.* (1999) Dimensional stability of cement-bonded particleboard. *Cement and Concrete Research* **29**,923–932.

Lin, X. *et al.* (1994) Approaches to improve properties of wood fiber reinforced cementitious composites. *Cement and Concrete Research* **24**, 1558–1566.

Parviz, S., Marikunte, S., and Won, J.-P. (1994) Wood fiber reinforced cement composites under wetting-drying and freezing-thawing cycles. *Journal of Materials in Civil Engineering* **6**, 595–611.

Simatupang, M.H. *et al.* (1995) Investigations on the influence of the addition of carbon dioxide on the production and properties of rapidly set wood-cement composites. *Cement and Concrete Composites* **17**, 187–197.

Sorushia, P., Narikunte, S., and Won, J. P. (1994) Wood fiber reinforced cement composites under wetting-drying and freezing-thawing cycles. *J. Nat. Civ. Eng.* **6**, 596–611.

27 Oil well cements

Oil well cements have been developed for use in the drilling of oil and gas wells. They are used mainly to restrict the movements of fluids between formations at different levels. In this way productive oil- and gas-bearing formations may be sealed from water-bearing layers, and – in the case of multilayer deposits – the production formations may be sealed off from each other. The hardened slurry is also intended to support and protect the casing. The cementing work is carried out by pumping the cement slurry down the steel casing of the well and up the annular space between it and the surrounding rock. More specialized techniques include squeeze cementing, in which the slurry is forced through a hole in the casing into a void or into a porous rock, and plugging, in which the casing is blocked at a specified depth.

Oil well cements must usually perform at elevated temperatures and pressures, both of which increase with increasing depth. The maximum temperature encountered at the bottom of deep wells may reach 250 °C, and may even exceed 300 °C in geothermal wells. Under these conditions the temperature of the slurry during pumping may reach 180 °C (bottom hole static temperature). The pressure to which the cement slurry is exposed is equal to the hydrostatic load plus the pumping pressure, and may reach 150 MPa.

The following are the main features desired in oil well cements:

- The viscosity of the cement slurry must be low enough to allow easy pumping. The slurry must stay sufficiently mobile for the whole time needed to complete the pumping procedure, which usually does not exceed 3–4 hours. In the oil industry the pumpability of the cement slurry is usually assessed by the thickening time test. In this the consistency is measured with a high-temperature, high-pressure consistometer, in which the temperature and pressure expected to occur during pumping are simulated.
- After setting, the slurry must attain a predetermined strength at an acceptable speed. The rate at which the material develops its strength is relevant, as it determines the time needed before drilling operations

can continue. A retrogression of strength must not occur under the conditions existing in the well.

- After being placed, the set cement slurry must exhibit sufficient resistance to the flow of liquids and gas. To achieve good performance in this respect, the volume changes that take place during setting and hardening must be negligible.
- The hardened cement slurry must be sufficiently resistant to chemical attack, especially by sulfates. In some fields corrosion by overheated CO_2-rich steam may also be critical, and must be taken into consideration.
- The available cement must be highly uniform, to ensure predictable and reproducible behavior of the slurry made from such a binder.

The requirements for oil well cements are laid down in the relevant specifications of the American Petroleum Institute (API), and are used as *de facto* standards all over the world. However, formal ISO international standards for oil well cementing are currently under development (Bensted, 1995). API Specification 10A, *Specifications for Well Cements*, divides oil well cements into several classes. Of these, classes G and H are the basic oil well cements, used most widely. Classes D, E, and F include retarded oil well cements, and Class C is a rapid-hardening cement. Class J includes special cements developed for high-temperature well cementing.

27.1   CLASS G AND CLASS H OIL WELL CEMENTS

Class G and Class H oil well cements are the most important ones in the API classification system, and are the most extensively used for cementing oil and gas wells. They can be tailored to cope with a wide range of well-cementing conditions by the addition of suitable additives to their slurries.

The chemical, physical and performance requirements for Class G and Class H oil well cements are summarized in Table 27.1. The requirements are identical for both classes, with one exception: the water/cement ratio to be used in the tests performed in determining the performance characteristics of these cements. The reasons for this distinction are historical (Bensted, 1992b), and will not be discussed in detail here.

In general, Class H cements are more coarsely ground than Class G cements; the specific surface area of Class H cements is typically in the range 220–300 m^2/kg, whereas that of Class G is in the range 270–350 m^2/kg. Class H cement was designated for lower water/cement ratios, and will hydrate more slowly than its Class G counterpart. Both Class G and Class H cements are available in moderately sulfate resistant (MSR) and high sulfate resistant (HSR) grades, which differ mainly in their C_3A content.

In terms of chemical composition, Class G and Class H oil well cements are low-C_3A Portland cements, and the specifications of both are typically

Table 27.1 API specifications for class G and class H oil well cements.

Chemical requirements Cement type:	MSR	HSR
MgO (max.)	6.0	6.0
SO_3 (max.)	3.0	3.0
Loss on ignition (max.)	3.0	3.0
Insoluble residue (max.)	0.75	0.75
C_3S	48–58	48–65
C_3A (max.)	8	3
$C_4AF + 2x\ C_3A$ (max.)	–	24
Na_2O equivalent (max.)	0.75	0.75
Physical and performance requirements		
Mix water (% by weight of cement)		
Class G		44
Class H		38
Schedule 5 thickening time (min.)		90–120
Max. consistency during first 15–30 min of schedule 5 test (Bc)		30
Compressive strength (MPa)		
8 h atmospheric pressure, 38 °C (min.)		2.1
8 h atmospheric pressure, 62 °C (min.)		10.3

met by high-iron, sulfate-resistant Portland cements (see also section 2.8). The free lime content in these cements is not limited by the API specifications, but it should be below 0.5 wt% to avoid difficulties with cement slurry rheology and retarder response. In producing Class G and Class H oil well cements, iron oxide (in the form of hematite or pyrites residues) must usually be added to the raw mix to produce more ferrite phase at the expense of tricalcium aluminate. The required amount of Fe_2O_3 is greater when a cement of HSR rather than MSR grade is produced.

Up to temperatures of about 100 °C the hydration products formed in the hydration process do not differ significantly from those formed at ambient temperature, but the hydration is accelerated unless retarders are added to the slurry. Of the two calcium silicate phases, dicalcium silicate is accelerated by elevated temperature more effectively than tricalcium silicate. In general, the C-S-H phase formed at elevated temperatures has a higher C/S ratio, and the silicate anion size is shifted upwards. It also contains higher amounts of incorporated aluminum and sulfate ions. The strength of the hardened slurry at shorter hydration times is obtained mainly from the hydration of C_3S, whereas after longer hydration times the contribution of C_2S to strength development increases significantly. Above about 70 °C the AFt phase becomes unstable, and the AFm phase is formed as the sole calcium aluminate (ferrite) sulfate hydrate.

The properties of slurries made with Class G and Class H cements may be modified by a variety of additives in order to cover a wide range of well depths and temperatures, pumping times, and other requirements.

To control the bulk density, low- or high-density materials may be added to the mix. To lowering the density, bentonite (sodium montmorrilonite based clay), typically in amounts between 2 and 12 wt%, may be used. Bentonite works by increasing the water demand of the slurry, as it can hold several times its own weight of water. The effectiveness of bentonite may be increased and its dosage lowered significantly if it is mixed with water and allowed to swell before being added to the slurry. Other materials that may be added to the slurry to lower its density include pulverized fly ash, diatomaceous earth, or expanded perlite. To increase the density of the slurry, finely ground hematite (Fe_2O_3, $d = 5.0$), baryte ($BaSO_4$, $d = 4.2$), ilmenite ($FeTiO_3$, $d = 4.7$), galena (PbS, $d = 7.5$), and similar high-density materials may be added to the system.

The flow properties of the slurry may be improved by adding suitable dispersing agents, such as sulfonated naphthalene formaldehyde condensate (SNFC), sulfonated melamine formaldehyde condensate (SMFC), or lignin-based dispersants.

Defoamers and deaerators are added to oil well cement slurries to prevent excessive foaming and air entrapment. They act by lowering the surface tension of the liquid phase, if introduced in very low amounts. They include lauryl alcohol, dibutylphthalate, and some silicone compounds, such as polydimethyl siloxane.

The thickening time and progress of hydration may be controlled by suitable accelerators, such as $CaCl_2$ or NaCl, and retarders, such as different sugars, citric acid, and calcium or sodium lignosulfonate, gluconate, and glucoheptanate. These last are especially indicated in deep drilling at high bottom-hole temperatures. Additions of NaCl in high concentrations may also be used in applications where a reduction of damage to salt strata is sought.

Gypsum or other expansive agents, such as calcium or magnesium oxide (Ghofrani and Planck, 1993), may be added to the mix to generate expansion and thus improve the sealing properties of the hardened material.

Radioactive tracers or coloring agents (such as methylene blue or phenolphthalein) may be used to help to trace the movement of the slurry.

At temperatures above 100 °C, initially coarse crystalline a-dicalcium silicate hydrate [a-C_2SH or $CA_2(HSiO_4)OH$] is formed as the main hydration product of pure Portland-type cements, such as plain Class G and Class H oil well cements. This orthorhombic phase possesses low compressive strength and high permeability. At higher temperatures (above about 200 °C) tricalcium silicate hydrate [jaffeite, $C_6S_2H_3$ or $CA_6Si_2O_7(OH)_6$], a similarly deleterious crystalline phase, may also be formed. As the intrinsic strength properties of these phases are rather low, their formation results in a low strength of the hardened material.

To avoid strength retrogression occurring, typically above about 130 °C, appropriate amounts of SiO_2 in the form of silica sand or silica flour (very finely ground quartz) have to be added to the slurry. Under these conditions the main calcium silicate hydrate formed below 200 °C is 11A tobermorite [$Ca_5(H_2Si_6O_{18}).4H_2O$]. The formation of this phase is associated with good strength and low permeability. At even higher temperatures (above about 150 °C) tobermorite is gradually transformed to xonotlite [C_6S_6H or $CA_6Si_6O_{17}(OH)_2$] and gyrolite [$C_8S_4H_8$ or $Ca_8(Si_4O_{10})_3(OH)_4.6H_2O$]. Both these phases possess a lower compressive strength and a somewhat higher permeability than tobermorite. Finally, above about 250 °C truscotite [$C_7S_{12}H_3$ or $Ca_7(Si_4O_{10})(Si_8O_{19})(OH)_4.H_2O$] may also be formed, which is weaker and less permeable than gyrolite. No aluminoferrite-derived phases have been independently established, and it must be assumed that all or most of the Al^{3+}, Fe^{3+}, and SO_4^{2-} ions become incorporated into the crystalline lattices of the existing calcium silicate hydrates.

27.2 OTHER OIL WELL CEMENTS

API Class J cement is basically a combination of β-dicalcium silicate and quartz interground in an approximate proportion of 60:40 (belite-silica cement). This type of cement is especially suitable for applications in deep oil wells and high-temperature geothermal wells, owing to its very low reactivity. At temperatures below 100 °C very little hydration occurs, even after several hours of curing: an amorphous C-S-H phase and calcium hydroxide are formed in the hydration of dicalcium silicate, whereas the quartz that is present does not participate in the hydration reaction. At temperatures above 100 °C quartz becomes gradually involved in the hydration process, even though the reaction still proceeds much more slowly than with Class G and Class H cements. Amorphous C-S-H and crystalline tobermorite are formed as the main products of reaction, as well as some xonotlite and gyrolite. If needed, the progress of the reaction can be slowed down further by adding suitable retarders to the mix.

Calcium aluminate cement may be considered for low-temperature cementing, especially in arctic conditions.

A calcium phosphate binder to be used as a lost circulation control material for geothermal wells may be produced by combining high-alumina cement with sodium phosphate and with borax serving as retarder. The high-alumina cement may be partially replaced by ground granulated blast furnace slag (Allan and Kukacka, 1995).

A sodium metasilicate modified high-alumina cement may be used in geothermal wells at temperatures up to 300 °C (Sugama and Garciello, 1996). Sodium calcium silicate hydrate and boehmite are formed as products of reaction in the hydrothermal reaction.

Portland cements extended with class F fly ash from coal burning power plants may be used for filler applications (Reeves, 1991) and for lightweight cement slurries. The fly ash content may reach up to 50 wt%.

Portland cement combined with granulated blast furnace slag may be used for general applications.

Combinations of granulated blast furnace slag with quartz sand have been used for high-temperature cementing (above 110 °C).

REFERENCES

Allan, M.L., and Kukacka, L.E. (1995) Calcium phosphate cements for lost circulation control in geothermal drilling. *Geothermics* **29**, 269–289.

Bensted, J. (1985) Oil well cements: a general review. *Chemistry in Industry (London)* **4**, 100–105.

Bensted, J. (1989) Oil well Cements. *World Cement* **20**, 346–357.

Bensted, J. (1991) API class C rapid-hardening oil well cement. *World Cement* **23**, 38–41.

Bensted, J. (1992a) Oil well cement standards: an update. *World Cement* **23**, 38–44.

Bensted, J. (1992b) Class G and H basic oil well cements. *World Cement* **23**, 44–50.

Bensted, J. (1995) Development of international standards for well cementing. *World Cement* **26**, 40–42.

Bensted, J. (1998) Special cements, in *Lea's Chemistry of Cement and Concrete* (ed. P.C. Hawlett), Arnold, London, pp. 779–835.

Brylicky, W., Malolepszy, J., and Stryczek, S. (1992) Alkali activated slag cementitious material for drilling operation, in *Proceedings 9th ICCC, New Delhi*, Vol.3, pp.312–318.

Ghofrani, R., and Planck, H. (1993) Prerequisites for successful use of calcia and magnesia expansive cements in cementing of annular space (in German). *Berichte der Deutschen Wissenschaftlichen Gesellschaft für Erdöl, Erdgas und Kohle, Tagungsberichte* **1993**, 87–100 [ref. CA 121/16265].

Lota, J.S. *et al.* (1992) The hydration of class G oil well cement at low and ambient temperatures with additions of sodium aluminate, in *Proceedings 9th ICCC, New Delhi*, Vol. 4, pp. 10–16.

Mohan, L. (1991) Advances in some special and newer cements, in *Cement and Concrete Science and Technology* (ed. S.N. Ghosh), Vol. 1, Part 1, ABI Books, New Delhi, pp. 253–313.

Reeves, N.K. (1991) Investigations of fly ash characteristics and behaviours for use in extended oil well cements, in *Proceedings 13th International Conference on Cement Microscopy, Tampa*, pp. 86–100.

Sugama, T., and Garciello, N. (1996) Sodium metasilicate-modified lightweight high alumina cements for use as geothermal well-cementing materials. *Advanced Cement Based Materials* **3**, 45–53.

28 Inorganic binders for toxic and radioactive waste disposal

Inorganic binders may be used to immobilize toxic or radioactive waste, with the aim of encapsulating it and converting it into an integral form of waste that can be conveniently and safely disposed. In this process, toxic and/or radioactive liquids, sludges, or dust are combined with the binder (and water) and converted into a solid form in the course of setting and hardening.

Several conditions must be fulfilled in using inorganic binders for waste immobilization. First, the waste material must not interfere with the setting and hardening of the inorganic binder. Some substances present in the waste to be processed may delay or even prevent setting and hardening of the binder that is used. In such instances it may be necessary to select another type of cement, or to reduce the concentration of the interfering substance in the mix to levels at which its action is reduced to an acceptable level.

The toxic or radioactive waste must be sufficiently well bound within the hardened cement paste to ensure extremely low leachability from the hardened material. Hardened cement pastes have an intrinsically microporous texture, and thus they are potentially more susceptible to leaching than may be apparent from their external geometry. The leachability of constituents bound within a cementitious system is significantly affected by the porosity of the cement matrix, which may be controlled by varying the initial water/cement ratio. Also important for controlling the rate of material transport in the existing pore system is the pore size distribution of the hardened cement paste, which depends on but is not a direct function of overall porosity.

The constituents of the toxic or radioactive waste may be bound within the cementitious matrix in several ways:

- by sorption to the surface of the cement hydration products that are present, particularly those with a high specific surface area. The ions or neutral species that are to be sorbed have to compete for sorption sites with species that are regular constituents of the pore solution, and which may be present in significantly higher concentrations. Thus to be effectively sorbed even at low concentrations the species in question must have very favorable distribution coefficients.

- by crystallochemical substitution: that is, by incorporation into the structure of the existing hydrate phases. A wide range of cations and anions may partially substitute for regular occupants in a range of structural sites of the existing reaction products.
- by precipitation and conversion into a non-soluble form, in the environment existing within the cement paste. Precipitation is the single most effective chemical mechanism of fixation, as a wide range of metal ions precipitate and convert to a compound of extremely low solubility in the high-pH environment existing in most inorganic cement systems. Some amphoteric elements, which have a minimum solubility at near-neutral pHs, may also be converted into non-soluble phases in a reaction with other constituents of the pore solution.

The cementitious system that is selected must perform well for a sufficiently long period of time, which in some applications may span hundreds or even thousands of years. The materials most likely to perform for long periods of time are those that are thermodynamically stable or near-stable, both intrinsically and under the environmental effects of the geologic repository. Hydrated products of silicate-, aluminate-, and phosphate-based binders appear to meet this requirement very closely. The high degree of thermodynamic stability of the formed reaction products represents a significant potential advantage of inorganic cements as encapsulant host materials for chemical and radioactive wastes, as compared with some other existing alternatives.

Also considered must be the amount of heat released in the setting and hardening process, and the kinetics of the heat release. This determines the temperature development within the hydrating paste, which in turn may affect the rate of hydration, the structure of the hydrates formed, the volume changes taking place in the paste, and ultimately the leachability of the species bound within the hardened material.

A significant fraction of toxic waste requiring disposal consists of inorganic species in which the interaction with the inorganic binder is relatively well understood. By contrast, the interaction between organic substances and inorganic binders is insufficiently understood, and in most instances the waste material must be evaluated case by case. Incineration of this type of waste may be a meaningful alternative.

The binding of radionuclides by a cement matrix is determined by their chemical nature, and is not dependent on their radiation characteristics. In most nuclear wastes the radionuclides that are present constitute only a small fraction of the total mass, and are intermixed with non-radioactive species. If some of them possess chemical toxicity, then such products are classified as mixed wastes.

In handling radioactive wastes, the increase of temperature caused by partial or complete conversion of the generated radiation energy into heat must also be taken into consideration. Based on its radioactivity, we can distinguish between low-level, intermediate-level and high-level radioactive waste. Immobilization with inorganic binders is limited to low-level and, to a far lesser degree, intermediate-level wastes.

The binder most commonly used for immobilization of toxic and radioactive waste is **Portland cement** (Atkins and Glasser, 1992; Glasser, 1992; Roy, 1992; Hills *et al.*, 1993; Macphee and Glasser, 1993; Bhatty *et al.*, 1996; Miller and Klemm, 1997). Some species that may be constituents of toxic wastes, such as Zn^{2+}, Pb^{2+}, borates, phosphates, citrates, or some saccharides, may slow down or even prevent the hydration process if present in concentrations that are too high. Conversely some substances, such as $CaCl_2$ and other chlorides, may accelerate hydration and make handling of the fresh mix difficult. To ensure proper setting and hardening it may be necessary to reduce the concentration of such interfering substances in the system to an acceptable level.

The immobilization of heavy metals, such as As, Ba, Cd, Cr, Hg, Pb, or Zn, is generally attributed to the precipitation of insoluble hydroxides in the high-pH environment of the Portland cement paste, or to adsorption (Glasser, 1992; Roy, 1992; Macphee and Glasser, 1993; Miller and Klemm, 1997). Experiments performed on pure calcium silicate hydrates indicate that such phases formed at ambient temperature in the presence of heavy metal ions are poorly crystallized, and their degree of condensation is increased. Zn, Pd, and Cd ions are immobilized by C-S-H very effectively, but Cr^{6+} is not (Nocun-Wczelik and Malolepszy, 1997). Some ions may occupy sites in the structure of the C-S-H phase: for example, it has been reported that Cr^{3+} may replace Si^{4+} in this phase (Miller and Klemm, 1997).

A particularly difficult problem is posed by the confinement of cesium, which in the form of a long-half-life isotope is a common constituent of radioactive waste. In the system C-S-A a matrix composition low in Al_2O_3 and nearly equimolar in its SiO_2 and CaO contents gives the best Cs retention (Bagosi and Cseteny, 1998).

The AFt phase present in hydrated Portland cement pastes possesses a range of structural sites that can be occupied by both cations and anions. Ions that can be incorporated into the structure of this phase include borate, chromate, selenate, and lead (Kumaratsan *et al.*, 1990; Glasser, 1992; McCarthy *et al.*, 1992; Miller and Klemm, 1997).

The leaching of heavy metals from a hardened Portland cement paste with water or acids also depends greatly on the existing porosity, and thus low leaching rates can be obtained only at very low water/cement ratios (van Eijk and Bronwers, 1998). The rate of leaching is also associated with calcium hydroxide removal from the hardened paste. A significant leaching of the heavy metals gets under way only after a substantial fraction of this paste constituent has been dissolved and the resultant increased porosity facilitates the transport of species by diffusion through the leached shell (van Eijk and Bronwers, 1998).

The way in which organic compounds are bound within the hardened Portland cement paste, and their effect on hydration and the structure of the hydrates formed, may vary greatly. Compounds that are insoluble in water usually have no effect, whereas those that are at least partially soluble may or may not slow down the rate of hydration, alter the structure of the hydrated paste, and decrease its strength. Compounds that exhibit an adverse effect on Portland cement hydration include various phenols, chlorophenols, and ethylene glycol, whereas methanol or ethanol have almost no effect (Grutzek, 1992).

It has been reported that the immobilization of waste in hardened Portland cement may be improved by adding sodium silicate to the system, as this additive accelerates the hydration reaction and lowers the amount of portlandite in the hydrated material (Scheetz and Hoffer, 1995).

In addition to plain Portland cement a variety of **blended cements** have also been used for waste disposal applications. In addition to Portland clinker or cement they may also contain granulated blast furnace slag, fly ash, natural pozzolanas, microsilica, or clays. It has been reported that the addition of fly ash or silica fume reduces the leachability of heavy metals, mainly by reducing the free calcium hydroxide content of the hardened paste (van Eijk and Bronwers, 1998). The presence of slag in the cementitious system reduces the leachability of technetium and chromium (Roy, 1992). The leachability of both cesium and strontium may be reduced by producing zeolitic phases in systems containing slag and fly ash, through reactions taking place at elevated temperatures (Roy, 1992). It is also possible to reduce the leachability of Cs by combining Portland cement with natural zeolites (Bagosi and Csetenyi, 1997)

Risch *et al.* (1997) studied the possibility of using **calcium aluminate** cement alone or in combination with Portland cement for the immobilization of inorganic wastes. They observed that the rate of leaching slowed down as the specimens became carbonated, producing a less permeable layer at the surface. However, delayed ettringite formation caused cracking of the test specimens, associated with an enlargement of the leachable surface and a decrease of physical stability.

A separate approach consists in precipitating ettringite from a solution that contains toxic ions, which become incorporated into the crystalline lattice of the produced solid phase (McCarthy *et al.*, 1992).

REFERENCES

Atkins, M., and Glasser, F.P. (1992) Application of cement based materials to radioactive waste immobilization. *Waste Management* **12**, 105–131.
Bagosi, S., and Csetenyi, L.J. (1997) Immobilization of Cs-loaded ion exchange

resins in zeolite-cement blends, in *Proceedings 10th ICCC, Göteborg*, paper 4iv045.

Bagosi, S., and Csetenyi, L.J. (1998) Caesium immobilisation in hydrated calcium-silicate-aluminate systems. *Cement and Concrete Research* **28**, 1753–1759.

Bhatty, J.I. *et al.* (1996) *Stabilization of heavy metals in Portland cement, silica fume/Portland cement and masonry cement matrices.* PCA R&P No 2067, Portland Cement Association, Skokie, IL, USA.

Glasser, F.P. (1992) Application of cements to the treatment and conditioning of toxic waste, in *Proceedings 9th ICCC, New Delhi*, Vol. 6, pp. 114–118.

Grutzek, M.W. (1992) Hazardous waste (organic, heavy metal): in cement disposal, in *Proceedings 9th ICCC, New Delhi*, Vol. 6, pp. 119–125.

Hills, C.D., Sollars, C.J., and Perry, R. (1993) Ordinary Portland cement based solidification of toxic wastes: the role of OPC reviewed. *Cement and Concrete Research* **23**, 196–212.

Kumaratsan, P. *et al.* (1990) Oxyanion substituted ettringites: synthesis and characterization and their potential role in immobilization of As, B, Cr, Se and V. *Materials Research Society Symposium Proceedings* **278**, 83–104.

Macphee, D.E., and Glasser, F.P. (1993) Immobilization science in cement systems. *MRS Bulletin* **18**, 66–71.

McCarthy, G.J., Hassett, D.J., and Bender, J.A. (1992) Synthesis, crystal chemistry and stability of ettringite, a material with potential applications in hazardous waste immobilization. *Materials Research Society Symposium Proceedings* **245**, 129–140.

Miller, F.M., and Klemm, W.A. (1997) Mechanistic considerations for the efficiency of Portland cement in waste stabilization, in *Proceedings 10th ICCC, Göteborg*, paper 4iv041.

Nocun-Wczelik, W., and Malolepszy, J. (1997) Studies on immobilization of heavy metals in cement paste: C-S-H leaching behaviour, in *Proceedings 10th ICCC, Göteborg*, paper 4iv 043.

Risch, A., Poellmann, H., and Ecker, M. (1997) Application of Portland cement and high alumina cement for immobilization/solidification of waste model composition, in *Proceedings 10th ICCC, Göteborg*, paper 4iv 044.

Roy, D.M. (1992) Cementing materials in nuclear waste management, in *Proceedings 9th ICCC, New Delhi*, Vol. 6, pp. 88–113.

Scheetz, B.E., and Hoffer, J.P. (1995) Characterization of sodium silicate-activated Portland cement: I. Matrixes for low-level radioactive waste forms. American Concrete Institute SP-158, pp. 91–110.

van Eijk, R.J., and Bronwers, H.J.H. (1998) Study of the relation between hydrated Portland Cement composition and leaching resistance. *Cement and Concrete Research* **28**, 815–828.

29 Cements for dental and medical applications

Inorganic binders are routinely used in dentistry, whereas their use in medical applications is limited, and is still in an exploratory state. In general, such cements must exhibit a short and easily controllable setting time, rapid strength development, a high ultimate strength, and very good volume stability. Acceptable biocompatibility and stability under exposure to saliva or body fluids are also obvious requirements.

In all cements used in dentistry the hardening process is based on an acid-base reaction. The cementing mix is obtained by mixing a powder of the basic component with a liquid phase containing the acid component. The basic component may be a metal oxide, such as ZnO or MgO, a calcium silicate, or a glass phase able to release free ions into the solution. The liquid phase consists of monomers, or more often polymers of a carboxylic acid or phosphoric acid.

Cements that contain a glass as the basic component are called glass ionomer cements. Here, the solid constituent is a very finely ground calcium-alumino-fluorosilicate glass with a typical composition $SiO_2 = 43$ wt%, $Al_2O_3 = 33.1$ wt%, $CaO = 11.4$ wt%, and $F = 7.7$ wt% (Sakai $et\ al.$, 1992). It may be produced by melting and subsequent rapid quenching of a blend of pertinent starting components. The liquid is typically a concentrated solution of polyacrylic acid or – less commonly – of other polymerized carboxylic acids, such as metacrylic acid $[CH_2{=}C(CH_3).COOH]$ or itaconic acid $[CH_2{=}C(COOH).CH_2.COOH]$:

$$n\ CH_2{=}CH{-}COOH\ \rightarrow\ {-}CH_2{-}CH{-}CH_2{-}CH{-}CH_2{-}CH{-}CH_2{-}\quad (29.1)$$

$$\underset{COOH\quad COOH}{}$$

    acrylic acid   polyacrylic acid

If both components are mixed together, an ion-exchange process takes place between the H^+ ions liberated by the dissociation of the carboxylic group and the Ca^{2+} and Al^{3+} ions of the glass. These ions subsequently form bridges between the carboxylic groups of neighboring polyacrylate chains:

$$-COOH \rightarrow -COO- + H^+ \tag{29.2}$$

$$2 -COO- + Ca^{2+} \rightarrow -COO\text{-}Ca\text{-}OOC- \tag{29.3}$$

In this way a three-dimensional structure develops, in which residua of glass particles, with an amorphous hydro-silica layer at the surface, are embedded in an organic-inorganic matrix consisting of metal-interlinked –C–C– chains. The reaction products are amorphous, and no crystalline phases can be detected by X-ray diffraction (Sakai *et al.*, 1992). At the same time the ratio of –COO– to –COOH (as determined by IR spectroscopy) increases, as the polymerization continues and the hardening of the cementitious system progresses. The Al/Ca ratio in the reaction product is significantly lower than that of the starting glass, as Al^{3+} ions react with the polyacrylic acid less readily than do Ca^{2+} ions (Sakai *et al.*, 1992). The bending strength of the hardened material increases with increasing molecular weight of the polyacrylate used, and with increasing fineness of the glass powder.

In a later development of glass ionomer cements, glass phosphonate cements were introduced, in which the glass powder is combined with polyvinyl phosphonic acid. These cements exhibit very favorable setting and working characteristics, and attain final compressive strengths of up to 200 MPa.

In a reaction of a polymerized carboxylic acid (such as polyacrylic acid) with metal oxides (such as ZnO) metallic bridges are again formed between pairs of carboxyl groups present in neighboring chains, with a simultaneous liberation of water:

$$2 -COOH + ZnO \rightarrow -COO\text{-}Zn\text{-}OOC- + H_2O \tag{29.4}$$

Such cements exhibit fast setting and hardening, and may attain compressive strengths of up to 100 MPa.

Combinations of zinc oxide with concentrated solutions of phosphoric acids also exhibit cementing properties. These zinc phosphate cements can be formulated to set within a few minutes, and develop strength rapidly. Zinc phosphate is formed as a product of the hardening reaction (see also section 12.4).

In calcium phosphate cements the hardening process is based on a reaction between dicalcium phosphate (or dicalcium phosphate dihydrate) and tetracalcium phosphate, and the product of reaction is hydroxyapatite $[Ca_5.(PO_4)_3.OH]$. The mixing liquid is usually a diluted aqueous solution of phosphoric acid, which accelerates the whole reaction. Small amounts of hydroxyapatite may also be added to the mix to promote the reaction and to adjust the setting time. A great advantage of calcium phosphate cement is its excellent biocompatibility, as hydroxyapatite, the product of the hardening reaction, is also a natural constituent of bone tissues. For this reason calcium phosphate cement may be considered for applications in both dental medicine and general surgery. It has also been suggested (Vanis and Odler, 1997) that compacts of the starting constituents should be produced, which may be implanted, to convert to hydroxyapatite after implantation. The chemistry of calcium phosphate cements is discussed in more detail in section 12.2.

Kokubo and Yoshihara (1991) produced a $CaO–SiO_2–P_2O_5–CaF_2$ glass that, if mixed with an ammonium phosphate solution, sets within minutes, yielding hydroxyapatite and $CaNH_4PO_4.H_2O$, together with amorphous SiO_2, as products of hydration.

REFERENCES

Brown, P.W., and Chow, L.C. (1986) A new calcium phosphate water-setting cement, in *Cement Research Progress* (ed. P.W. Brown), American Ceramic Society, Westerville, OH, USA, pp. 353–374.

Chow, L.C. (1991) Self setting calcium phosphate cement. *Materials Research Society Symposium Proceedings* **179**, 3–24.

Jiang, W., and Yang, N. (1992) Phosphate bonded dental cement, in *Proceedings 9th ICCC, New Delhi*, Vol. 3, pp. 338–344.

Kokubo, T., and Yoshihara, S. (1991) Bioactive bone cement based on $CaO-SiO_2-P_2O_5$ glass. *Journal of the American Ceramic Society* **74**, 173–174.

Sakai, E. *et al.* (1992) Mechanism of glass-ionomer hardening, in *Proceedings 9th ICCC, New Delhi*, Vol. 3, pp. 413–418.

Vanis, P., and Odler, I. (1997) Investigations on calcium phosphate cements, in *Proceedings 10th ICCC, Göteborg*, paper 2ii063.

Wiliams, D.F. (ed.) (1992) Medical and dental materials, in *Medical Science and Technology*, Vol. 14, VHC, Weinheim, Germany.

Xie, L., and Monroe, E.A. (1991) Calcium phosphate dental cements. *Materials Research Society Symposium Proceedings* **179**, 25–39.

30 Cements for miscellaneous special applications

30.1 CEMENTS FOR APPLICATIONS IN WHICH ELECTRICAL CONDUCTIVITY OF CONCRETE IS REQUIRED

Both fresh and hardened cement pastes exhibit electrical conductivity, which is brought about by the movement of ions in the liquid phase. The limiting factors are the size and tortuosity of the existing pore system, rather than the mobility of the ions involved and their concentration in the liquid phase (Tumidajski *et al.*, 1996). The electrical conductivity increases with increasing water/cement ratio and declines with progressing hydration, as the volume of the pores declines. The following conductivity values [in $(\Omega.m)^{-1}$] were reported for Portland cement pastes made with different water/cement ratios and hydrated between one and 28 days (Tumidajski, 1996):

w/c	1 d	14 d	28 d
0.35	0.285	0.085	0.070
0.40	0.402	0.113	0.095
0.45	0.470	0.152	0.113

Another factor that greatly affects the electrical conductivity is the presence of free, non-combined water in the hardened paste. In general, the conductivity declines as the paste loses free water by drying.

In mortars and concretes the electrical conductivity is affected adversely by the presence of the aggregate, which possesses a low conductivity, and positively by the development of the transition zone between the aggregate and the hardened cement paste, whose conductivity is particularly high (Garboczi *et al.*, 1995; Xie and Tang, 1988). Thus the resultant conductivity will represent a compromise between the insulating effect of the aggregate and the enhancing effect on conductivity brought about by the existence of the transitional zone. As a result of this interaction, the conductivity may increase with increasing aggregate content at very low aggregate volume fractions, but declines consistently if the volume fraction of the aggregate is higher. The following conductivity values [in $(\Omega.m)^{-1}$] were reported for mortars made from a Portland cement with variable volume fractions of the aggregate ($w/s = 0.45$) (Tumidajski 1996):

$V_{aggregate}/V_{mortar}$	1 d	14 d	28 d
0	0.285	0.085	0.070
0.006	0.308	0.092	0.072
0.030	0.304	0.098	0.078
0.093	0.309	0.088	0.067
0.323	0.192	0.051	0.046

If all other variables remain constant, the electrical conductivity will also depend indirectly on the cement employed, and directly on the concentration of individual ions in the pore solution.

In Portland slag cements the electrical conductivity was found to decline with increasing slag content. The following are average values [in $(\Omega.m)^{-1}$] found in cements with different slag contents ($w/c = 0.50$, water-saturated pastes) (Hinrichs 1987):

Slag (%)	2 d	25 d	365 d
0	0.42	0.10	0.07
25	0.45	0.12	0.07
50	0.15	0.08	0.03
70	0.10	0.04	0.01

The electrical conductivity of cement pastes, mortars and concrete may be increased by incorporating electrically conductive constituents into the mix. Especially effective in this respect are graphite powders or graphite fibers (Fu and Chung, 1996; Solotov *et al.*, 1997). In laboratory experiments the electrical resistivity of a Portland cement paste ($w/c = 0.55$, cured for 28 d at RH $= 30\%$) declined from 1.62×10^5 to 1.93×10^4 by adding 0.5% carbon filaments (per weight of the cement) to the mix (Fu and Chung, 1996).

Applications of electrically conductive concrete include conductive floors and cathodic protection of reinforced bridge decks.

30.2 CEMENT FOR STRUCTURES DESIGNED TO SHIELD AGAINST ELECTROMAGNETIC RADIATION

The ability of structures to shield against electromagnetic radiation may be important in buildings that house electronic equipment, or in electric power plants. As inorganic cements possess only a very limited capability to shield electromagnetic radiation, suitable additives have to be added to the concrete mix to achieve such an ability.

Effective shielding capability may be achieved by incorporating electrically conductive materials, especially carbon fibers or carbon filaments, into the concrete mix. Owing to the existence of a skin effect, the electromagnetic radiation at high frequencies interacts with a conductor (such as carbon fiber) only near its surface. As a result, at a constant volume fraction of fibers the shielding effectiveness will increase with decreasing fiber diameter. Especially effective appears to be the incorporation of carbon filaments, rather than conventional carbon fibers, because of their smaller diameter (Fu and Chung, 1996). In laboratory experiments the attenuation was increased from 0.4 dB to 24.4 dB at 1.0 GHz by adding 0.5% of carbon filaments (per weight of cement) to a plain Portland cement paste ($w/c = 0.50$, cured for 28 d at RH $= 30\%$).

30.3 RADIO WAVE–REFLECTING CONCRETE

Normal Portland cement based concrete exhibits a low reflectibility and a high transmissivity to radio waves. This ratio may be reversed by incorporating small amounts (0.5 vol.%) of carbon filaments in the mix. In such mixes the absorbed radiation is negligible, compared with the reflected radiation (Fu and Chung, 1998). Concretes of this type may be potentially useful for lateral guidance in automatic highways.

30.4 CEMENTS AND CONCRETES FOR USE IN STRUCTURES DESIGNED TO SHIELD AGAINST IONIZING RADIATION

Concrete walls with the capacity to shield against ionizing radiation may be required in atomic power plants and facilities using radioisotopes, X-ray equipment, and the like.

The shielding capability of a concrete wall may be effectively controlled by its thickness, and by the selection of the aggregate used. In general, the shielding capacity (at a constant thickness of the wall) increases with increasing density of the aggregate. Aggregates that exhibit a high shielding capability include baryte ($BaSO_4$), hematite (Fe_2O_3), and magnetite

(Fe_3O_4). To increase the density even further, metallic iron in the form of powder or chips may also be added to the fresh concrete mix. In this way concrete densities of $d \leq 4.0$ kg/L or more may be achieved.

In most instances Portland cement is used as the binder, but the shielding capacity may be increased even further by using a cement in which a fraction of CaO has been substituted by BaO (see also section 15.8).

REFERENCES

Fu, X., and Chung, D.D.L. (1996) Submicron carbon filament cement-matrix composites for electromagnetic interference shielding. *Cement and Concrete Research* **26**, 1467–1472.

Fu, X., and Chung, D.D.L. (1998) Radio-wave-reflecting concrete for lateral guidance in automatic highways. *Cement and Concrete Research* **28**, 795–801.

Garboczi, E.J., Schwartz, L.M., and Bentz, D.P. (1995) Modelling the influence of the interfacial zone on the D.C. electrical conductivity of mortar. *Advanced Cement Based Materials* **2**, 169–181.

Hansson, I.L.H. (1983) Electrical resistivity measurements of Portland cement based materials. *Cement and Concrete Research* **13**, 675–683.

Hansson, I.L.H., and Hansson, C.M. (1985) Ion conduction in cement-based materials. *Cement and Concrete Research* **15**, 201–212.

Hinrichs, W. (1987) Investigations on the hydration of blastfurnace slag Portland cements (in German), PhD thesis, University of Clausthal.

Solotov, A. *et al.* (1997) Properties of electrically conductive concretes (in Russian). *Izvestiya Vysshikh Uchebnykh Zavednii, Stroitel'stvo* (6), 38–44 [ref. CA 127/247533].

Tamas, F.D. (1982) Electrical conductivity of cement pastes. *Cement and Concrete Research* **12**, 115–120.

Tamas, F.D., Farkas, E., and Roy, D.N. (1984) Electrical conductivity of pastes made from clinker and gypsum. *British Ceramic Proceedings*, pp. 237–248.

Tumidajski, P.J. (1996) Electrical conductivity of Portland cement mortars. *Cement and Concrete Research* **26**, 529–534.

Tumidajski, P.J. *et al.* (1996) On the relationship between porosity and electrical resistivity in cementitious systems. *Cement and Concrete Research* **26**, 539–544.

Xie, P., and Tang, M. (1988) Effect of Portland cement paste – aggregate interface on electrical conductivity and chemical corrosion resistance of mortar. *Il Cemento* **85**, 33–42.

Appendix

Phases common in cement chemistry

AFm phase
A phase of the general formula $[Ca_2(Al,Fe)(OH)_6]_2.X.xH_2O$ or $3CaO.(Al,Fe)_2O_3.CaX.yH_2O$ [abbreviation $C_4(A,F).X.H_y$], where X represents one formula unit of a double charged, or two formula units of a singly charged anion.

AFt phase
A phase of the general formula $[Ca3(Al,Fe)(OH)_6.12H_2O]_2.X_3.xH_2O$ or $3CaO.(Al,Fe)_2O_3.3CX.32H_2O$ [abbreviation $C_6(A,F)X_3H_{32}$], where X represents one formula unit of a doubly charged ion, or two formula units of a singly charged anion. Normally $x \leq 2$.

alinite
A phase that – in a simplified form – may be described by the chemical formula $Ca_{11}[3(SiO_4).(AlO_4).O_2Cl]$. Doped with foreign ions, it is a constituent of alinite cement.

alite
A form of tricalcium silicate, $3CaO.SiO_2$ (abbreviation C_3S), doped with foreign ions. Main constituent of Portland clinker.

anhydrite
$CaSO_4$ (abbreviation $C\bar{S}$).

apthitalite
$K_3Na(SO_4)_2$ (abbreviation $K_3N\bar{S}_4$).

arkanite
K_2SO_4 (abbreviation $K\bar{S}$).

belite
A form of β-dicalcium silicate, β-2CaO.SiO$_2$ (abbreviation β-C$_2$S), doped with foreign ions. A regular constituent of Portland clinker.

bregidite
α'-C$_2$S; α'-modification of dicalcium silicate.

brownmillerite
Tetracalcium aluminate ferrite, 4CaO.Al$_2$O$_3$.Fe$_2$O$_3$ (abbreviation C$_4$AF). Ferrite phase with $x = 0.5$.

brucite
Magnesium oxide, Mg(OH)$_2$ (abbreviation MH).

calcium fluoroaluminate
11CaO.7Al$_2$O$_3$.CaF$_2$ (abbreviation C$_{11}$A$_7$.CaF$_2$).

calcium langbeinite
K$_2$Ca$_2$(SO$_4$)$_3$ (abbreviation KC$_2$$\bar{\text{S}}$$_3$).

C-S-H phase
Product of hydration of dicalcium and tricalcium silicate of the general formula CaO$_x$.SiO$_2$.H$_2$O$_y$, where x and y may vary over a wide range.

ettringite
6CaO.Al$_2$O$_3$.3SO$_3$.32H$_2$O, abbreviation C$_6$A$\bar{\text{S}}$$_3H_{32}$. An AFt phase.

ferrite phase
A phase of the composition Ca$_2$(Al$_x$,Fe$_{(1-x)}$)$_2$O$_5$, where $0 < x < 0.7$ [abbreviation C$_2$(A$_1$F)].

fluorellastidite
9CaO.3SiO$_2$.3SO$_3$.CaF$_2$ (abbreviation 3C$_2$S.3C$\bar{\text{S}}$.CaF$_2$ or C$_9$S$_3$$\bar{\text{S}}$$_3$$\bar{\text{F}}$).

Friedel's salt
Ca$_2$Al(OH)$_6$]Cl.2H$_2$O (abbreviation C$_3$A.CaCl$_2$.10H$_2$O).

gehlenite
Ca$_2$Al$_2$SiO$_7$ (abbreviation C$_2$AS).

gehlenite hydrate (= strätlingite)
2CaO.Al$_2$O$_3$.SiO$_2$.8H$_2$O (abbreviation C$_2$ASH$_8$).

gibbsite
γ–Al(OH)$_3$ (abbreviation γ–AH$_3$).

gypsum
$CaSO_4.2H_2O$ (abbreviation $C\bar{S}H_2$).

hemihydrate
$CaSO_4.\frac{1}{2}H_2O$ (abbreviation $C\bar{S}H_{0.5}$).

hydrogarnet
A solid solution within the compositional region C_3AH_6, C_3FH_6, C_3AS_3, and C_3FS_3.

hydrotalcite
$[Mg_{0.75}.Al_{0.25}.(OH)_2].(CO_3)_{0.125}.(H_2O)_{0.5}$. Hydrotalcite-type phases are derived from hydrotalcite by replacing some of the Mg^{2+} by Al^{3+} or Fe^{3+}, and by balancing the charge with some anions (such as CO_3^{2-}) occupying interlayer sites.

hydroxyapatite
$Ca_5(PO_4)_3OH$. Main reaction product in the hardening of calcium phosphate cements.

kleinite = Klein's compound
$4CaO.3Al_2O_3.SO_3$ or $Ca_4.(Al_6O_{12}).(SO_4)$ (abbreviation $C_4A_3\bar{S}$).

larnite
β-$2CaO.SiO_2$ (abbreviation β-C_2S); β modification of dicalcium silicate.

lime
CaO, abbreviation C.

monosulfate
$4CaO.Al_2O_3.SO_3.12H_2O$ (abbreviation $C_4A\bar{S}H_{12}$). An AFm phase.

periclase
MgO (abbreviation M).

spurrite
Calcium silicocarbonate, $Ca_5(SiO_4)_2.(CO_3)$ (abbreviation $C_5S_2\bar{C}$).

strätlingite (gehlenite hydrate)
$2CaO.Al_2O_3.SiO_2.8H_2O$ (abbreviation C_2ASH_8).

struvite
$NH_4MgPO_4.6H_2O$. Main reaction product in the hardening of magnesium phosphate cements.

sulfospurrite
Calcium silicosulfate, $Ca_5(SiO_4)_2.(SO_4)$ (abbreviation $C_5S_2\bar{S}$).

syngienite
$CaK_2(SO_4)_2.H_2O$ (abbreviation $CK\bar{S}_2H$).

tricalcium aluminate
$3CaO.Al_2O_3$ or $Ca_3Al_2O_6$ (abbreviation C_3A).

trisulfate
An alternative designation for ettringite.

Index